# Contents

# Preface

Categories and functors, it has often been repeated, were introduced thirty years ago by Eilenberg and MacLane [11] to understand and study certain constructions in algebraic topology. It was soon realized that they provided a useful language in which to treat large tracts of mathematics, ranging as far afield as algebraic geometry on the one hand [19] and automata theory [10] on the other. Thus category theory was developed with the specific needs of certain of these fields in mind. Indeed, it is fair to say that many of the most significant contributions came from mathematicians, expert in one or another area, who forged the new theory to their own use. But as the discipline gained momentum, it started generating internal problems of its own, and an ever increasing band of mathematicians who worked on them became known as categorists. In this respect the situation resembles that of group theory. After people had been working with permutation groups, substitution groups, transformation groups for decades, the notion of 'abstract groups' evolved during the third quarter of last century. This general concept rapidly made clear why the older theories had many features in common. In time, however, questions began to be asked in pure group theory which, as everyone knows, were not always easy to answer.

In this book we do not lose sight of the origins of the subject: categories are there to make different topics more transparent by revealing common underlying patterns. This is particularly true of the notion of adjoint functor which is introduced at an early stage and remains a central theme throughout the book. In view of applications, we have also stuck to the traditional description of a category as consisting of objects and morphisms, rather than as just morphisms with certain operations, sometimes favoured by 'pure' categorists.

The material in the first two chapters is mostly standard, but the arrangement perhaps is not. In chapter 1 representable and adjoint functors straightaway take the stage and are used in our treatment of products and limits. The latter owes much to Lambek [23]. Throughout, many examples and exercises should convince the reader that he is doing 'real' mathematics – albeit a rather superficial part of it – and not just building castles in the air. Only in chapter 2 do we get acquainted with monomorphisms and epimorphisms, kernels and cokernels, which feature right at the beginning of most texts, and we gradually introduce more structure in our categories. Thus we pass from additive and exact categories to abelian and Grothendieck categories. We could not resist presenting the pretty juggling of axioms defining an abelian category, mainly following D. Puppe [34]. Our treatment of Grothendieck categories is frankly utilitarian, geared to the needs of homological algebra; our account has benefited from Popescu [33].

Homological algebra also arose out of algebraic topology when its practitioners began to consider homology groups rather than just Betti numbers. Essentially it deals with derived functors; the treatise by Cartan–Eilenberg [9] was followed by those of MacLane [25] and Hilton–Stammbach [20]. Chapter 3 presents the elements of that theory, but without going at all into applications. As opposed to these books, where the theory is set up for modules and it is then remarked as an afterthought that it also carries through for abelian categories, we work with these straightaway as in Mitchell's book [29]. We first present the theory of Yoneda extensions in a given abelian category. From these we build a large new category and by extending a given functor from the original category to the new category we obtain its sequence of satellites in one fell swoop. Thus the Kan extension theorem yields an existence theorem for satellites. This elegant method was suggested by P. Gabriel in his review of Mitchell's book [15]. I am grateful to him for telling me about it back in 1968. The large category involving the Ext's has the additional advantage that the additivity of these functors follows very easily, a fact also noticed by Brinkmann [7] in a similar setting. Our treatment of derived functors is more conventional; it follows the lines laid down by Grothendieck [18]. We do not discuss spectral sequences.

The fourth and final chapter deals with sheaves and their cohomology. This is an important topic in its own right, but also one in which adjoint functors are employed to great advantage. The cohomology of sheaves of modules displays the techniques developed in the chapter on homological algebra. Applications of sheaf cohomology are manifold, in various fields of mathematics. They are not touched upon here; only the elements of the theory are presented. For a more extensive treatment the reader is referred to the monographs [17], [42] and [6].

In some recent books on categories, the author explains in his preface that he intends to write a textbook as well as a work of reference, for students as well as mature mathematicians. This makes four objectives in all, which seems a tall order to fill. So let me state explicitly that this book is meant as a textbook, not a monograph, treatise or work of reference. I have had in mind students rather than mature mathematicians, learners rather than experts. Even so, the wisdom of publishing as a book notes from a course given seven years ago is legitimately open to question. The subject has rapidly developed in the meantime, but I believe that most of the material in this book should still be considered basic. To my mind, the most important developments have been nonabelian homological algebra and the theory of topoi. In both fields, an authoritative treatise still remains to be written; see however [35], [1], and [43] respectively. For both subjects, certain parts of this book form a useful if not absolutely necessary preliminary.

As already mentioned, the book arose out of a course, given at Utrecht University during the first semester of 1968/1969, followed by a seminar. Notes of the course were taken by A. G. van Asch and W. L. J. van der Kallen. In the seminar, S. H. Nienhuys-Cheng and J. W. Nienhuys exposed sheaves and their cohomology. Notes of their lectures were taken by W. H. Hesselink. To all these people the original Dutch notes, put out by Utrecht University in 1970, owe much. Dr Hesselink moreover has helped considerably with the revision of the fourth chapter for the present edition. My colleague Dr C. J. Penning of the University of Amsterdam undertook the translation. However, his contribution has been far greater than the rendering of the text into English. He made many suggestions and revisions and the final form was decided upon during frequent discussions, in which he often managed to boost my flagging morale. Finally I wish to thank J. Lambek for urging me to publish these notes in the first place and for insisting when I remained reluctant; F. Oort for discouraging and P. Gabriel for encouraging the project. All these people share in the merits, if any, of the final result; but only the author is to blame for its shortcomings. And now, gentle reader, bring along an open mind and judge for yourself.

J.R.S.
Utrecht

# 1

# General concepts

## 1.1 Categories

**1.1.1 Definition** A *category* **C** is a system of morphisms and objects. We say that $f$ is a *morphism in* **C** *from the object* $A$ *to the object* $B$ and write $f: A \to B$ or $A \xrightarrow{f} B$. The following conditions should be satisfied.

(i) For each morphism $f$ in **C** there are unique objects $A$ and $B$ in **C** such that $f: A \to B$.

(ii) For each pair of objects $A$ and $B$ in **C** the class of morphisms $f$ such that $f: A \to B$ is a set $(A, B)$. This set may be empty.

(iii) For all objects $A$, $B$ and $C$ in **C** there is a mapping (called *composition* or *product*) $(B, C) \times (A, B) \to (A, C)$ which assigns to a pair $\ulcorner g, f \urcorner$, with $g \in (B, C)$ and $f \in (A, B)$, the product $gf \in (A, C)$.

(iv) Existence of identities: for every object $A$ in **C** there is a morphism $1_A: A \to A$ with the property that for every object $C$ in **C** and for every couple of morphisms $f: A \to C$ and $g: C \to A$ we have $f1_A = f$ and $1_A g = g$.

(v) Associativity: for objects $A$, $B$, $C$ and $D$ and morphisms $f: A \to B$, $g: B \to C$ and $h: C \to D$ in **C** we have $(hg)f = h(gf)$.

**Comments** This definition is abstracted from the case that objects are sets and morphisms are mappings. In the abstraction objects are not necessarily sets, nor are morphisms necessarily mappings. The morphisms are the essential ingredients of the theory; the objects are of minor importance.

ad (i). Whether the statement $f: A \to B$ is true or untrue is given together with the category. It is a statement within this category.

ad (ii). Set theory is taken to be known, in particular the difference between class and set. We review this point briefly. A class is not a set if it is bigger than every set. For instance the class of all sets is larger than every set and hence is not a set. If a class $K$ is not larger than a given set $X$, then $K$ itself is a set. Big sets can be constructed by taking the cartesian product $\prod_{i \in I} A_i$ of sets $A_i$, where the index set $I$ is also a set. In this book we take a rather naive view of these matters since we are not dealing with foundations. A more careful discussion may be found in [26].

ad (iii). When there is danger of confusion we sometimes specify the category in which morphisms are considered and write $(A, B)_{\mathbf{C}}$. For similar reasons we occasionally write the composition of $f$ and $g$ as $g \circ f$.

ad (iv). The identity morphisms are unique. For the existence of two such identity morphisms $1_A$ and $e_A$ for an object $A$ implies $1_A = 1_A e_A = e_A$.

**Notation** Instead of '$A$ is an object in the category $\mathbf{C}$' we write $A \in \mathbf{C}$. This does not therefore have the meaning it has in set theory, since $\mathbf{C}$ need not be a set.

**1.1.2 Examples of categories** (*a*) **Sets.** This is the system consisting of sets and mappings. We agree that for each set $X$ there is a unique map going from the void set $\emptyset$ to $X$. Caution: for $Y \subset Z$ we distinguish between $f: X \to Y$ and $g: X \to Z$, even when $f(x) = g(x)$ for all elements $x$ of $X$, in order to comply with axiom (i).

(*b*) $\mathbf{Sets}_*$. The category of sets with base-points. Objects are nonvoid sets $V$ with a given point $*_V$. Morphisms are mappings that map base-points to base-points.

(*c*) **Top.** This is the category of topological spaces. Objects are topological spaces and morphisms are continuous mappings. **Hausd** is the category of Hausdorff spaces.

(*d*) $\mathbf{Top}_*$. As above, with base-points.

(*e*) **Gr.** Groups with group-homomorphisms.

(*f*) **Ab.** Abelian groups with group-homomorphisms.

(*g*) $\mathbf{V}_k$. The category of vector spaces over a given field $k$ with linear mappings.

(*h*) **Rg**. This is the category of rings with ring-homomorphisms. Rings are supposed to have an identity element and ring-homomorphisms are supposed to map the identity element to the identity element. The smallest ring consists of only one element.

(*i*) **CRg.** Commutative rings and ring-homomorphisms.

(*j*) $\mathbf{M}_R$. $R$ is a fixed ring. This is the category of right $R$-modules. The morphisms are $R$-linear mappings. The modules should be right unitary: $x1 = x$ for all elements $x$ of $M \in \mathbf{M}_R$. Analogously, ${}_R\mathbf{M}$ is the category of left $R$-modules.

(*k*) **CR-alg.** This is the category of commutative $R$-algebras with algebra-homomorphisms. The algebras are supposed to have an identity element and the algebra-homomorphisms should transform identity element to identity element.

(*l*) **TopGr.** Topological groups and continuous group-homomorphisms.

(*m*) Let $I$ be a preordered class. A category **I** is constructed by taking the elements of $I$ as objects. The set of morphisms $(i, j)$ from $i$ to $j$ is empty unless $i \leq j$ in which case $(i, j)$ is the set consisting of one element.

(*n*) Let $G$ be a group. The category **G** is the category with single object $G$ and with morphisms all left multiplications.

(*o*) Let **C** be a category. The *dual category* $\mathbf{C}^\circ$ is defined as follows: the objects and morphisms of **C** and $\mathbf{C}^\circ$ are the same but the morphisms of $\mathbf{C}^\circ$ run 'in the opposite direction' (arrows are reversed); in other words, for every pair of objects $A$ and $B$ we have $(A, B)_{\mathbf{C}} = (B, A)_{\mathbf{C}^\circ}$. For $f \in (A, B)_{\mathbf{C}}$ we write $f^\circ \in (B, A)_{\mathbf{C}^\circ}$. In the case when $g \in (B, C)_{\mathbf{C}}$, composition in $\mathbf{C}^\circ$ is defined by $f^\circ g^\circ = (gf)^\circ$.

(*p*) Let **A** and **B** be categories. The *product category* $\mathbf{A} \times \mathbf{B}$ is defined as follows. Objects are pairs $\ulcorner A, B \urcorner$ of objects with $A \in \mathbf{A}$ and $B \in \mathbf{B}$. Morphisms are pairs $\ulcorner f, g \urcorner$ of morphisms with $f$ a morphism in **A** and $g$ in **B**. The product of any number of categories is defined similarly.

(*q*) Let **C** be a category. One defines a category $\mathbf{C}^2$ in the following way. Objects of $\mathbf{C}^2$ are the morphisms of **C**. Morphisms of $\mathbf{C}^2$ are certain pairs of morphisms of **C**. For $f: A \to B$ and $g: C \to D$ in **C**, the pair $\ulcorner \alpha, \beta \urcorner$ is a morphism from $f$ to $g$ in $\mathbf{C}^2$ if and only if $\alpha: A \to C$ and $\beta: B \to D$ make the following diagram commutative (i.e. $\beta f = g\alpha$):

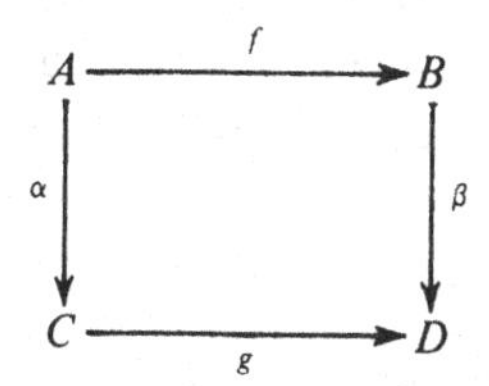


Note that $\mathbf{C}^2$ is not the same category as $\mathbf{C} \times \mathbf{C}$.

**1.1.3 Terminology** For $f: A \to B$ in a category **C**, $A$ is called the *domain* of $f$ and $B$ the *range* of $f$. We call $f$ an *isomorphism* (notation $f: A \xrightarrow{\simeq} B$) and $A$ and $B$ are called *isomorphic* (notation $A \simeq B$) provided there is a morphism $g: B \to A$ in **C** such that $fg = 1_B$ and $gf = 1_A$. Given $f$, such a morphism $g$ is necessarily unique.

A category is called *small* provided the class of objects is a set. In this case the class of all morphisms is also a set since this class equals $\bigcup_{A,B \in \mathbf{C}} (A, B)$ which is a set.

A category is called *concrete* provided its objects are sets endowed with a certain structure which is conserved by morphisms. Examples (*a*) to (*l*) are concrete categories. A precise definition will be given in the next section.

$\mathbf{C}'$ is called a *subcategory* of a category $\mathbf{C}$ provided:

(i) $C' \in \mathbf{C}' \Rightarrow C' \in \mathbf{C}$ for all $C'$;

(ii) $(A, B)_{\mathbf{C}'} \subset (A, B)_{\mathbf{C}}$ for all $A, B \in \mathbf{C}'$;

(iii) $(1_{C'})_{\mathbf{C}'} = (1_{C'})_{\mathbf{C}}$.

$\mathbf{C}'$ is called a *full subcategory* of $\mathbf{C}$ provided it is a subcategory with the stronger condition:

(ii)′ $(A, B)_{\mathbf{C}'} = (A, B)_{\mathbf{C}}$ for all $A$ and $B \in \mathbf{C}'$.

For example **Ab** is a full subcategory of **Gr**. The category of all metric spaces with isometries is not a full subcategory of **Top** but **Hausd** is.

## 1.2 Functors

**1.2.1 Definition** A *covariant functor* $T$ from a category $\mathbf{C}$ to a category $\mathbf{D}$ is a prescription which assigns to each object $C \in \mathbf{C}$ an object $TC \in \mathbf{D}$ and to each morphism $f \in (A, B)_{\mathbf{C}}$ a morphism $Tf \in (TA, TB)_{\mathbf{D}}$ such that the following conditions are satisfied:

(i) for all $C \in \mathbf{C}$, $T1_C = 1_{TC}$;

(ii) $T(gf) = TgTf$ for all $f \in (A, B)_{\mathbf{C}}$ and all $g \in (B, C)_{\mathbf{C}}$.

Notation $T: \mathbf{C} \to \mathbf{D}$ or $\mathbf{C} \xrightarrow{T} \mathbf{D}$.

**1.2.2 Examples** (*a*) $T: \mathbf{Top} \to \mathbf{Sets}$. $T$ is the functor that forgets the topological structure (the *forgetful functor*). Continuous maps between topological spaces are now considered just as maps between the underlying sets.

(*b*) $T: \mathbf{Gr} \to \mathbf{Sets}_*$. Similar to (*a*). In the underlying set $TG$ for $G \in \mathbf{Gr}$, take the identity element as the base-point $*$.

(*c*) Let $G$ and $H$ be groups and let $f \in (G, H)_{\mathbf{Gr}}$. Consider the categories $\mathbf{G}$ and $\mathbf{H}$ as described in 1.1.2(*n*). Then one may define $T: \mathbf{G} \to \mathbf{H}$ by $TG = H$ and $T(\lambda_a) = \lambda_{f(a)}$ ($\lambda_a$: left multiplication by $a$ in $G$).

(d) $T: \mathbf{Gr} \to \mathbf{Gr}$. For $G \in \mathbf{Gr}$ let $TG = [G, G]$ (commutator subgroup of $G$) and for $f \in (G, H)_{\mathbf{Gr}}$ let $Tf$ be the restriction of $f$ to $[G, G]$.

(*e*) $T: \mathbf{Gr} \to \mathbf{Ab}$. For $G \in \mathbf{Gr}$ let $TG = G/[G, G]$ and for $f \in (G, H)_{\mathbf{Gr}}$ let $Tf$ be defined by $Tf(a[G, G]) = f(a)[H, H]$, $a \in G$.

(*f*) $T: \mathbf{Top}_* \to \mathbf{Gr}$. For $(X, *) \in \mathbf{Top}_*$, $T(X, *) = \pi(X, *)$ (fundamental group of $X$ with respect to the base-point $*$). See [39, 1.8].

(*g*) $T: \mathbf{Top} \to \mathbf{Ab}$. $T = H_n$ ($n^{\text{th}}$-singular homology functor). See [39, 4.4].

(*h*) Let $R \in \mathbf{Rg}$ and $N \in \mathbf{M}_R$. $T: {}_R\mathbf{M} \to \mathbf{Ab}$ is defined by $TM = N \otimes_R M$ and for $f: M \to L$ in ${}_R\mathbf{M}$ by $Tf = 1 \otimes f: N \otimes_R M \to N \otimes_R L$. If $R$ is commutative there is no distinction between left and right modules. In that case $N \otimes_R M$ is an $R$-module and one may consider $T$ as a functor from ${}_R\mathbf{M}$ to ${}_R\mathbf{M}$.

**1.2.3 Terminology** A functor $T: \mathbf{C} \to \mathbf{D}$ is called *faithful* provided $Tf = Tg$ implies $f = g$; i.e. for all $A, B \in \mathbf{C}$ the mapping $T: (A, B)_\mathbf{C} \to (TA, TB)_\mathbf{D}$ is injective. If all these mappings are surjective, the functor is called *full*. $T$ is called an *embedding* provided $T$ is faithful and $TA = TB$ implies $A = B$. A category $\mathbf{C}$ is called *concrete* provided there is a faithful functor $T: \mathbf{C} \to \mathbf{Sets}$. This makes precise the description of a concrete category in 1.1.3.

Let $\mathbf{C}$ be any category and $C \in \mathbf{C}$. Consider the covariant functor $h_C: \mathbf{C} \to \mathbf{Sets}$ defined by $h_C A = (C, A)$ for $A \in \mathbf{C}$ and $h_C f(u) = fu$ for $f \in (A, B)_\mathbf{C}$, $u \in (C, A)_\mathbf{C}$. This functor is basic in category theory, since it describes the composition of morphisms in the given category.

**1.2.4 Definition** A *contravariant functor* $T$ from a category $\mathbf{C}$ to a category $\mathbf{D}$ is defined as in 1.2.1 except that now $Tf \in (TB, TA)_\mathbf{D}$ and condition (ii) reads $T(gf) = Tf \circ Tg$.

An important example is the contravariant functor (for $C \in \mathbf{C}$) $h^C: \mathbf{C} \to \mathbf{Sets}$ defined by $h^C A = (A, C)$ for $A \in \mathbf{C}$ and $h^C f(u) = uf$ for $f \in (A, B)_\mathbf{C}$, $u \in (B, C)_\mathbf{C}$.

The contravariant functor $^\circ: \mathbf{C} \to \mathbf{C}^\circ$ introduced in 1.1.2 (*o*) will be denoted by $\Delta$ (the dualizing functor). Thus $\Delta$ is defined by $\Delta C = C$ and for $f: C \to C'$ by putting $\Delta f = f^\circ: C' \to C$. For any contravariant functor $T: \mathbf{C} \to \mathbf{D}$ one can consider the compositions $T\Delta: \mathbf{C}^\circ \to \mathbf{D}$ and $\Delta T: \mathbf{C} \to \mathbf{D}^\circ$ which are covariant. In this way we often identify a contravariant functor $T: \mathbf{C} \to \mathbf{D}$ with its covariant counterpart $T\Delta: \mathbf{C}^\circ \to \mathbf{D}$. When this does not give rise to confusion we sometimes drop the $\Delta$.

A contravariant functor $T$ is called full, faithful or an embedding functor provided $T\Delta$ is such.

Some more examples of contravariant functors are:

(*a*) $H^n: \mathbf{Top} \to \mathbf{Ab}$ with $H^n$ the $n^{\text{th}}$-cohomology functor. See [28, ch. 5], [39, ch. 6].

(*b*) Spec: $\mathbf{CRg} \to \mathbf{Top}$ (the spectrum of a commutative ring, see [19]). 'Functor' on its own usually means a convariant one.

**1.2.5 Multifunctors** One can also consider functors of more than one variable. Such a functor may be covariant in all variables, contravariant in all variables, or covariant in some and contravariant in the other variables. The reader should write out the formulas for this situation for himself. As a typical example we give the following special case.

Let **C** be any category. For any two objects $A$ and $B$ in **C** denote $(A, B)_{\mathbf{C}}$ by $H\ulcorner A, B\urcorner$. Then $H$ is a *bifunctor* (functor of two variables), contravariant in the first and covariant in the second variable. We consider this as a covariant functor from $\mathbf{C}^{\circ}\times\mathbf{C}$ to **Sets**. Explicitly, if $f^{\circ}: A'\to A$ (i.e. $f: A\to A'$) and $g: B\to B'$, then $H\ulcorner f, g\urcorner: H\ulcorner A, B\urcorner\to H\ulcorner A', B'\urcorner$ is given by $H\ulcorner f, g\urcorner(u)=g\circ u\circ f^{\circ}$ for $u\in H\ulcorner A, B\urcorner$. $H\ulcorner f, g\urcorner$ is also denoted by $(f, g)_{\mathbf{C}}$. We often denote $H$ by $(-, -)_{\mathbf{C}}$.

## 1.3 Morphisms of functors

**1.3.1 Example** Let $V$ be a vector space over a field $k$ and $V^{**}$ its double dual which in our language would be denoted by $((V, k)_{\mathbf{V}_k}, k)_{\mathbf{V}_k}$. There is a particular linear mapping from $V$ to $V^{**}$ that has some remarkable properties which we shall describe now. First let the linear mapping $\hat{}: V\to V^{**}$ be defined by $\hat{v}(f)=f(v)$ for $v\in V$ and $f\in V^{*}$ (the dual of $V$). Now let $\phi: V\to W$ be a linear map. Let $\phi^{**}: V^{**}\to W^{**}$ be the corresponding map between the double duals. The following diagram is then commutative:

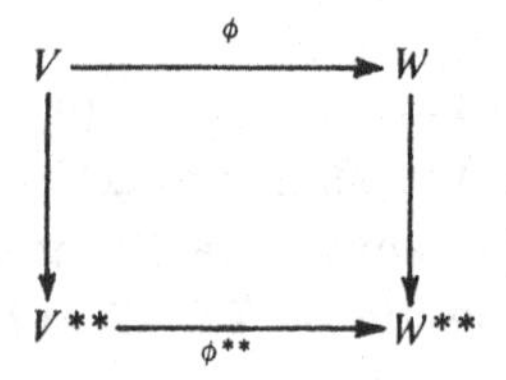


i.e. $\hat{}\circ\phi=\phi^{**}\circ\hat{}$, as the reader may easily check.

**1.3.2 Definition** Let **C** and **D** be categories and $S$ and $T$ functors from **C** to **D**:

$$\mathbf{C}\ \underset{T}{\overset{S}{\rightrightarrows}}\ \mathbf{D}$$

A *morphism* $\eta$ *from* $S$ *to* $T$ is a class of morphisms $\eta(C)$ in **D**, indexed by the objects $C$ of **C**, such that

(i) $\eta(C): SC\to TC$ in **D** for all $C\in\mathbf{C}$;

(ii) for every $f: A\to B$ in **C** the following diagram is commutative:

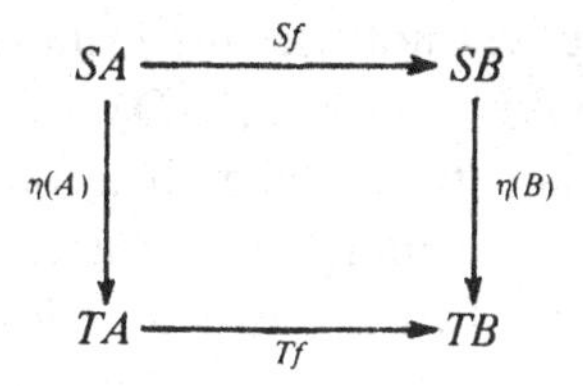


Example 1.3.1 shows a morphism ^ from the identity functor $I: \mathbf{V}_k \to \mathbf{V}_k$ to the 'double dual' functor $**: \mathbf{V}_k \to \mathbf{V}_k$. If one tries to define such a morphism of functors between the identity functor and the 'single dual' functor * one runs into the difficulty that there is no candidate to replace ^ for this situation. Even if one restricts to the category of finite dimensional vector spaces over $k$, so that one knows that $V$ and $V^*$ are isomorphic, these isomorphisms depend on the choice of bases and we cannot choose linear maps $\eta(V)$: $V \to V^*$ such that condition (ii) of definition 1.3.2 is satisfied. It is this difference between the mappings $\hat{}(V)$: $V \to V^{**}$ and $\eta(V)$: $V \to V^*$ that motivates the name of 'natural transformation' for the first. It is in this sense that the mysterious word 'canonical' is mostly used. Thus a morphism $\eta$ of functors is also called a *natural transformation of functors*. As such, this important notion was introduced in the paper of Eilenberg and MacLane [11], which is at the origin of category theory. Indeed, the example of vector spaces and their double duals is theirs.

If for all the objects $C$ of the category $\mathbf{C}$ the morphisms $\eta(C)$ are isomorphisms, the morphism $\eta$ of functors is called a *natural equivalence* (see 1.3.4).

**1.3.3** Let $\mathbf{C}$ and $\mathbf{D}$ be categories. Consider the 'category' $(\mathbf{C}, \mathbf{D})$ whose objects are the covariant functors from $\mathbf{C}$ to $\mathbf{D}$ and whose morphisms are the functor morphisms defined in 1.3.2. Is $(\mathbf{C}, \mathbf{D})$ a category? It is easily verified that with the obvious composition of morphisms of functors the axioms for a category are satisfied except perhaps axiom (ii). If the category $\mathbf{C}$ is small this axiom is satisfied since for functor morphisms $S, T: \mathbf{C} \to \mathbf{D}$ we have

$$(S, T) \in \prod_{C \in \mathbf{C}} (S(C), T(C))$$

and since the right hand side is a set; see 1.1.1 comment (ii).

If the category $\mathbf{C}$ is not small one sometimes may get around this difficulty as will be seen later (1.9.8 and following).

Although not always justified we still use notation and terminology for $(\mathbf{C}, \mathbf{D})$ as if it were a category, except, of course, in those cases where the missing axiom is essential. Thus we will not use, for instance, the notation $h^T$ for $T \in (\mathbf{C}, \mathbf{D})$ since this would mean $h^T: (\mathbf{C}, \mathbf{D}) \to \mathbf{Sets}$.

Categories allow one to define functors between them. Natural transformations between functors then occur as the morphisms in a 'category' where objects are the functors between two given categories. This 'closure of category theory within itself' is of fundamental importance, both from a foundational point of view, and in the more sophisticated branches of the theory. Brashly, one sometimes speaks of the 'category' **Cat**: its objects are all categories (or possibly all small categories) while $(\mathbf{C}, \mathbf{D})_{\mathbf{Cat}}$ consists of all functors from **C** to **D**.

**1.3.4 Remark** If a morphism $\eta$ of functors

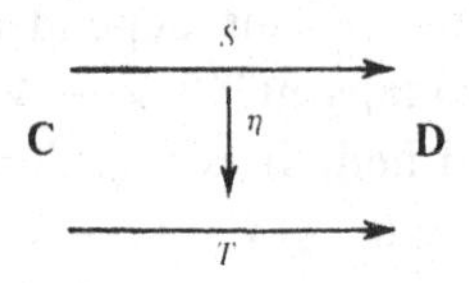


is such that for every $C \in \mathbf{C}$ the morphism $\eta(C)$ is an isomorphism in the category **D**, the reader may easily establish the fact that then the morphism $\eta$ has the usual properties required for an isomorphism in the 'category' $(\mathbf{C}, \mathbf{D})$, and vice versa. Therefore a natural equivalence $\eta: S \to T$ is also called an isomorphism from $S$ to $T$. The functors $S$ and $T$ are called *naturally equivalent* or *isomorphic*. We denote this by $S \simeq T$.

**1.3.5 Definition** Let $S$ and $T$ be contravariant functors from **C** to **D**. A morphism $\eta$ from $S$ to $T$ assigns to each $C \in \mathbf{C}$ a morphism $\eta(C)$ such that

(i) $\eta(C): SC \to TC$ for all $C \in \mathbf{C}$;

(ii) for $f: A \to B$ in **C** the following diagram is commutative:

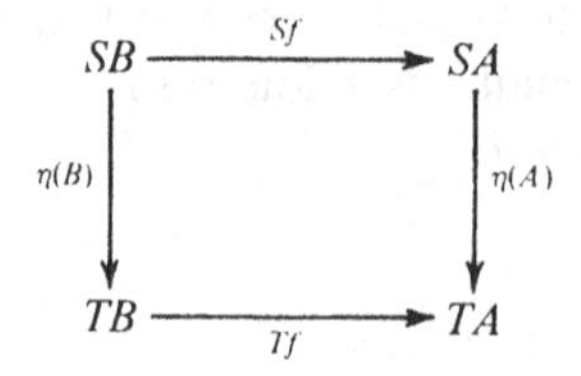


In other words, $\eta$ is a morphism between the covariant functors $S\Delta$ and $T\Delta$ from $\mathbf{C}^\circ$ to **D**.

**1.3.6 Some more examples** ($a$) There is a completely analogous morphism of functors $\hat{}: I \to {**}$ from the identity functor to the double dual functor for the case $\mathbf{M}_R$ instead of $\mathbf{V}_k$. As the ring $R$ need not be commutative one has to distinguish between $\mathbf{M}_R$ and ${}_R\mathbf{M}$. For any object $M \in \mathbf{M}_R$ the single dual $M^*$ can be made in the evident and usual way into an object of ${}_R\mathbf{M}$. Thus the double dual $M^{**}$ is again an object of $\mathbf{M}_R$.

(*b*) Let $F: {}_R\mathbf{M} \to {}_R\mathbf{M}$ be the functor that assigns to each left $R$-module $M$ the free left $R$-module $FM$ with a base consisting of all nonnull elements of $M$. This yields a morphism $\varepsilon: F \to I$ from the functor $F$ to the identity functor by mapping each basis element to its underlying element in $M$.

(*c*) The so-called Hurewicz homomorphism $\tau: \pi_n \to H_n$ (see [28, p. 322], [39, p. 390]) is a morphism of functors:

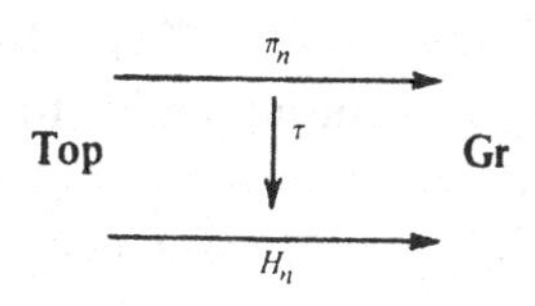


(*d*) Let $R$ be a commutative ring and let $T$ be the bifunctor $\overset{\circ}{\mathbf{M}}_R \times \mathbf{M}_R \to \mathbf{M}_R$: $T\ulcorner M, N \urcorner = (M, R) \otimes_R N$, and $T\ulcorner f, g \urcorner (u \otimes y) = uf^\circ \otimes g(y)$. Let $H$ be the bifunctor $\overset{\circ}{\mathbf{M}}_R \times \mathbf{M}_R \to \mathbf{M}_R$ defined by $H\ulcorner M, N \urcorner = (M, N)$ and $H\ulcorner f, g \urcorner h = ghf^\circ$ as in 1.2.5. Define $\theta: T \to H$ by $\theta\ulcorner M, N \urcorner (u \otimes y)(m) = u(m)y$ for $u \in (M, R)$, $y \in N$ and $m \in M$. It is left to the reader to verify that $\theta$ is well defined and is indeed a morphism from $T$ to $H$. This is an example of a morphism between bifunctors. More generally one defines morphisms between multifunctors. Keeping certain objects constant one obtains by restriction of the variables new (multi-) functors.

(*e*) Let $T: \mathbf{A} \to \mathbf{B}$ be a functor. Then $1_T$ is a frequently occurring functor isomorphism from $T$ to $T$ given by $1_T(A) = 1_{TA}: TA \to TA$.

## 1.4 Representable functors

**1.4.1 Example** Let $S$: **Gr** → **Sets** be the forgetful functor which assigns to each group $G$ its underlying set $SG$. Let $\eta: h_{\mathbb{Z}} \to S$ be the morphism defined by $\eta(G)(f) = f(1)$ for $f \in (\mathbb{Z}, G)_{\mathbf{Gr}}$, where 1 is the generator of the infinite cyclic group $\mathbb{Z}$. (For typographical reasons we write $h_{\mathbb{Z}}$ for the functor $(\mathbb{Z}, -)_{\mathbf{Gr}}$.) Define $\lambda: S \to h_{\mathbb{Z}}$ by putting, for $x \in SG$, $\lambda(G)(x) = f$ if $f$ is the group-homomorphism $\mathbb{Z} \to G$ defined by $f(1) = x$. The reader may easily verify that $\eta$ and $\lambda$ are morphisms of functors and that $\eta\lambda = 1_S$ while $\lambda\eta = 1_{h_{\mathbb{Z}}}$. In other words, the functors $S$ and $h_{\mathbb{Z}}$ are isomorphic or naturally equivalent as discussed in 1.3.4.

**1.4.2 The Yoneda lemma** This situation may be generalized. However we first establish the following fact. Let **C** be any category and $S$ and $T$ functors from **C** to **Sets**. As we have already pointed out, $(S, T)$ need not be a set. But, for $C \in \mathbf{C}$, $(h_C, T)$ is a set, which can be seen as follows. Since $\eta: h_C \to T$ is a morphism of functors we have for each $f: C \to X$ the commutative square

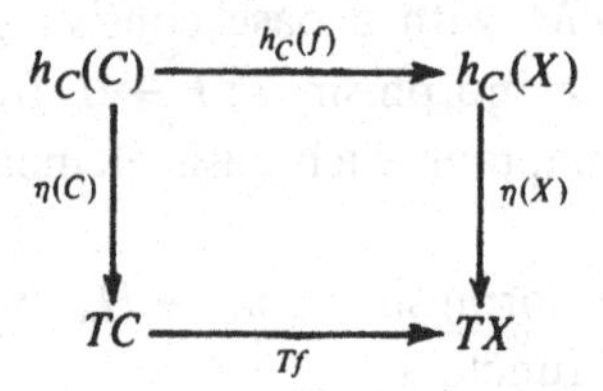


Analysing these maps for $1_C \in h_C C$ we find $Tf(\eta(C)(1_C)) = (Tf \circ \eta(C))(1_C) = (\eta(X) \circ h_C f(1_C) = \eta(X)(f)$. Consequently for each $X$ the mapping $\eta(X)$ is entirely determined by the element $\eta(C)(1_C) \in TC$. The latter being a set, so is $(h_C, T)$. This point having been settled we proceed with the main discussion.

**Yoneda's lemma** There are mappings

$$\Lambda: (h_C, T) \to TC$$

and

$$\Theta: TC \to (h_C, T)$$

such that

$$\Theta\Lambda = 1_{(h_C, T)}$$

and

$$\Lambda\Theta = 1_{TC}.$$

*Proof.* For $\eta \in (h_C, T)$ let $\Lambda(\eta) = \eta(C)(1_C)$ and for $c \in TC$ let $\Theta(c)(X)(f) = Tf(c)$, for $X \in \mathbf{C}$ and $f \in (C, X)$. We have to check that $\Theta(c) \in (h_C, T)$. It follows from the definition of $\Theta$ that $\Theta(c)(X)$: $(C, X) \to TX$; for if $f \in (C, X)$, then $Tf \in (TC, TX)$ and hence $Tf(c) \in TX$. Now let $g: X \to Y$ in **C**, then the following diagram commutes:

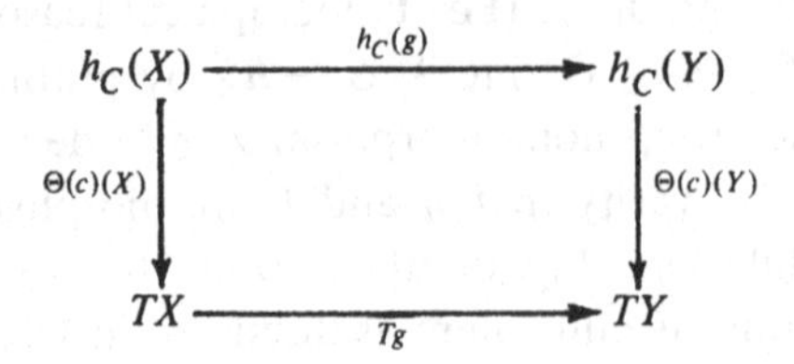

so that $\Theta(c)$ is indeed a morphism from $h_C$ to $T$. Since $(\Theta\Lambda)(\eta) = \Theta(\Lambda(\eta)) = \Theta(\eta(C)(1_C))$, we have $(\Theta\Lambda)(\eta)(X)(f) = \Theta(\eta(C)(1_C))(X)(f) = Tf(\eta(C)(1_C)) = \eta(X)(f)$ for $X \in \mathbf{C}$ and $f \in (C, X)$. Thus $(\Theta\Lambda)(\eta) = \eta$ or $\Theta\Lambda = 1_{(h_C, T)}$. Also $(\Lambda\Theta)(c) = \Theta(c)(C)(1_C) = T1_C(c) = c$ for all $c \in TC$ and thus $\Lambda\Theta = 1_{TC}$.

The above says that for each $C \in \mathbf{C}$ and $T: \mathbf{C} \to \mathbf{Sets}$ the sets $(h_C, T)$ and $TC$ are isomorphic by means of the mappings $\Theta$ and $\Lambda$, which are inverses of each other:

$$(h_C, T) \underset{\Theta}{\overset{\Lambda}{\rightleftarrows}} TC.$$

It would have been more accurate to indicate by $\Lambda\ulcorner C, T\urcorner$ and $\Theta\ulcorner C, T\urcorner$ that these mappings depend on $C$ and $T$. It is left to the reader to show that $\Lambda$ and $\Theta$ are 'natural' in $C$ and $T$ (or canonical!). By this is meant that the following diagram commutes

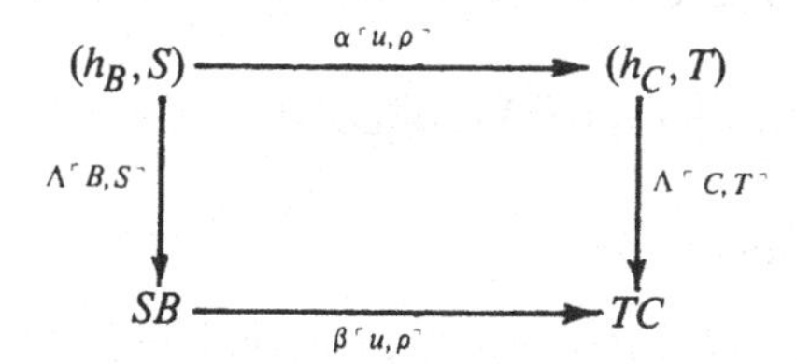


for any $B, C \in \mathbf{C}$, $S, T: \mathbf{C} \to \mathbf{Sets}$, $u: B \to C$ in $\mathbf{C}$ and $\rho: S \to T$ in $(\mathbf{C}, \mathbf{Sets})$, where $\alpha$ and $\beta$ are defined by $\alpha\ulcorner u, \rho\urcorner(\eta)(X)(f) = (\rho(X)\eta(X))(f \circ u)$, $\beta\ulcorner u, \rho\urcorner = Tu \circ \rho(B)$, $\eta \in (h_B, S)$, $X \in \mathbf{C}$ and $f \in h_C(X)$. Since $\Theta$ is the inverse of $\Lambda$, the mapping $\Theta$ is thus also natural in $C$ and $T$, so that the Yoneda lemma states that $\Lambda: (h-, -) \to -(-)$ and $\Theta: -(-) \to (h-, -)$ are inverse isomorphisms. This fact is at the root of most developments in categories of functors.

**1.4.3 Examples** (*a*) Let $T = h_A$ for a certain object $A \in \mathbf{C}$. Then the lemma states $(h_C, h_A) \simeq h_A C = (A, C)$.

(*b*) Let $S$: $\mathbf{Gr} \to \mathbf{Sets}$ be the forgetful functor. According to 1.4.1 this functor is isomorphic to $h_{\mathbb{Z}}$. Thus $(S, S)_{(\mathbf{Gr}, \mathbf{Sets})} \simeq (h_{\mathbb{Z}}, h_{\mathbb{Z}})_{(\mathbf{Gr}, \mathbf{Sets})}$. But this is again isomorphic to $(\mathbb{Z}, \mathbb{Z})_{\mathbf{Gr}}$. Since $(\mathbb{Z}, \mathbb{Z})_{\mathbf{Gr}} \xrightarrow{\sim} \mathbb{Z}$ (assigns to each $f \in (\mathbb{Z}, \mathbb{Z})_{\mathbf{Gr}}$ the element $f(1) \in \mathbb{Z}$), we see that $(S, S)_{(\mathbf{Gr}, \mathbf{Sets})} \simeq \mathbb{Z}$.

(*c*) Suppose $T: \mathbf{C} \to \mathbf{Sets}$ is a contravariant functor. Then (1.2.4) $T^\circ: \mathbf{C}^\circ \to \mathbf{Sets}$ where $T^\circ = T\Delta$ is covariant. The Yoneda lemma gives $(h_C, T^\circ) \simeq T^\circ C = TC$, and one may also check that, since $(h_C, T^\circ) \simeq (h^C, T)$, we have $(h^C, T) \simeq TC$. In particular, for $T = h^A$ (for a certain object $A \in \mathbf{C}$), $(h^C, h^A) \simeq h^A C = (C, A)$.

**1.4.4 Definition** Consider the situation of the Yoneda lemma:

$$(h_C, T) \underset{\Theta}{\overset{\Lambda}{\rightleftarrows}} TC.$$

An element $c \in TC$ is called *universal* provided the morphism $\Theta(c)$: $h_C \to T$ is an isomorphism of functors. The functor $T$ is called *representable* in this case and the couple $\ulcorner C, c \urcorner$ is said to represent the functor. We sometimes also call $\ulcorner C, c \urcorner$ the *representation* of the functor $T$.

**1.4.5 Lemma** (Notation same as above.) The couple $\ulcorner C, c \urcorner$ represents the functor $T$ if and only if for every $C' \in \mathbf{C}$ and $c' \in TC'$ there is a unique morphism $f$: $C \to C'$ such that $Tf(c) = c'$. If in addition $\ulcorner C', c' \urcorner$ also represents the functor $T$ this morphism is an isomorphism.

*Proof.* The following statements are equivalent.

(i) For every $C' \in \mathbf{C}$ and $c' \in TC'$ there is a unique morphism $f$: $C \to C'$ with $Tf(c) = c'$.

(ii) For every $C' \in \mathbf{C}$ the morphism $\Theta(c)$ is bijective.

(iii) $\Theta(c)$ is an isomorphism.

(iv) $\ulcorner C, c \urcorner$ represents the functor $T$.

Suppose $\ulcorner C', c' \urcorner$ also represents $T$. Then there are unique morphisms $f$: $C \to C'$ and $g$: $C' \to C$ with $Tf(c) = c'$ and $Tg(c') = c$, so $T(gf)(c) = TgTf(c) = Tg(c') = c$. But also $T1_C(c) = 1_{TC}(c) = c$. Uniqueness now implies $gf = 1_C$ and the morphisms $f$ and $g$ are isomorphisms. This lemma states the 'unicity up to isomorphism' of universal constructions which abound in mathematics (see 1.4.8).

**1.4.6** For a contravariant functor $T$: $\mathbf{C} \to \mathbf{Sets}$ the couple $\ulcorner C, c \urcorner$ is said to represent $T$ provided $\Theta(c)$ is an isomorphism from $h^C$ to $T$. In this case $\ulcorner C, c \urcorner$ represents the functor if and only if for every $C' \in \mathbf{C}$ and $c' \in TC'$ there is a unique $f$: $C' \to C$ with $Tf(c) = c'$. If the couple $\ulcorner C', c' \urcorner$ also represents $T$, the morphism is an isomorphism.

**1.4.7 Yoneda embedding** If we define functors $h_{\cdot}$: $\mathbf{C}^\circ \to (\mathbf{C}, \mathbf{Sets})$ and $h^{\cdot}$: $\mathbf{C} \to (\mathbf{C}^\circ, \mathbf{Sets})$ by putting $h_{\cdot}(C) = h_C$, $h_{\cdot}(g)(A) = h^A g$, $h^{\cdot}(A) = h^A$, $h^{\cdot}(f)(C) = h_C f$, examples $(a)$ and $(c)$ of 1.4.3 can also be described in the following way. $h_{\cdot}$ is a faithful contravariant functor embedding $\mathbf{C}$ dually as a full subcategory of $(\mathbf{C}, \mathbf{Sets})$. Representable functors from $\mathbf{C}$ to $\mathbf{Sets}$ are those functors that are isomorphic to an object of this subcategory. The functor $h_{\cdot}$ is faithful and full because the mapping $(B, C) \to (h_C, h_B)$ is an isomorphism. The functor $h^{\cdot}$ is a faithful covariant functor embedding **C** as a full subcategory of $(\mathbf{C}^\circ, \mathbf{Sets})$. Contravariant representable functors are functors isomorphic to an object of this subcategory.

**1.4.8 Examples** (*a*) For any category **C** and any object $C \in \mathbf{C}$ the functors $h_C$ and $h^C$ are both represented by the couple $\ulcorner C, 1_C \urcorner$.

(*b*) The forgetful functor $S$; $\mathbf{Gr} \to \mathbf{Sets}$ is represented by the couple $\ulcorner \mathbb{Z}, 1 \urcorner$ (see 1.4.1).

(*c*) Let $G \in \mathbf{Gr}$ be a fixed group, and $V$ a subset of $G$ (not necessarily a subgroup). Define, for $X \in \mathbf{Gr}$,

$$TX = \{f \in (G, X)_{\mathbf{Gr}} | f(v) = 1 \in X \text{ for all } v \in V\}.$$

For morphisms, define $Tg(f) = gf$ for $g: X \to Y$ in **Gr** and $f \in TX$. Since $Tg(f)(v) = gf(v) = g(1) = 1 \in Y$ we have $Tg(f) \in TY$. This makes $T$ a covariant functor from **Gr** to **Sets** which is represented by the pair $\ulcorner G/N, c \urcorner$ where $N$ stands for the smallest normal subgroup of $G$ containing $V$ and $c$ is the canonical homomorphism $G \to G/N$ as the reader may easily check for himself.

(*d*) For $C \in \mathbf{Top}$ and $\sim$ an equivalence relation on $C$ we define $T$ by

$$TX = \{f \in (C, X) | c_1 \sim c_2 \Rightarrow f(c_1) = f(c_2)\}$$

and $Tg(f) = gf$ for $g \in (X, Y)$ and $f \in TX$.

Let $C/\sim$ be the set of equivalence classes in $C$ and $c: C \to C/\sim$ the canonical map. Make $C/\sim$ into a topological space by defining $U \subset C/\sim$ to be open if and only if $c^{-1}(U)$ is open in $C$. Prove that $\ulcorner C/\sim, c \urcorner$ represents $T$.

(*e*) Same as above for $C \in \mathbf{Hausd}$ and $T$: $\mathbf{Hausd} \to \mathbf{Sets}$. Show that $T$ is representable; this amounts to proving that $c: C \to C/\sim$ is a continuous mapping between Hausdorff spaces.

(*f*) Let $R$ be a commutative ring, $M$ and $N$ both $R$-modules. Define $T: \mathbf{M}_R \to \mathbf{Sets}$ by putting $TX$ equal to the set of all $R$-bilinear maps of $M \times N$ into $X$ and $Tf(\phi) = f\phi$ for $\phi: M \times N \to X$ and $f; X \to Y$. Prove that $T$ is represented by $\ulcorner M \otimes_R N, c \urcorner$ with $c: M \times N \to M \otimes_R N$ the canonical bilinear map of the cartesian product into the tensor product.

(*g*) Let $E$ be a fixed group. Define $T$: $\mathbf{Gr} \to \mathbf{Sets}$ by $TG = (E, G)_{\mathbf{Sets}}$ and $T\phi(f) = \phi f$ for $f \in (E, G)_{\mathbf{Sets}}$ and $\phi \in (G, H)_{\mathbf{Gr}}$. Denote the free group generated by the elements $\neq 1$ of $E$ by $F_E$. Show that $T$ is represented by $\ulcorner F_E, c \urcorner$, where $c$ is the canonical map $c: E \to F_E$.

Analogous constructions pertain to the categories $\mathbf{M}_R$ and **CR-alg**.

($h$) Define the functors $P_*$: **Sets** $\to$ **Sets** and $P^*$: **Sets**$^\circ \to$ **Sets** as follows. For $A \in$ **Sets** let $P^*(A) = P_*(A)$ be equal to the collection of all subsets of $A$. For $f: A \to B$ let $P^*f: P^*B \to P^*A$ be defined by $(P^*f)(Y) = f^{-1}(Y)$, i.e. the inverse image of $Y$ under the mapping $f$, and let $P_*f: P_*A \to P_*B$ be defined by $(P_*f)(X) = f(X)$, i.e. the image of $X$ under the mapping $f$. Show that the contravariant functor $P^*$ is represented by $\ulcorner C, \{1\} \urcorner$ where $C = \{0, 1\}$, the idea being that of a characteristic function. Also show that $P_*$ is not representable.

($i$) Let $*$ be a set consisting of one element only, which we also denote by $*$. Let $T$: $\mathbf{C} \to$ **Sets** be defined by $TC = *$ for all $C \in \mathbf{C}$ and $Tf = 1_*$ for all $f: A \to B$. It is obvious that $T$ can be considered both as a covariant and as a contravariant functor. Show that the contravariant functor $T$ is represented by $\ulcorner C, * \urcorner$ if and only if for every $X \in \mathbf{C}$ there is precisely one morphism from $X$ to $C$. Such a $C$ is called a *terminal object* of **C**. Similarly the *covariant functor* $T$ is represented by $\ulcorner C, * \urcorner$ if and only if for every $X \in \mathbf{C}$ there is precisely one morphism from $C$ to $X$. The object $C$ is called an *initial object* in this case. For example in the category **Sets** terminal objects are the sets consisting of one element and the only initial object is the empty set. The same holds for **Top**, whereas in **Gr** and **Ab** the group consisting of just one element is the only initial as well as the only terminal object. In a category without a terminal or an initial object it is always possible to adjoin such an object.

($j$) Let $T$ be a covariant functor from the category **C** to **Sets** and let $\mathbf{C}_T$ be the category whose objects are pairs $\ulcorner C, c \urcorner$ with $C \in \mathbf{C}$ and $c \in TC$ and whose morphisms from $\ulcorner A, a \urcorner$ to $\ulcorner B, b \urcorner$ are those morphisms $f \in (A, B)_\mathbf{C}$ for which $Tf(a) = b$. The initial objects in this so-called 'comma category' $\mathbf{C}_T$ are precisely those pairs $\ulcorner C, c \urcorner$ which represent the functor $T$.

For a contravariant functor $T$: $\mathbf{C} \to$ **Sets** one defines in a similar way a category ${}_T\mathbf{C}$, where the morphisms $\ulcorner A, a \urcorner \to \ulcorner B, b \urcorner$ are those $f \in (B, A)_\mathbf{C}$ with $Tf(a) = b$.

All these examples are rather elementary. This is not always the case. In algebraic geometry, for instance, it is the content of serious mathematical theorems that certain functors are representable. The difficulty there lies in constructing the representing couple.

### 1.5 Products and sums

**1.5.1** We now come to a special case of a representable functor which is of fundamental importance in mathematics.

Let **C** be a category and let $I$ be some index set. Consider a set of objects of **C** indexed by $I$: $\{C_i | i \in I\}$. Define a contravariant functor

$T: \mathbf{C} \to \mathbf{Sets}$ by $TX = \prod_{i \in I}(X, C_i)$ and, for $s: Y \to X$, $Ts: TX \to TY$ given by $Ts(f_i) = f_i s$. Then the pair $\ulcorner C, c \urcorner$ represents the contravariant functor $T$ if and only if for every $X \in \mathbf{C}$ and $x = (x_i) \in TX$ there is exactly one morphism $s: X \to C$ such that $Ts(c) = x$. In other words, for every $X \in \mathbf{C}$ and every set of morphisms $x_i: X \to C_i$ there should be precisely one morphism $s: X \to C$ such that the following diagram commutes:

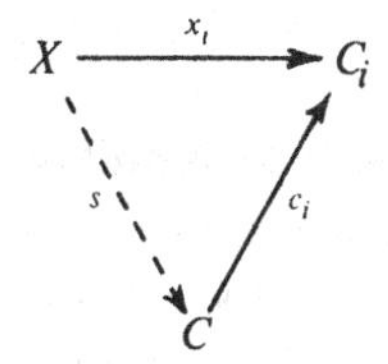


In this case $\ulcorner C, (c_i) \urcorner$ is called a *product* (or *direct product*) of the objects $C_i$ and is denoted by $\prod_{i \in I} C_i$. This product may as well be characterized by the fact that the functors $(\text{-}, \prod_{i \in I} C_i)_{\mathbf{C}}$ and $\prod_{i \in I}(\text{-}, C_i)_{\mathbf{C}}$ from $\mathbf{C}$ to **Sets** are isomorphic. Also the point of view of 1.4.8 ($j$) can be used: consider the category ${}_T\mathbf{C}$ whose objects are couples $\ulcorner X, (x_i) \urcorner$ where $x_i: X \to C_i$ and where a morphism $\alpha$ from $\ulcorner Y, (y_i) \urcorner$ to $\ulcorner X, (x_i) \urcorner$ is given by $\alpha: X \to Y$ such that $y_i \alpha = x_i$. The existence of a product for the objects $C_i$, or the representability of the contravariant functor $T$, is equivalent to the existence of an initial object in ${}_T\mathbf{C}$.

If the product $\ulcorner C, (c_i) \urcorner$ for the objects $C_i \in \mathbf{C}$ exists, the morphisms $c_i: C \to C_i$ are mostly denoted by $p_i$ and are called the *projections* of the product $C$ onto the factors $C_i$.

There are two facts about products that should at least be mentioned, but whose proofs are left (as tedious formal arguments) to the reader: ($a$) $C_1 \prod (C_2 \prod C_3) \simeq (C_1 \prod C_2) \prod C_3$ and ($b$) $C_1 \prod C_2 \simeq C_2 \prod C_1$.

Finally we note that when the category $\mathbf{C}$ has a product for every pair of objects, $\text{-}\prod\text{-}$ may be considered as a covariant functor from $\mathbf{C} \times \mathbf{C}$ to $\mathbf{C}$. The situation may be completely described by the following diagram:

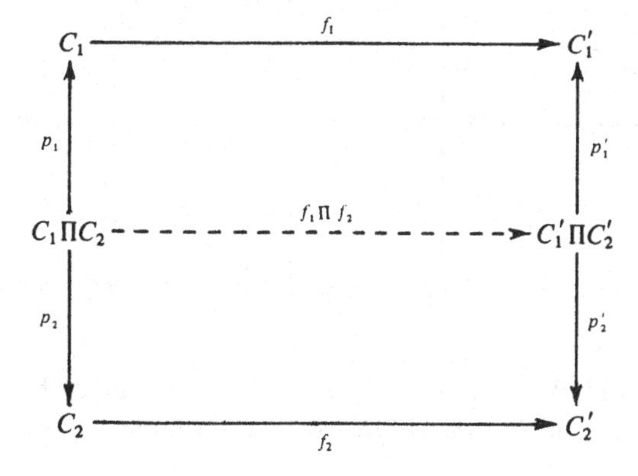

All this can be dualized to obtain the notion of *coproduct* or *sum* (reverse all arrows), sometimes called *direct sum.* The sum is denoted $C=\coprod_{i\in I} C_i$ and the morphisms $c_i$ or $q_i$: $C_i\to C$ are called *injections.*

Warning. The terms 'projection' and 'injection', in this context, are misleading. Although in specific examples (see below) these morphisms do have the properties one expects from projections and injections, in general they do not.

**1.5.2 Proposition** Let **C** be a category with sums and products. Let $\{A_j\}_{j\in J}$ and $\{B_i\}_{i\in I}$ be two collections of objects in **C** with morphisms $f_{ij}$: $A_j\to B_i$ for all $i\in I$ and $j\in J$. Then there is exactly one morphism $f$: $\coprod_j A_j\to\prod_i B_i$ with $p_i f q_j=f_{ij}$ for all $i\in I$ and $j\in J$. Notation: $f=(f_{ij})$.

*Proof.* For every $i\in I$ there is exactly one morphism $f_i$: $\coprod_j A_j\to B_i$ with $f_i q_j=f_{ij}$ for all $j\in J$. For these morphisms $(f_i)$ there is again exactly one morphism $f$: $\coprod_j A_j\to\prod_i B_i$ with $p_i f=f_i$ for all $i\in I$. This morphism $f$ satisfies the relations $p_i f q_j=f_{ij}$. Suppose also $g$ satisfies these relations, i.e. $p_i g q_j=f_{ij}$. Then the unicity of the morphisms $f_i$ implies $p_i g=f_i$, so that the unicity of $f$ in turn implies $f=g$.

Conversely, for a given $f$: $\coprod_j A_j\to\prod_i B_i$ we may choose $f_{ij}=p_i f q_j$. Thus there is a bijection between the collection of systems $(f_{ij})$, $i\in I$, $j\in I$, and the set of morphisms $\coprod_j A_j\to\prod_i B_i$. When $I$ and $J$ are finite the system of morphisms $(f_{ij})$ is often written as a matrix with entry $f_{ij}$ in row number $i$ and column number $j$. In the case $J=\{1\}$ and $I=\{1,2\}$ we henceforth use the notation $f=\begin{pmatrix}f_1\\ f_2\end{pmatrix}$. Note that $q_1=1_A$ and $p_i\begin{pmatrix}f_1\\ f_2\end{pmatrix}=f_i$ $(i=1,2)$. Similarly, for the case $J=\{1,2\}$ and $I=\{1\}$, we write $g=(g_1, g_2)$.

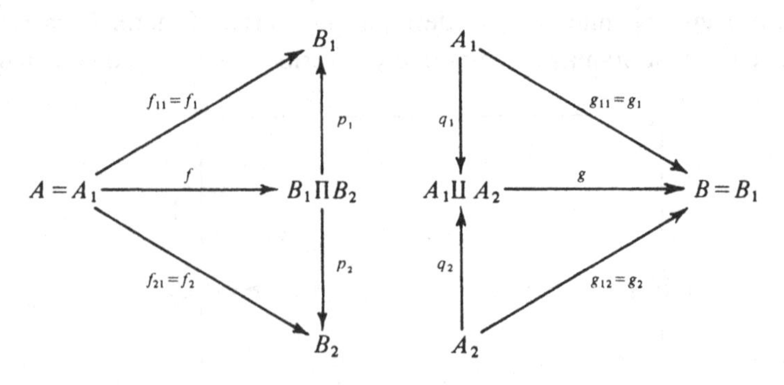


In this notation, if $f$: $\coprod_j A_j\to\prod_i B_i$ then $p_i f$: $\coprod_j A_j\to B_i$ is given by the $i^{\text{th}}$ row of the matrix $(f_{ij})$ and $fq_j$: $A_j\to\prod B_i$ by its $j^{\text{th}}$ column.

**1.5.3 Examples** Since these notions are important we list a number of examples, several of which will already be familiar to the reader.

(*a*) Although we assume that the reader is conversant with the elementary notions of set theory, such as intersection, union, cartesian product and disjoint union, we make a few comments on these. Consider the category **C** = **Sets**. In this category the product $\prod_{i\in I} C_i$ of any collection of sets $C_i$, $i \in I$, coincides with the cartesian product, and the sum $\coprod_{i\in I} C_i$ of these sets equals the so-called disjoint union, by which is meant that $c_i \in C_i$ is never considered equal to an element $c_j \in C_j$ if $i \neq j$. To stress this fact one sometimes identifies the set $C_i$ with the set $C_i \times \{i\}$ in such a disjoint union. The more elementary notions of intersection and union of sets usually considered then turn out to be what are, categorically speaking, called fibred products and sums. (See 1.5.5(*a*).)

(*b*) For **Top** the product $\prod_{i\in I} C_i$ is the cartesian product provided with the topology whose open sets are the sets of the form $\prod_{i\in I} U_i$, with $U_i$ open in $C_i$ and $U_i \neq C_i$ for only finitely many $i \in I$. For, inspection of the diagram

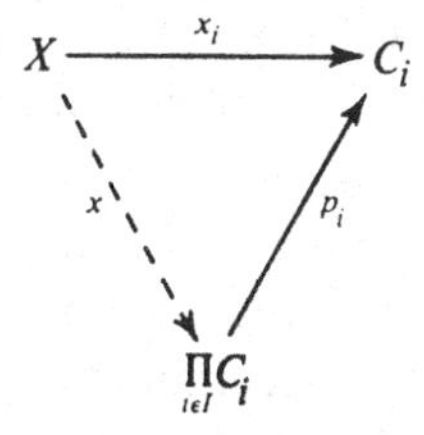


shows that sets of the form $\prod_{i\in I} U_i$, with $U_i = C_i$ for all but one index $j$ for which $U_j$ is open in $C_j$, must necessarily be taken as open sets to make the $p_j$ continuous. The topology thus generated is exactly the topology referred to above and this topology still guarantees the continuity of the map $x$ in the above diagram, whose existence and uniqueness already follows from the set-theoretical situation, described in example (*a*).

The sum $\coprod_{i\in I} C_i$ in the category **Top** is obtained by taking the disjoint union provided with the topology whose open sets are of the form $\bigcup_{i\in I} U_i$ with $U_i$ open in $C_i$. Also here, inspection of the diagram

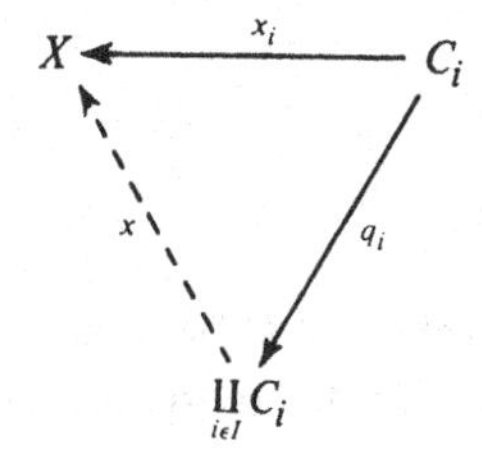

shows that open sets of the above form are necessary to ensure the continuity of the map $x$ (whose existence and uniqueness again follows from the set-theoretical aspect) and this topology also still makes the injections $q_i$ continuous.

(*c*) For $\mathbf{C} = \mathbf{Sets}_*$ the product is given by the cartesian product, taking as base-point the point all the components of which are $*$. The sum is given by the disjoint union with, however, all base-points identified.

(*d*) For $\mathbf{C} = \mathbf{M}_R$ the product is given by the cartesian product whose operations (addition and scalar multiplication) are given componentwise. Restriction to the submodule consisting of all elements with finitely many nonnull components yields the sum in this category. When the index set $I$ is finite, product and sum coincide.

(*e*) For $\mathbf{C} = \mathbf{Gr}$ the product is the cartesian product with componentwise multiplication, whereas the sum is given by the so-called free product. This and other clashes of nomenclature of course are due to the fact that such notions have been studied in many instances before the general concept of sums and products in a category came into being.

(*f*) Let $R$ be a commutative ring and let $\mathbf{C} = \mathbf{CR\text{-}alg}$. The product in this category is again the cartesian product with operations defined componentwise. The tensor product, with componentwise multiplication, yields the sum for this category.

(*g*) Let $\mathbf{C} = \mathbf{Ban}_1$ be the category of real or complex Banach spaces whose morphisms are the bounded linear maps with norm $\leqslant 1$. To obtain a product in this category for an arbitrary collection of objects $B_i$ in $\mathbf{Ban}_1$, we first form the cartesian product $B'$ of the sets $B_i$. With componentwise defined addition and scalar multiplication $B'$ becomes a linear space. Next consider

$$B = \{(b_i) \in B' | \sup_{i \in I} \|b_i\| < \infty\}.$$

Then $B$ is a linear subspace of $B'$ which turns out to be a Banach space if one defines $\|(b_i)\| = \sup_{i \in I} \|b_i\|$. The projections $p_j((b_i)) = b_j$ are indeed morphisms in the category since

$$\frac{\|p_j((b_i))\|}{\|(b_i)\|} = \frac{\|b_j\|}{\|(b_j)\|} \leqslant 1;$$

hence $\|p_j\| \leqslant 1$.

**1.5.4 Cartesian and cocartesian squares** Let $\mathbf{C}$ be a category in which two fixed objects and morphisms into a $C \in \mathbf{C}$ are given:

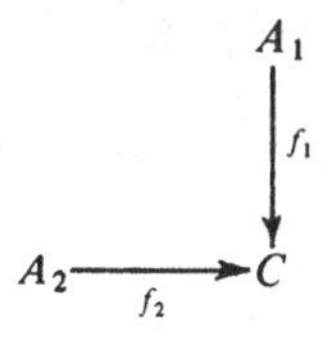


Define the contravariant functor $T: \mathbf{C} \to \mathbf{Sets}$ as follows:

$$TX = \{h_1, h_2 \in (X, A_1) \times (X, A_2) | f_1 h_1 = f_2 h_2\}$$

and for $g: Y \to X$ let $Tg: TX \to TY$ be given by $Tg \ulcorner h_1, h_2 \urcorner = \ulcorner h_1 g, h_2 g \urcorner$. If $T$ is represented by $\ulcorner D, \ulcorner p_1, p_2 \urcorner \urcorner$ this means that for every $X \in \mathbf{C}$ and $\ulcorner h_1, h_2 \urcorner \in TX$ there is exactly one morphism $g: X \to D$ with $Tg \ulcorner p_1, p_2 \urcorner = \ulcorner h_1, h_2 \urcorner$, i.e. $p_1 g = h_1$ and $p_2 g = h_2$.

In this case the following commutative diagram is called a *fibred product of $A_1$ and $A_2$ over $C$*:

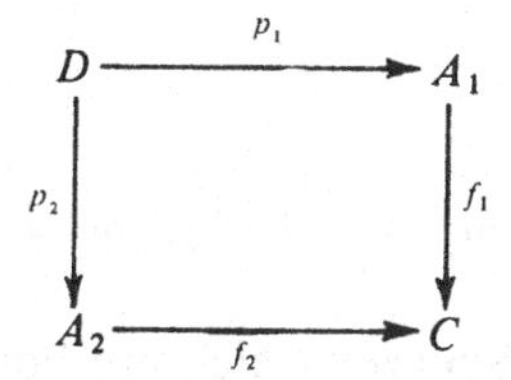


Such a fibred product diagram is called a *cartesian square.*

Notation: $D = A_1 \prod_C A_2$ ($f_1$ and $f_2$ being understood as obvious from the context).

Now consider all commutative squares of the form (*).

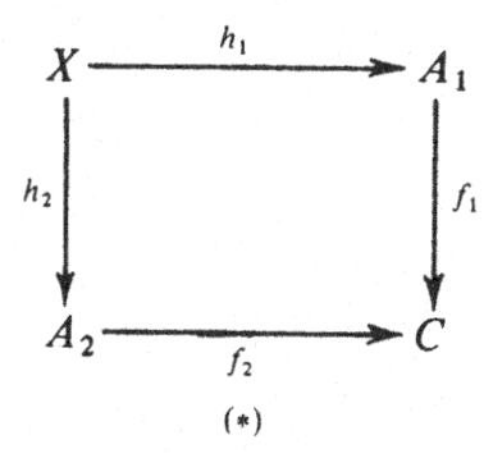


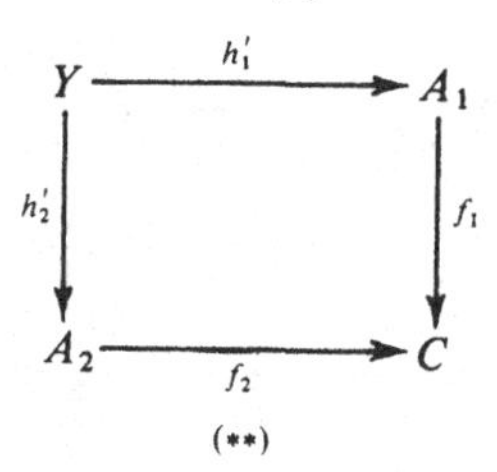


Define a morphism from the square (*) to the square (**) to be a morphism $\alpha: X \to Y$ such that $h_1' \alpha = h_1$ and $h_2' \alpha = h_2$.

In this way one obtains a category whose objects are commutative squares and whose morphisms are morphisms between such squares. The terminal objects in this category are precisely the fibred products of $A_1$ and $A_2$ over $C$.

Remark. In the case that $\mathbf{C}$ has a terminal object *pt* one can easily check that $A_1 \prod A_2 \simeq A_1 \prod_{pt} A_2$.

**Exercise** Prove that if in the diagram

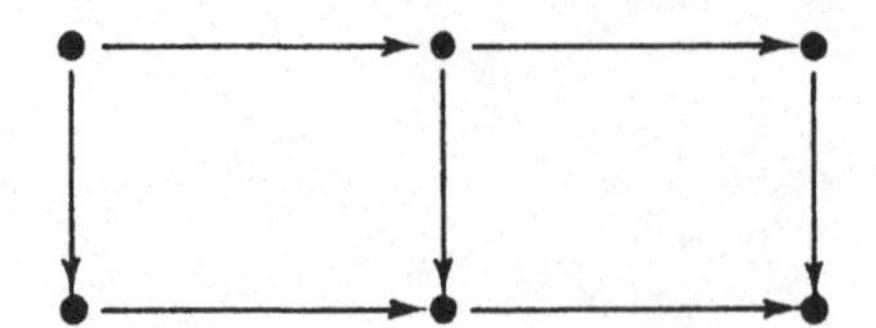

both squares are cartesian then so is the composite rectangle. Here and elsewhere we sometimes allow objects and morphisms to remain nameless.

All the above may be dualized, in which case we obtain the *fibred sum of $A_1$ and $A_2$ over $C$*:

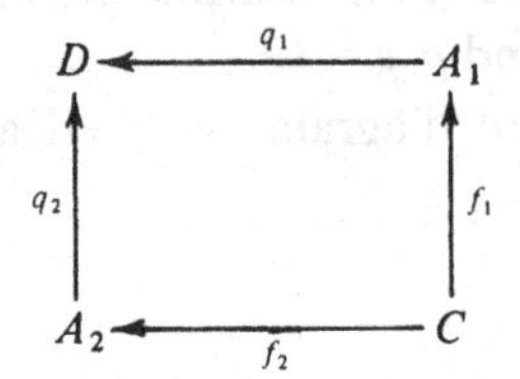


Notation: $D = A_1 \coprod_C A_2$ (again the morphisms $f_1$ and $f_2$ mostly being understood from the context).

Here we speak of *cocartesian squares*. Squares which are both cartesian and cocartesian are called *bicartesian*.

### 1.5.5 Examples

(*a*) **C** = **Sets.**

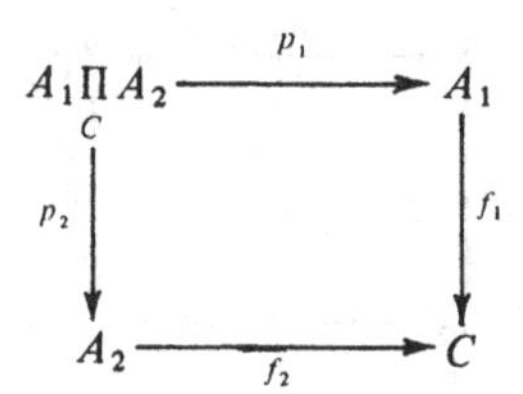


$$A_1 \prod_C A_2 = \{a_1, a_2 \in A_1 \times A_2 | f_1(a_1) = f_2(a_2)\}$$

Remark. If $A_1$ and $A_2$ are subsets of $C$ with inclusions $f_1$ and $f_2$, then $A_1 \prod_C A_2 = A_1 \cap A_2$, the intersection of $A_1$ and $A_2$.

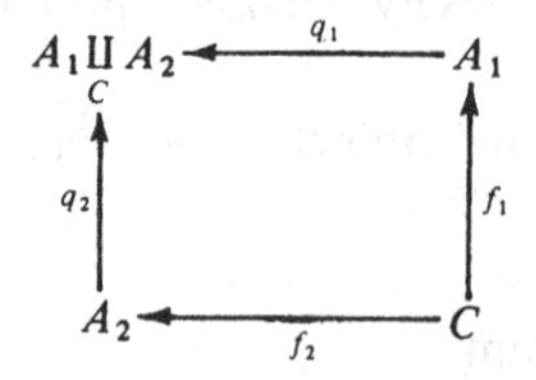

First form the disjoint union of $A_1$ and $A_2$: $A_1 \coprod A_2$. Then define the equivalence relation $\sim$ on $A_1 \coprod A_2$ by $a_1 \sim a_2$ ($a_1 \in A_1$ and $a_2 \in A_2$) if and only if there is an element $c \in C$ such that $a_1 = f_1(c)$ and $a_2 = f_2(c)$. Then $A_1 \coprod_C A_2 = A_1 \coprod A_2/\sim$.

Remark. If $C$ equals the intersection of $A_1$ and $A_2$ with inclusions $f_1$ and $f_2$, then $A_1 \coprod_C A_2$ becomes the usual union $A_1 \cup A_2$.

(*b*) $\mathbf{C} = \mathbf{Top}$. $A_1 \prod_C A_2$ is, as a set, equal to the fibred product of $A_1$ and $A_2$ over $C$ and is given the topology induced by the topology defined in 1.5.3(*b*).

$A_1 \coprod_C A_2$ is, again just as a set, equal to the fibred sum of $A_1$ and $A_2$ over $C$ and it is given the identification topology of the canonical mapping

$$c: A_1 \coprod A_2 \to A_1 \coprod A_2/\sim = A_1 \coprod_C A_2,$$

i.e. a subset $U$ *of* $A_1 \coprod_C A_2$ is open if and only if $c^{-1}(U)$ is open in $A_1 \coprod A_2$. The topology of $A_1 \coprod A_2$ is described in 1.5.3(*b*).

(*c*) $\mathbf{C} = \mathbf{M}_R$. $A_1 \prod_C A_2$ is as in example (*a*).

For $A_1 \coprod_C A_2$, first consider the sum $A_1 \coprod A_2$ and let $B$ be the submodule consisting of all pairs $\ulcorner a_1, a_2 \urcorner$ for which there is an element $c \in C$ with $a_1 = f_1(c)$ and $a_2 = -f_2(c)$. Then $A_1 \coprod_C A_2 = (A_1 \coprod A_2)/B$.

(*d*) $\mathbf{C} = \mathbf{Gr}$. $G_1 \prod_G G_2$ is the subgroup of $G_1 \prod G_2$ consisting of all pairs $\ulcorner g_1, g_2 \urcorner$ with $f_1(g_1) = f_2(g_2)$. $G_1 \coprod G_2$ equals $G_1 \coprod G_2/H$ where $H$ is the smallest normal subgroup in $G_1 \coprod G_2$ (which is the free product of $G_1$ and $G_2$) that contains the set of all elements of the from $f_1(g)f_2(g)^{-1}$ with $g \in G$.

## 1.6 Limits

**1.6.1 Definitions** (*a*) Let $I$ be a preordered set. $I$ is called *directed* (*directed from above*, more precisely) provided for all $i$ and $j \in I$ there is an element $k \in I$ with $i \leq k$ and $j \leq k$.

(*b*) Let $\mathbf{C}$ be a category and $I$ a preordered set. As pointed out in 1.1.2 example (*m*) $I$ can be made into a category $\mathbf{I}$. Objects are the elements $i \in I$ and morphisms are the pairs $\ulcorner i, j \urcorner$ with $i \leq j$. Let $F: \mathbf{I} \to \mathbf{C}$ be a covariant functor, where $I$ is directed. Put $f_{ij} = F\ulcorner i, j \urcorner$ and $C_i = Fi$. Then $\{C_i, f_{ij}\}$ is called an *inductive system in* $\mathbf{C}$. In other words such a system is characterized by:

(i) for $i \leq j$ there is a single morphism $f_{ij}: C_i \to C_j$ in the system; otherwise there is no morphism $C_i \to C_j$;

(ii) $f_{ii} = 1_{C_i}$ for all $i \in I$;
(iii) $f_{jk}f_{ij} = f_{ik}$ for $i \leq j \leq k$;
(iv) for every $C_i$ and $C_j$ in the system there is a $C_k$ in the system which allows morphisms $f_{ik}: C_i \to C_k$ and $f_{jk}: C_j \to C_k$ in the system.

(*c*) Finally we define a covariant functor $T: \mathbf{C} \to \mathbf{Sets}$. For $X \in \mathbf{C}$ let

$$TX = \{(g_i) \in \prod_{i \in I} (C_i, X) | \text{ for } i \leq j \ g_j f_{ij} = g_i\}.$$

For $s: X \to Y$ in $\mathbf{C}$ let $Ts: TX \to TY$ in $\mathbf{Sets}$ be defined by $Ts(g_i) = (sg_i)$.

(*d*) If this covariant functor $T$ is representable and if it is represented by the couple $\ulcorner L, (l_i) \urcorner$, then this pair is called the *direct limit* of the inductive system $\{C_i, f_{ij}\}$. Referring to 1.4.5 we may also describe the direct limit as follows. The couple $\ulcorner L, l(i) \urcorner$ is a direct limit provided:
(i) for all $i \leq j$ the triangle (*) commutes;
(ii) if $\ulcorner X, (g_i) \urcorner$ also yields a commutative triangle for all $i \leq j$, then there is exactly one $u: L \to X$ such that $ul_i = g_i$ for all $i$; see (**).

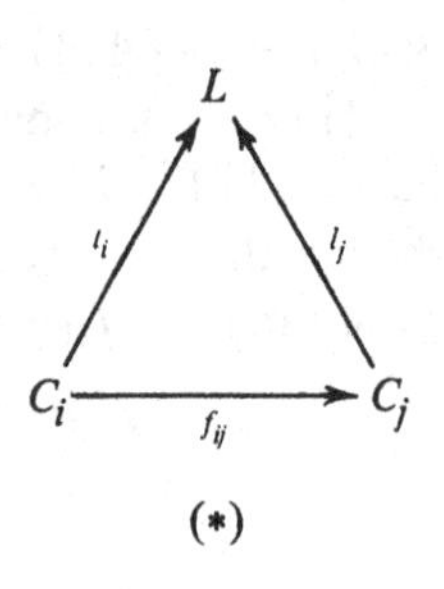


(*)

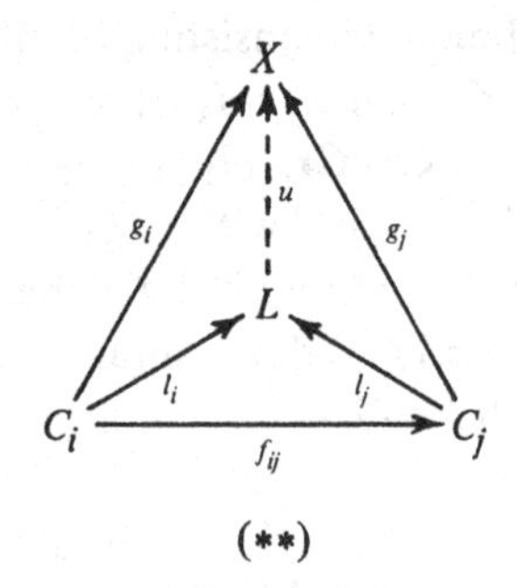


(**)

In still another way the above may be expressed using the category whose objects are commutative triangles as in the picture equipped with the appropriate morphisms. A direct limit $\ulcorner L, l(i) \urcorner$ then corresponds to an initial object in this category.

(*e*) A direct limit is sometimes called an *inductive limit* and also denoted by $\ulcorner \varinjlim C_i, (l_i) \urcorner$ or merely as $\varinjlim C_i$, the morphisms $(l_i)$ being understood as obvious from the context.

All the above machinery may be dualized. In that case the system $\{C_i, f_{ij}\}$ is called a *projective system*. If the corresponding contravariant functor $F: \mathbf{I} \to \mathbf{C}$ is representable, a representing couple is called an *inverse* or *projective limit*. It is denoted by $\ulcorner \varprojlim C_i, (l_i) \urcorner$ or, again, merely by $\varprojlim C_i$.

**1.6.2 Examples** (*a*) In **Sets** every projective system has an inverse limit. For let $\{C_i, f_{ij}\}$ be such a projective system. Take

$$\varprojlim C_i = L = \{(c_k) \in \prod_{k \in I} C_k \mid \text{if } i \leqslant j \text{ then } f_{ij}c_j = c_i\}$$

and take $l_i = p_i|L$.

Also every inductive system has a direct limit. Let $\{C_i, f_{ij}\}$ be such a system in **Sets**. First consider the disjoint union $C = \coprod_{i \in I} C_i$. Then consider the relation $\sim$ on $C$ defined by $c_i \sim c_j$ ($c_i \in C_i$ and $c_j \in C_j$) if and only if there is an index $k \in I$ with $i \leqslant k$ and $j \leqslant k$ and if there are morphisms $f_{ik}$ and $f_{jk}$ in the inductive system such that $f_{ij}(c_i) = f_{jk}(c_j)$.

One may easily check that this relation is an equivalence relation (the fact that the preordered set is directed is needed to prove the transitivity law). The set $L = C/\sim$ now satisfies the requirements for a direct limit $\varinjlim C_i$.

(*b*) In the category $\mathbf{M}_R$ direct and inverse limits exist for inductive and projective systems and their construction is quite analogous to the constructions in the preceding example. For the inverse limit first take the product $\prod M_i$ and then take the submodule $L$ consisting of those elements $(m_i)$ with the property that $f_{ij}m_j = m_i$ for $i \leqslant j$. Choose $l_i = p_i|L$.

For the direct limit first take $\coprod M_i = M$. Instead of the equivalence relation above, one takes the submodule $N$ generated by the elements of type $q_jf_{ij}m_i + q_im_i$ for all $m_i \in M_i$ and $i \leqslant j$ and puts $L = M \bmod N$.

(*c*) In **Gr** the constructions are performed similarly using normal subgroups instead of submodules.

(*d*) Let **C** be the category of the commutative rings **CRg** and let $p \in \mathbb{Z}$ be a prime. Take $C_i = \mathbb{Z}/(p^i)$ for $i \in \mathbb{Z}, i \leqslant 0$. For $i \leqslant j$ let $f_{ij}$: $\mathbb{Z}/(p^j) \to \mathbb{Z}/(p^i)$ be the ring-homomorphism given by $f_{ij}(m \bmod p^j) = m \bmod p^i$. This makes $\{C_i, f_{ij}\}$ into a projective system. It has an inverse limit denoted by $\mathbb{Z}_p = \varprojlim \mathbb{Z}/(p^i)$ and this commutative ring is called the ring of the $p$-adic integers. It can be described as follows. The elements may be considered as formal power series $\alpha = \sum_{n=0}^{\infty} a_n p^n$, with $a_n \in \mathbb{Z}/(p)$. Finite sums represent an integer developed in the number system with base $p$. Addition and multiplication are performed according to the rules in this system. Successive terms of infinite series are to be considered approximations for the element up to successive powers of $p$. The mappings $l_i$: $\mathbb{Z}_p \to \mathbb{Z}/(p^i)$ are given by

$$l_i\left(\sum_{n=0}^{\infty} a_n p^n\right) = \sum_{n=0}^{i-1} a_n p^n \bmod p^i$$

(*e*) $\mathbf{C} = \mathbf{Gr}$ and $G$ is a fixed infinite group. Consider the set $I$ of all normal subgroups $N$ of $G$ such that $G/N$ has finite order. The set $I$ will be partially ordered by $M \leqslant N \Leftrightarrow N \subset M$. Moreover $I$ is directed, which can be seen as follows.

$$\begin{aligned}|G/M\cap N| &= |N/M\cap N|\,.\,|(G/M\cap N)/(N/M\cap N)|\\ &= |N/M\cap N|\,.\,|G/N| = |MN/M|\,.\,|G/N|\\ &\leqslant |G/M|\,.\,|G/N| < \infty\end{aligned}$$

if both $M$ and $N$ belong to $I$. Since $M\cap N \geqslant M$ and $M\cap N \geqslant N$ and $M\cap N\in I$, $I$ is directed. Define $C_N = G/N$, $N\in I$. For $N\leqslant M$ let $f_{NM}$ be the homomorphism $G/M\to G/N$ with $f_{NM}(aM) = aN$. Then $\{C_N, f_{MN}\}$ is a projective system. Its inverse limit, which exists according to example ($c$), is called $G$. We may also work in the category **TopGr**, furnishing the finite groups $C_N$ with the discrete topology. The group $\hat{G}$ then is compact and totally disconnected; it is called the profinite group associated with $G$. The projections $p_N: G\to C_N$ induce a morphism $u: G\to\hat{G}$. If $G$ is the Galois group of an infinite field extension, $u$ is an isomorphism [5, app. 2].

($f$) $\mathbf{C} = \mathbf{Ab}$. Consider $\mathbb{Q}$ (the rational numbers) as an object of **Ab**. Let $I$ be the set $\mathbb{N}$ of natural numbers preordered by the relation $n\leqslant m$ if and only if $n$ divides $m$. Then $I$ is directed. Take $A_n$ as the subgroup of $\mathbb{Q}$ generated by $1/n$. For $n\leqslant m$ define $f_{nm}: A_n\to A_m$ to be the embedding of $A_n$ in $A_m$. The system $\{A_n, f_{nm}\}$ is an inductive system with direct limit $\ulcorner\mathbb{Q}, (l_n)\urcorner$ where $l_n$ is the embedding of $A_n$ in $\mathbb{Q}$. Since the groups $A_n$ are free abelian groups on one generator we may also look for the direct limit of the system $\{A_n, f_{nm}\}$ in the category **B** of free abelian groups. The group $\mathbb{Q}$ is not an object of this category since it is not free. It is left to the reader to show that in **B** the inductive limit is the null group. Let $T$: **B** $\to$ **Ab** be the functor that embeds the objects of **B** in **Ab**. Then the above example shows that the direct limit of the system $\{A_n, f_{nm}\}$ is not taken into the direct limit of the system $\{TA_n, Tf_{nm}\} = \{A_n, f_{nm}\}$ in **Ab**.

### 1.7 Adjoint functors

**1.7.1** Let $S$: **A** $\to$ **B** and $T$: **B** $\to$ **A** be covariant functors. There are many examples in mathematics in which the sets $(SA, B)_{\mathbf{B}}$ and $(A, TB)_{\mathbf{A}}$ are isomorphic in a natural way (to be made precise below). In such cases the functor $S$ is called a left adjoint of $T$ and $T$ is called a right adjoint of $S$ (the term left and right referring to the position of the functor with respect to the comma). Before launching into the pertinent details we give an example which may serve as motivation for the general treatment. Further examples will be postponed until 1.7.12.

Let $E$ be a fixed set. Consider the covariant functor $S$: **Sets** $\to$ **Sets** given by $SA = A \times E$ (cartesian product of $A$ and $E$) and $Su = \ulcorner u, 1 \urcorner$ for $u: A \to A'$. Also consider the covariant functor $T$: **Sets** $\to$ **Sets** given by $TB = (E, B)$ and $Tv = v \circ$ - (composition with $v$) for $v: B \to B'$.

We want to show that the sets $(SA, B) = (A \times E, B)$ and $(A, TB) = (A, (E, B))$ are isomorphic. A convenient way to do so is to construct a mapping $\phi: (SA, B) \to (A, TB)$ and show that it is a bijection or show that it has an inverse $\psi: (A, TB) \to (SA, B)$. This should be done for every pair of objects $A$ and $B$, so $\phi$ and $\psi$ are more accurately denoted by $\phi \ulcorner A, B \urcorner$ and $\psi \ulcorner A, B \urcorner$.

Furthermore we want these mappings $\phi$ and $\psi$ to be natural in $A$ and $B$. This means that we require the following diagram to be commutative for every pair $\ulcorner A, B \urcorner$ and $\ulcorner A', B' \urcorner$ where, in order to achieve that

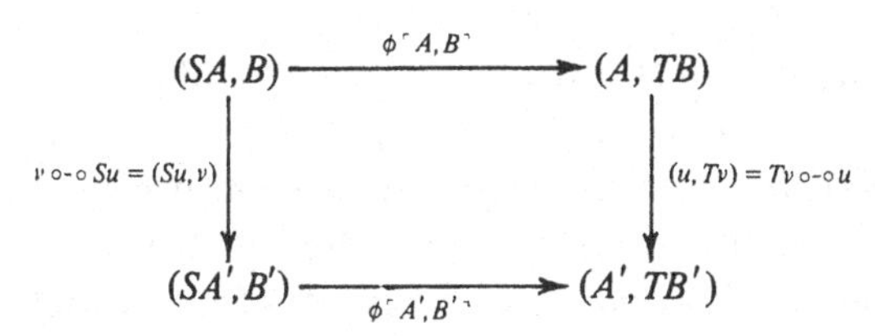


$(Su, v)$ is a mapping from $(SA, B)$ to $(SA', B')$, $u$ should be a map from $A'$ to $A$, and thus $Su: SA' \to SA$; for $v$ the requirement is $v: B \to B'$. Since it is more convenient to have all arrows run in the direction from nonprimed to primed objects we usually regard $(S\text{-}, \text{-})$ and $(\text{-}, T\text{-})$ as bifunctors from **Sets**$^\circ \times$ **Sets** to **Sets**. We conclude by defining $\phi$ and $\psi$, leaving it to the reader to check that they determine morphisms between these bifunctors.

Let $g$ be an element of the set $(SA, B) = (A \times E, B)$. Then define $\phi g \in (A, (E, B))$ by $(\phi g)(a)(e) = g \ulcorner a, e \urcorner$. Let $f$ be an element of $(A, TB) = (A, (E, B))$. Define $\psi f \in (A \times E, B)$ by $(\psi f) \ulcorner a, e \urcorner = (fa)(e)$.

**1.7.2 Definitions** Let $S$: $\mathbf{A} \to \mathbf{B}$ and $T$: $\mathbf{B} \to \mathbf{A}$ be covariant functors. Consider the bifunctors

$$(S\text{-}, \text{-})_{\mathbf{B}}: \mathbf{A}^\circ \times \mathbf{B} \to \mathbf{Sets} \quad \text{and} \quad (\text{-}, T\text{-})_{\mathbf{A}}: \mathbf{A}^\circ \times \mathbf{B} \to \mathbf{Sets}$$

defined by

$$(S\text{-}, \text{-})_{\mathbf{B}} \ulcorner A, B \urcorner = (SA, B)_{\mathbf{B}} \quad \text{and} \quad (\text{-}, T\text{-})_{\mathbf{A}} \ulcorner A, B \urcorner = (A, TB)_{\mathbf{A}}$$

as far as objects are concerned. For $u^\circ: A \to A'$ in $\mathbf{A}^\circ$ (and thus $u: A' \to A$ in $\mathbf{A}$) and $v: B \to B'$ in $\mathbf{B}$ we define $(SA, B)_{\mathbf{B}} \to (SA', B')_{\mathbf{B}}$ by the map $g \mapsto v \circ g \circ Su$. Similarly $(A, TB)_{\mathbf{A}} \to (A', TB')_{\mathbf{A}}$ is given by $f \mapsto Tv \circ f \circ u$.

$S$ is called a *left adjoint of* $T$ and $T$ a *right adjoint of* $S$ provided the functors $(S\text{-}, \text{-})_{\mathbf{B}}$ and $(\text{-}, T\text{-})_{\mathbf{A}}$ are isomorphic.

Note that the definition is not symmetric in $S$ and $T$.

We want to obtain criteria under which $S$ is a left adjoint of $T$. Suppose, to begin with, that $\phi: (S\text{-}, \text{-})_{\mathbf{B}} \to (\text{-}, T\text{-})_{\mathbf{A}}$ is a morphism between these two bifunctors from $\mathbf{A}^{\circ} \times \mathbf{B}$ to **Sets**. This means that for every pair of objects $A \in \mathbf{A}$ and $B \in \mathbf{B}$ the following diagram commutes

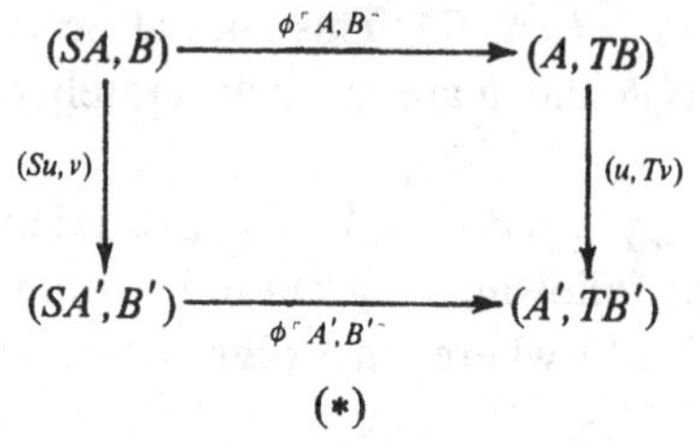


for $u^{\circ}: A \to A'$ and $v: B \to B'$. We may then consider $\phi \ulcorner A, SA \urcorner$: $(SA, A)_{\mathbf{B}} \to (A, TSA)_{\mathbf{A}}$ and we define $\Phi(A) = \phi \ulcorner A, SA \urcorner (1_{SA})$. Thus for every $A \in \mathbf{A}$, $\Phi(A)$ is a morphism from $A$ to $TSA$. It may easily be checked that $\Phi$ is a morphism from the identity functor $I_{\mathbf{A}}$ (from $\mathbf{A}$ to $\mathbf{A}$) to the composite functor $TS$ (which is also a functor from $\mathbf{A}$ to $\mathbf{A}$).

Thus we have shown that to any functor morphism $\phi$ there can be assigned a functor morphism $\Phi = \alpha\phi$.

Conversely, let a morphism of functors $\Phi: I_{\mathbf{A}} \to TS$ be given. Then we define a morphism of functors $\phi = \beta\Phi: (S\text{-}, \text{-})_{\mathbf{B}} \to (\text{-}, T\text{-})_{\mathbf{A}}$ by $\phi \ulcorner A, B \urcorner (g) = Tg \circ \Phi(A)$ for $A \in \mathbf{A}$, $B \in \mathbf{B}$ and $g: SA \to B$. Again we have to check that this is a functor morphism. This means that for the $\phi$ just defined the square (*) above should commute.

Inspection of what happens to elements yields for $f: SA \to B$

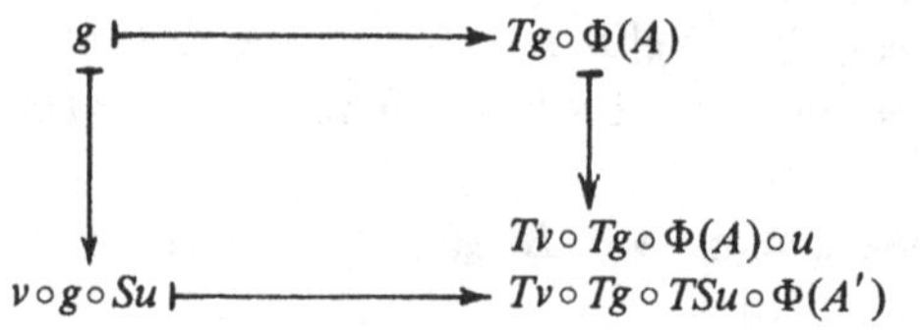


But the two items in the lower right hand corner are equal since $\Phi(A) \circ u = TSu \circ \Phi(A')$ which expresses the fact that $\Phi$ is a functor morphism so that the square

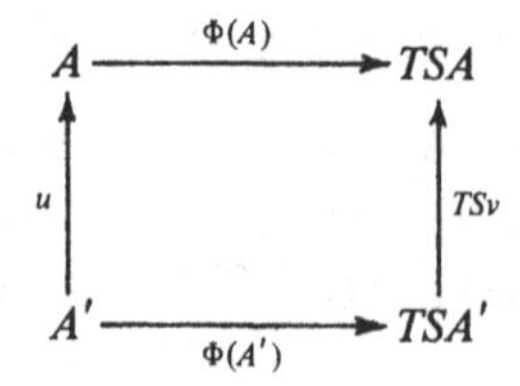


must be commutative.

In a similar way one shows that to any morphism $\psi: (\text{-}, T\text{-})_{\mathbf{A}} \to (S\text{-}, \text{-})_{\mathbf{B}}$ there can be associated a morphism of functors $\Psi: ST \to I_{\mathbf{B}}$ defined by $\Psi(B) = \psi\ulcorner TB, B\urcorner(1_{TB})$ and, conversely, for a given functor morphism $\Psi: ST \to I_{\mathbf{B}}$ one may define a morphism of functors $\psi: (\text{-}, T\text{-})_{\mathbf{A}} \to (S\text{-}, \text{-})_{\mathbf{B}}$ by $\psi\ulcorner A, B\urcorner(f) = \Psi(B) \circ Sf$ for $A \in \mathbf{A}$, $B \in \mathbf{B}$ and $f: A \to TB$.

For the associations $\alpha: \phi \mapsto \Phi$ and $\beta: \Phi\beta \mapsto \phi$ we have been considering there is the following lemma.

**1.7.3 Lemma** $\alpha\beta = 1_{(I_{\mathbf{A}}, TS)}$ and $\beta\alpha = 1_{((S\text{-},\text{-})_{\mathbf{B}}, (\text{-},T\text{-})_{\mathbf{A}})}$.
A similar assertion holds for $\psi$ and $\Psi$.

*Proof.* To show that $\alpha\beta = 1_{(I_{\mathbf{A}}, TS)}$, let $A \in \mathbf{A}$ and $\Phi: I_{\mathbf{A}} \to TS$. Then we have $(\alpha(\beta\Phi))A = (\beta\Phi)\ulcorner A, SA\urcorner(1_{SA}) = T1_{SA} \circ \Phi(A) = \Phi(A)$. This holds for all $A \in \mathbf{A}$, hence $\alpha(\beta\Phi) = \Phi$ or $\alpha\beta = 1_{(I_{\mathbf{A}}, TS)}$.

To show that $\beta\alpha = 1_{((S\text{-},\text{-}),(\text{-},T\text{-}))}$, let $A \in \mathbf{A}$, $B \in \mathbf{B}$, $f \in (SA, B)_{\mathbf{B}}$ and $\phi: (S\text{-}, \text{-})_{\mathbf{B}} \to (\text{-}, T\text{-})_{\mathbf{A}}$. Then

$$(\beta(\alpha\phi))\ulcorner A, B\urcorner(f) = Tf \circ \phi\ulcorner A, SA\urcorner(1_{SA}) = Tf \circ \Phi(A)$$
$$= \phi\ulcorner A, B\urcorner(f).$$

We are now in a position to state necessary and sufficient conditions for two functors $S$ and $T$ as above to be adjoint ($S$ a left adjoint of $T$, or $T$ a right adjoint of $S$). With the notation as in the last section we have

**1.7.4 Proposition**
(i) $\phi\psi = 1_{(\text{-},T\text{-})_{\mathbf{A}}}$ if and only if

$$T \xrightarrow{\Phi T} TST \xrightarrow{T\Psi} T = T \xrightarrow{1_T} T;$$

(ii) $\psi\phi = 1_{(S\text{-},\text{-})_{\mathbf{B}}}$ if and only if

$$S \xrightarrow{S\Phi} STS \xrightarrow{\Psi S} S = S \xrightarrow{1_S} S.$$

The functor morphisms $\Phi T$, $T\psi$, $S\Phi$ and $\Psi S$ are defined in the obvious way, that is $(\Phi T)(B) = \Phi(TB): TB \to TSTB$ and $(T\Psi)(B) = T\Psi(B): TSTB \to TB$ etc.

In this notation we have followed the custom to write $TB$ and $Tf$ for values of a functor $T$ on objects and morphisms, but for a morphism of functors such as $\Phi$ to use parentheses, e.g. $\Phi(A)$, for its evaluations. However, we shall let intelligibility prevail over dogmatism.

*Proof.* We will only give a proof for the assertion (i), leaving the proof of (ii) to the reader. In the first place we remark that $\Phi T$ and $T\Psi$ are morphisms of functors as may easily be seen by writing down the pertinent squares. Next suppose that $\phi\psi = 1_{(-,T-)_{\mathbf{A}}}$. Let $B$ be an object of $\mathbf{B}$. Then we have $\phi\psi\ulcorner TB, B\urcorner = 1_{(TB,TB)_{\mathbf{A}}}$. Rewriting the left hand member of the equality we obtain $\phi\ulcorner TB, B\urcorner \circ \psi\ulcorner TB, B\urcorner$. Applying both members to the element $1_{TB}$ we find

$$(\phi\ulcorner TB, B\urcorner \circ \psi\ulcorner TB, B\urcorner)(1_{TB}) = 1_{TB}.$$

Rewriting the expression on the left we obtain

$$\phi\ulcorner TB, B\urcorner \psi(B) = T\Psi(B) \circ \Phi(TB) = (T\Psi \circ \Phi T)(B)$$

Thus $T\Psi \circ \Phi T = 1_T$.

Conversely let $T\Psi \circ \Phi T = 1_T$, $A \in \mathbf{A}$, $B \in \mathbf{B}$ and $f \in (A, TB)$. Then

$$\begin{aligned}
\phi\psi\ulcorner A, B\urcorner(f) &= (\phi\ulcorner A, B\urcorner \circ \psi\ulcorner A, B\urcorner)(f) \\
&= T(\psi\ulcorner A, B\urcorner(f)) \circ \phi(A) \\
&= T(\Psi(B) \circ Sf) \circ \Phi(A) \\
&= T\Psi(B) \circ TSF \circ \Phi(A) \\
&= T\Psi(B) \circ \Phi(TB) \circ f \\
&= 1_B \circ f = f
\end{aligned}$$

so that $\phi\psi = 1_{(-,T-)_{\mathbf{A}}}$.

In the situation described by the proposition above, it is customary to call the morphisms of functors $\phi$, $\psi$, $\Phi$, $\Psi$ *adjunctions*. We shall sometimes refer to the whole set of data as an *adjoint situation*.

As an aid to the reader we list the relevant notations in a table.

| | |
|---|---|
| $S: \mathbf{A} \to \mathbf{B}$ | $T: \mathbf{B} \to \mathbf{A}$ |
| $\phi: (S-,-)_{\mathbf{B}} \to (-,T-)_{\mathbf{A}}$ | $\psi: (-,T-)_{\mathbf{B}} \to (S-,-)_{\mathbf{B}}$ |
| $\Phi: I_{\mathbf{A}} \to TS$ | $\Psi: ST \to I_{\mathbf{B}}$ |
| $\Phi(A) = \phi\ulcorner A, SA\urcorner(1_{SA})$ | $\Psi(B) = \psi\ulcorner TB, B\urcorner(1_{TB})$ |
| $\phi\ulcorner A, B\urcorner(g) = Tg \circ \Phi(A)$ | $\psi\ulcorner A, B\urcorner(f) = \Psi(B) \circ Sf$ |
| for $g: SA \to B$<br>$Tg: TSA \to TB$ | for $f: A \to TB$<br>$Sf: SA \to STB$ |
| $\Phi T: T \to TST$ | $T\Psi: TST \to T$ |
| $S\Phi: S \to STS$ | $\Psi S: STS \to S$ |

**1.7.5 Proposition** If both $T_1$ and $T_2$ are right adjoints of a functor $S$: $\mathbf{A} \to \mathbf{B}$ they are isomorphic.

*Proof.* We have

$$A \underset{T_1}{\overset{S}{\rightleftarrows}} B \quad \text{and} \quad A \underset{T_2}{\overset{S}{\rightleftarrows}} B$$

with $(S\text{-}, \text{-})_{\mathbf{B}} \simeq (\text{-}, T_1\text{-})_{\mathbf{A}}$ and $(S\text{-}, \text{-})_{\mathbf{B}} \simeq (\text{-}, T_2\text{-})_{\mathbf{A}}$. Then there is an isomorphism $\chi$: $(\text{-}, T_1\text{-})_{\mathbf{A}} \simeq (\text{-}, T_2\text{-})_{\mathbf{A}}$. For any $v$: $B \to B'$ in $\mathbf{B}$ the following square commutes:

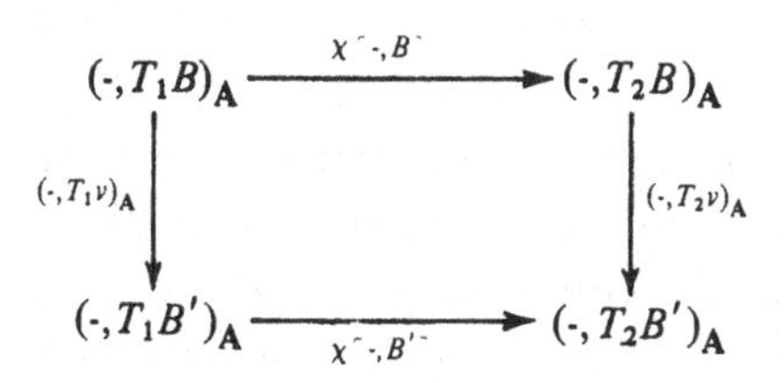

The entry $(\text{-}, T_1B)_{\mathbf{A}}$ in the above square was denoted earlier by $h^{T_1B}$ (see 1.2.4) and $(\text{-}, T_1v)_{\mathbf{A}}$ by $h^{T_1v}$ (see 1.4.7). We know that the functor $h^{\cdot}$ is a covariant functor $\mathbf{A} \to (\mathbf{A}^\circ, \mathbf{Sets})$ embedding $\mathbf{A}$ as a full subcategory of $(\mathbf{A}, \mathbf{Sets})$ (see 1.4.7). This implies that for the functor morphism $\chi\ulcorner\text{-}, B\urcorner$ there must exist $\mathrm{X}(B)$: $T_1B \to T_2B$ such that $h^{\mathrm{X}(B)} = \chi\ulcorner\text{-}, B\urcorner$. Since $h^{\cdot}$ is faithful the following square is commutative.

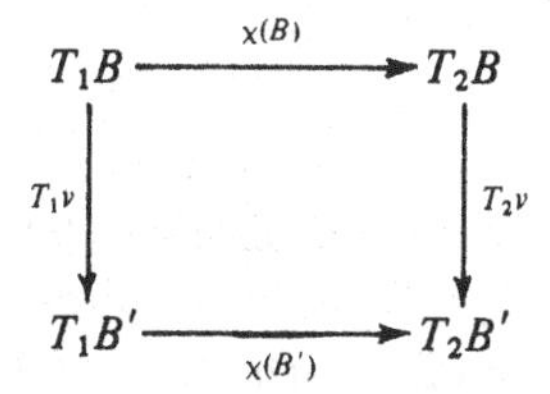

Therefore X is a morphism of functors. Since $\chi$ is an isomorphism, so must be $\chi\ulcorner\text{-}, B\urcorner$ and because $h^{\cdot}$ is fully faithful the same must hold for $\mathrm{X}(B)$. Thus $\mathrm{X}$: $T_1 \to T_2$ is an isomorphism between the functors $T_1$ and $T_2$.

Dually one may prove that if two functors $S_1$ and $S_2$ are both left adjoint to a functor $T$: $\mathbf{B} \to \mathbf{A}$, the functors $S_1$ and $S_2$ must be isomorphic. The reader should moreover note that if in the situation

$$\mathbf{A} \underset{T}{\overset{S}{\rightleftarrows}} \mathbf{B} \underset{T'}{\overset{S'}{\rightleftarrows}} \mathbf{C}$$

the functor $T$ is a left adjoint of $S$ and $T'$ of $S'$, the composite $TT'$ is a left adjoint of $S'S$. The proof is straightforward.

It is worth observing that the notions left–right are not dual in the definition of adjoint functors. Nevertheless, to each statement about left adjoints there is a related statement about right adjoints, which is loosely referred to as its 'dual'. It is possible to set up a formal scheme of translation so as to derive one from the other. But it is often more instructive to work out the dual statement for oneself.

**1.7.6 Definition** Now suppose that $S$ and $T$ are contravariant functors:

$$\mathbf{A} \underset{T}{\overset{S}{\rightleftarrows}} \mathbf{B}.$$

In that case $(S\text{-}, \text{-})_{\mathbf{B}}$ and $(T\text{-}, \text{-})_{\mathbf{A}}$ are covariant functors respectively from $\mathbf{A}\times\mathbf{B}$ to **Sets** and $\mathbf{B}\times\mathbf{A}$ to **Sets**. For convenience, however, we will also consider $(T\text{-}, \text{-})_{\mathbf{A}}$ as a covariant functor from $\mathbf{A}\times\mathbf{B}$ to **Sets**.

The pair of functors $\ulcorner S, T\urcorner$ is called *adjoint on the left* provided the functors $(S\text{-}, \text{-})_{\mathbf{B}}$ and $(T\text{-}, \text{-})_{\mathbf{A}}$ are isomorphic.

In the same situation one may similarly check that $(\text{-}, S\text{-})_{\mathbf{B}}$ and $(\text{-}, T\text{-})_{\mathbf{A}}$ are covariant functors from $\mathbf{A}^{\circ}\times\mathbf{B}^{\circ}$ to **Sets**. If these functors are isomorphic we call $S$ and $T$ *adjoint on the right.*

The following remark perhaps clarifies these definitions. We may regard $S$ and $T$ as covariant functors between $\mathbf{A}^{\circ}$ and $\mathbf{B}$ and assume $S$ is a left adjoint of $T$. As contravariant functors $S$ and $T$ are then adjoint on the left. If we work with $\mathbf{A}$ and $\mathbf{B}^{\circ}$ the functors $S$ and $T$ are adjoint on the right. The details are left as an exercise.

We want to know under what conditions a covariant functor $T: \mathbf{B}\rightarrow\mathbf{A}$ has a left adjoint $S: \mathbf{A}\rightarrow\mathbf{B}$. For this purpose we define the covariant functor $F_A$, for $A\in\mathbf{A}$, by $F_A = h_A \circ T$. In other words $F_AB = (A, TB)_{\mathbf{A}}$ for $B\in\mathbf{B}$ making $F_A$ a functor from $\mathbf{B}$ to **Sets**. In the following we often use lemma 1.4.5: a covariant functor $F: \mathbf{C}\rightarrow$ **Sets** is represented by the couple $\ulcorner C, c\urcorner$ if and only if for every $C'\in\mathbf{C}$ and $c'\in FC'$ there is a unique $f: C\rightarrow C'$ with $Ff(c) = c'$.

**1.7.7 Theorem** A functor $T: \mathbf{B}\rightarrow\mathbf{A}$ has a left adjoint $S: \mathbf{A}\rightarrow\mathbf{B}$ if and only if for every $A\in\mathbf{A}$ the covariant functor $F_A = h_A\circ T: \mathbf{B}\rightarrow$ **Sets** is representable.

*Proof.* First suppose that $S: \mathbf{A}\rightarrow\mathbf{B}$ is a left adjoint of $T$. We claim that the couple $\ulcorner SA, \Phi(A)\urcorner$ represents the functor $F_A$. To prove this let $B\in\mathbf{B}$ and let $f\in F_AB = (A, TB)_{\mathbf{A}}$. We have to show that there is exactly one morphism $g: SA\rightarrow B$ such that $(F_Ag)\Phi(A) = f$. But $(F_Ag)\Phi(A) = Tg\circ\Phi(A) = \phi\ulcorner A, B\urcorner(g)$. The fact that $\Phi: (S\text{-}, \text{-})_{\mathbf{B}}\rightarrow(\text{-}, T\text{-})_{\mathbf{A}}$ is an isomorphism yields the result. Conversely, suppose for every $A \in \mathbf{A}$ the covariant functor $F_A = h_A \circ T$ is representable. Thus for every $A$ we have a representing couple, which we shall denote by $\ulcorner SA, \Phi(A) \urcorner$. We will show that $S$ can be made into a functor from $\mathbf{A}$ to $\mathbf{B}$ which is a left adjoint of $T$. Suppose therefore that $u: A \to A'$ is a morphism in $\mathbf{A}$. We shall define $Su: SA \to SA'$. Since $SA' \in \mathbf{B}$ and $\Phi(A') \circ u \in F_A SA'$, the fact that $\ulcorner SA, \Phi(A) \urcorner$ represents $F_A$ means that there is exactly one morphism $g: SA \to SA'$, which we call $Su$, such that $(F_A g)\Phi(A) = \Phi(A') \circ u$. Thus $g = Su$ is uniquely determined by the condition $Tg \circ \Phi(A) = \Phi(A') \circ u$. For $u': A' \to A''$, $Su' = g': SA' \to SA''$ is the unique morphism defined by $(F_A g')\Phi(A') = \Phi(A'') \circ u'$, or, again, by the condition $Tg' \circ \Phi(A') = \Phi(A'') \circ u'$. To show that $S(u' \circ u) = Su' \circ Su$, we have

$$
\begin{aligned}
T(Su' \circ Su) \circ \Phi(A) &= TSu' \circ TSu \circ \Phi(A) \\
&= TSu' \circ \Phi(A') \circ u \\
&= \Phi(A'') \circ u' \circ u.
\end{aligned}
$$

Since also $T(S(u' \circ u)) \circ \Phi(A) = \Phi(A'') \circ (u' \circ u)$, it follows from uniqueness that

$$S(u' \circ u) = Su' \circ Su.$$

The fact that $S1_A = 1_{SA}$ follows readily from $T1_A \circ \Phi(A) = \Phi(A) \circ 1_A$.

Now we have proved that $S$ is indeed a functor from $\mathbf{A}$ to $\mathbf{B}$. Also $\Phi$ turns out to be a morphism from the functor $I_{\mathbf{A}}$ to $TS$. It remains to be shown that $S$ is a left adjoint of $T$. The Yoneda lemma states that $\Theta: F_A SA \to (h_{SA}, F_A)$ is a bijection of sets, natural in $A$. If however $\ulcorner SA, \Phi(A) \urcorner$ represents the functor $F_A$, this means that $\Theta(\Phi(A)): h_{SA} \to F_A$ is an isomorphism of functors, while $\Theta(\Phi(A))$ was defined for $B \in \mathbf{B}$ by $\Theta(\Phi(A))(B): h_{SA}B = (SA, B)_{\mathbf{B}} \to F_A B = (A, TB)_{\mathbf{A}}$ in such a way that $\Theta(\Phi(A))(B)(g) = (F_A g)\Phi(A)$ for $g \in h_{SA}B$. Denoting this isomorphism $\Theta(\Phi(A))(B)$ by $\phi \ulcorner A, B \urcorner$, we have constructed an isomorphism of functors $\phi: (S\text{-}, \text{-})_{\mathbf{B}} \to (\text{-}, T\text{-})_{\mathbf{A}}$, thus showing that $S$ is indeed a left adjoint of $T$. Since also $\phi \ulcorner A, B \urcorner (g) = (F_A g)\Phi(A) = Tg \circ \Phi(A)$, the notation $\Phi$ is justified.

We might as well have asked for the conditions for a given functor to have a right adjoint. The reader is invited to prove the following theorem.

**1.7.7° Theorem** A covariant functor $S: \mathbf{A} \to \mathbf{B}$ has a right adjoint $T: \mathbf{B} \to \mathbf{A}$ if and only if for every $B \in B$ the contravariant functor $G^B = h^B \circ S$: $\mathbf{A} \to \mathbf{Sets}$ is representable. We sometimes write $G^B$: $\mathbf{A}^\circ \to \mathbf{Sets}$.

Similar conditions can be given for two contravariant functors to be adjoint (on the left or right).

Using theorem 1.7.7° we can give an alternative proof for 1.7.5. Let $T_1$ and $T_2$ both be right adjoint to $S$: $\mathbf{A} \to \mathbf{B}$ with adjunctions $\Psi_1$ and $\Psi_2$. The couples $\ulcorner T_1B, \Psi_1(B) \urcorner$ and $\ulcorner T_2B, \Psi_2(B) \urcorner$ both represent the contravariant functor $G^B$. According to definition and lemma 1.4.6 this means that $T_1B$ and $T_2B$ are isomorphic. We denote the isomorphism by $\eta(B)$: $T_2B \to T_1B$. The relation between $\eta(B)$, $\Psi_1(B)$ and $\Psi_2(B)$ is then given by $(G^B\eta(B))\Psi_1(B) = \Psi_2(B)$. The following diagram commutes:

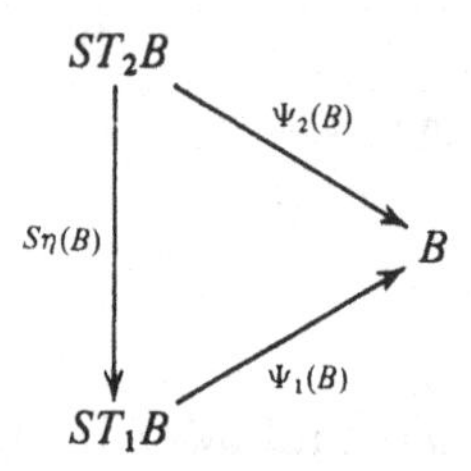


As soon as we have proved that $\eta$ is a morphism of functors, it has to be an isomorphism of functors and we are through. The reader will have no trouble in completing the proof.

**1.7.8 Theorem** Let $S$: $\mathbf{A} \to \mathbf{B}$ be a left adjoint of $T$: $\mathbf{B} \to \mathbf{A}$. Then the functor $T$ is fully faithful if and only if $\Psi$: $ST \to I_{\mathbf{B}}$ is an isomorphism.

*Proof.* We shall only prove one direction of the statement, leaving the converse as an exercise. Assume $T$ to be fully faithful. We want to show that $\Psi$ is an isomorphism of functors. Let $B$ be any object of $\mathbf{B}$. According to 1.7.4(i) we have $T\Psi(B) \circ \Phi(TB) = 1_{TB}$. Now $\Phi(TB)$: $TB \to TSTB$ and since $T$ is full there exists a morphism $g$: $B \to STB$ in $\mathbf{B}$ such that $Tg = \Phi(TB)$. But then $T\Psi(B) \circ Tg = 1_{TB} = T1_B$. Since $T$ is faithful $\Psi(B) \circ g = 1_B$. We also have

$$\begin{aligned}\phi \ulcorner TB, STB \urcorner (g \circ \Psi(B)) &= Tg \circ T\Psi(B) \circ \Phi(TB) \\ &= \Phi(TB) \circ T\psi(B) \circ \Phi(TB) \\ &= 1_{TB} \circ \Phi(TB) = \Phi(TB) \\ &= \phi \ulcorner TB, STB \urcorner (1_{STB})\end{aligned}$$

Since $\phi$ is an isomorphism of functors it follows that $g \circ \Psi(B) = 1_{STB}$.

Together with $\Psi(B) \circ g = 1_B$ this yields that $\Psi(B)$ is an isomorphism. This being the case of all objects $B$ we have the result that $\Psi$ is an isomorphism of functors.

Dually we have

**1.7.8° Theorem** Let $S$: $\mathbf{A} \to \mathbf{B}$ be a left adjoint of $T$: $\mathbf{B} \to \mathbf{A}$. Then the functor $S$ is fully faithful if and only if $\Phi$: $I_{\mathbf{A}} \to TS$ is an isomorphism.

**1.7.9** Let $T$: $\mathbf{B} \to \mathbf{A}$ be a functor and let $A \in \mathbf{A}$. If the covariant functor $F_A = h_A \circ T$ is representable, a representing couple $\ulcorner \bar{A}, e_A \urcorner$ is characterized by the following two properties:

(i) $\bar{A} \in \mathbf{B}$ and $e_A$: $A \to T\bar{A}$ in $\mathbf{A}$;

(ii) for every $B \in \mathbf{B}$ and $f$: $A \to TB$ in $\mathbf{A}$ there is exactly one $g$: $\bar{A} \to B$ in $\mathbf{B}$ such that $f = Tg \circ e_A$. In diagram:

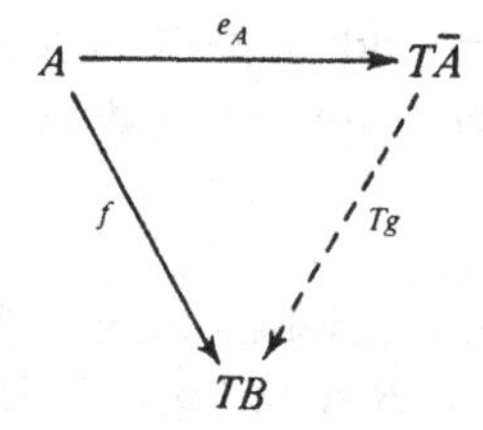


Suppose now that $T$ embeds $\mathbf{B}$ as a subcategory of $\mathbf{A}$. Conditions (i) and (ii) are then equivalent to:

(iii) given $A \in \mathbf{A}$ there is an object $\bar{A} \in \mathbf{B}$ and $e_A$: $A \to \bar{A}$ such that for every morphism $f$: $A \to B$, $B \in \mathbf{B}$, there is a unique $g$: $\bar{A} \to B$ in $\mathbf{B}$ such that $g \circ e_A = f$.

A couple $\ulcorner A, e_A \urcorner$ satisfying (iii) is called a *reflection* of $A$ in the subcategory $\mathbf{B}$.

The embedding $T$: $\mathbf{B} \to \mathbf{A}$ is full if and only if every object $B$ in the subcategory $\mathbf{B}$ is reflected by the couple $\ulcorner B, 1_B \urcorner$.

We say that $T$: $\mathbf{B} \to \mathbf{A}$ embeds $\mathbf{B}$ as a *reflective subcategory* of $\mathbf{A}$ if every $A \in \mathbf{A}$ has a reflection in $\mathbf{B}$.

According to theorem 1.7.7 and its proof this means $T$ has a left adjoint $S$ determined by putting $SA = \bar{A}$, $\Phi(A) = e_A$ for every $A \in \mathbf{A}$. Conversely, when the embedding $T$: $\mathbf{B} \to \mathbf{A}$ has a left adjoint, $\mathbf{B}$ must be a reflective subcategory of $\mathbf{A}$.

This frequently occurs for full embeddings $T$. Then $\Psi$: $ST \to I_{\mathbf{B}}$ is an isomorphism in virtue of theorem 1.7.8. For $B = TB \in \mathbf{B}$ one may choose the reflection $\ulcorner B, 1_B \urcorner$, leaving the couples $\ulcorner SA, \Phi(A) \urcorner$ unchanged for $A \notin \mathbf{B}$. By theorem 1.7.7 this determines a modified adjoint situation $S'$: $\mathbf{A} \to \mathbf{B}$, $\Phi'$: $I_{\mathbf{A}} \to TS$. By 1.7.5° (the dual of 1.7.5) we know that $S' \simeq S$. From the identity $T\Psi \circ \Phi' T = 1_T$ (proposition 1.7.4(i)) and the fact that $\Phi'(TB) = 1_{TB}$ we see that $T\Psi(B) = 1_{TB}$. Since $T$ is faithful this implies $\Psi(B) = 1_B$, i.e. $ST = I_{\mathbf{B}}$ and $\Psi = 1_{ST}$. From $T\Psi \circ \Phi T = 1_T$ it follows that also $\Phi T = 1_T$. From $\Psi S \circ S\Phi = 1_S$ (proposition 1.7.4(ii)) one now easily derives $S\Phi = 1_S$. On the other hand, $ST = I_{\mathbf{B}}$ certainly entails that $TB_1 = TB_2$ if and only if $B_1 = B_2 \in \mathbf{B}$. Combined with theorem 1.7.8 this shows

**1.7.10 Theorem** Let $T$: $\mathbf{B} \to \mathbf{A}$ have a left adjoint. Then $T$ embeds $\mathbf{B}$ as a full reflective subcategory of $\mathbf{A}$ if and only if $ST = I_{\mathbf{B}}$ and $\Psi = 1_{ST}$ for a certain left adjoint $S$: $\mathbf{A} \to \mathbf{B}$.

The following formulas hold in this case: $\Phi T = 1_T$ and $S\Phi = 1_S$; $\psi \ulcorner A, B \urcorner (f) = Sf$ for $f \in (A, TB)_{\mathbf{A}}$ (see table after proposition 1.7.4), and therefore $S\phi \ulcorner A, B \urcorner (g) = \psi \ulcorner A, B \urcorner \phi \ulcorner A, B \urcorner (g) = g$ for $g \in (SA, B)_{\mathbf{B}}$.

*Coreflections* are defined similarly. The embedding $T$: $\mathbf{B} \to \mathbf{A}$ is then required to have a right adjoint. The reader is invited to develop the appropriate formalism.

**1.7.10° Theorem** Let $S$: $\mathbf{A} \to \mathbf{B}$ have a right adjoint. Then $S$ embeds $\mathbf{A}$ as a full coreflective subcategory of $B$ if and only if $TS = I_A$ and $\Phi = 1_{TS}$ for a certain right adjoint $T$: $\mathbf{B} \to \mathbf{A}$.

The following formulas hold in this case: $\Psi S = 1_S$ and $T\Psi = 1_T$; $\phi \ulcorner A, B \urcorner (g) = Tg$ for $g \in (SA, B)_{\mathbf{B}}$ (see table after proposition 1.7.4), and therefore $T\psi \ulcorner A, B \urcorner (f) = \phi \ulcorner A, B \urcorner \psi \ulcorner A, B \urcorner (f) = f$ for $f \in (A, TB)_{\mathbf{A}}$.

A functor $S$: $\mathbf{A} \to \mathbf{B}$ is called an *equivalence* if there exists a functor $T$: $\mathbf{B} \to \mathbf{A}$ such that $TS \simeq I_{\mathbf{A}}$ and $ST \simeq I_{\mathbf{B}}$. The proof of the next result is left as an exercise.

**1.7.11 Theorem** Let $S$: $\mathbf{A} \to \mathbf{B}$ be a functor. The following conditions are equivalent:

(i) $S$ is an equivalence of categories.

(ii) There exists an adjoint situation $\ulcorner S, T, \Phi, \Psi \urcorner$ with $\Phi$ and $\Psi$ isomorphisms.

(iii) $S$ is full and faithful and every object $B$ of $\mathbf{B}$ is isomorphic to $SA$ for some object $A \in \mathbf{A}$.

The only point worth noting is that, in the proof that (iii) implies (ii), one puts $TB = A$ and shows that $T$ may be uniquely extended to a functor with the required properties.

If $S$ is an embedding, the result can be strengthened to read $\Psi = 1_{ST}$, as in 1.7.10. Equivalence between categories through contravariant functors are defined similarly and sometimes called *dualities*. An example is the functor $\Delta$ (see 1.2.4).

**Exercise** A full subcategory $\mathbf{B}$ of $\mathbf{A}$ is called *skeletal* or a *skeleton* if every object $A \in \mathbf{A}$ is isomorphic to exactly one $B \in \mathbf{B}$. Prove that the embedding $T: \mathbf{B} \to \mathbf{A}$ is an equivalence by constructing its pseudo-inverse $S: \mathbf{A} \to \mathbf{B}$ in the obvious and only possible manner. By pseudo-inverse is meant that $ST = I_{\mathbf{B}}$ and $TS \simeq I_{\mathbf{A}}$.

This fact allows one for many purposes to replace a category by one of its skeletons, which sometimes is small, while the original category is not.

As another application, consider a full subcategory $\mathbf{B}$ of $\mathbf{A}$, with embedding $T: \mathbf{B} \to \mathbf{A}$. Let $\mathbf{B}'$ be the full subcategory of $\mathbf{A}$ consisting of all objects $A \in \mathbf{A}$ which are isomorphic in $\mathbf{A}$ to an object $TB$, $B \in \mathbf{B}$. Then $\mathbf{B}'$ is equivalent with $\mathbf{B}$, since both are equivalent with one of their skeletons.

**1.7.12 Examples** The notion of adjoint functor, first made explicit in the classical paper of Kan [22], is basic in much of mathematics. The following examples, which range from the banal to the statement of important results, should convince the reader of its wide applicability. The advantage of the formalism is then to indicate, even in fairly complicated situations, which statements need proof and which follow automatically.

($a$) $\mathbf{A} = \mathbf{B} = \mathbf{Sets}$. Let $E$ be a fixed object of **Sets**. Define $SA = A \times E$ and define $T = h_E$. Section 1.7.1 showed that $S$ is a left adjoint of $T$.

($b$) Let $R$ be a commutative ring. $\mathbf{A} = \mathbf{B} = \mathbf{M}_R$. Choose a fixed object $M \in \mathbf{M}_R$. Let $SA = A \otimes_R M$ for $A \in \mathbf{A}$. For $f: A \to A'$ in $\mathbf{A}$, let $Sf = f \otimes_R 1$. Let $T = h_M$. Then $S$ is a left adjoint of $T$. If $R$ is not commutative we choose a fixed left module $M$ and we define $S$ and $T$ as above. But now we have $S: \mathbf{M}_R \to \mathbf{Ab}$ and $T: \mathbf{Ab} \to \mathbf{M}_R$. Again $S$ is a left adjoint of $T$. The maps are defined as in example ($a$), but it should of course be verified that they yield morphisms in the appropriate categories.

($c$) There is another variant of example ($a$) for topological spaces. Let $X$ be a fixed topological space. Then for any $B \in \mathbf{Top}$ the set $(X, B)$ may be equipped with the compact–open topology. This is the coarsest topology with the property that for any compact $C \subset X$ and any open $U \subset B$, the set $\{f \in (X, B) | f(C) \subset U\}$ is open. Then $(X, \text{-})_{\mathbf{Top}}$ is a functor from **Top** to **Top**, as the reader may check. For convenience we now consider $\mathbf{A} = \mathbf{B} = \mathbf{Hausd}$ and define for a fixed $X \in \mathbf{A}$ the functors $S = -\prod X$ and $T = (X, -)$. Both are functors from **A** to **A**. It is an exercise in point set topology to prove that, if $X$ is a compact space, $S$ is a left adjoint of $T$. One needs to show that the set maps involved are continuous. This still remains true if we only assume $X$ to be locally compact [21, ch. V, sect. 3.9].

(*d*) Let $\mathbf{A} = \mathbf{B}$ be a category with finite sums and finite products. Define $S: \mathbf{A} \to \mathbf{B}$ by $SA = A \coprod A$ and $Sf = f \coprod f$ (see 1.5.1) and define $T: \mathbf{B} \to \mathbf{A}$ by $TB = B \prod B$ and $Tg = g \prod g$. The reader may prove that $S$ is a left adjoint of $T$.

(*e*) $\mathbf{A} = \mathbf{Gr}$ and $\mathbf{B} = \mathbf{Ab}$, $T: \mathbf{B} \to \mathbf{A}$ is the embedding and $S: \mathbf{A} \to \mathbf{B}$ is given by $SG = G/[G, G]$ where $[G, G]$ is the commutator subgroup of $G$. Then $S$ is a left adjoint of $T$. The isomorphism $(SG, H)_{\mathbf{Ab}} \simeq (G, TH)_{\mathbf{Gr}}$ is clearly illustrated by the commutative diagram

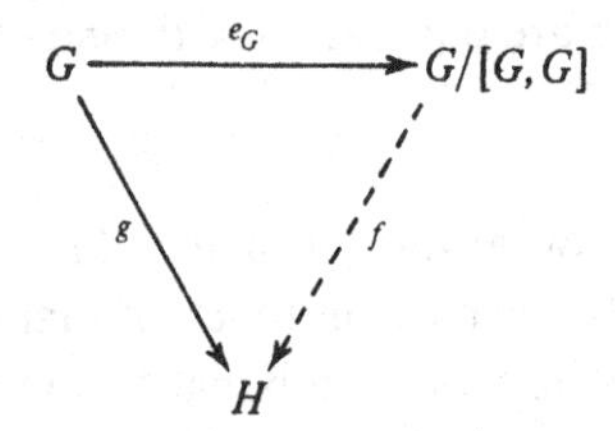


Every homomorphism from $G$ to an abelian group $H$ can be factored uniquely through $e_G$. Note that in fact we should have written

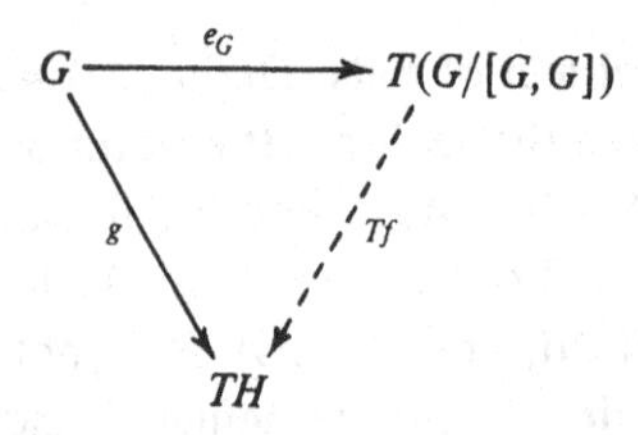


$T$ fully embeds the category of abelian groups as a reflective subcategory into the category of groups.

(*f*) Let **A** be the category of metric spaces whose morphisms are isometries. Let **B** be the category of complete metric spaces and isometries. Then $S: \mathbf{A} \to \mathbf{B}$ is defined by $SA =$ completion of $A$ and, for $f: A \to A_1$, $Sf$ is the unique isometry $Sf: SA \to SA_1$ whose restriction is $f$. Let $T: \mathbf{B} \to \mathbf{A}$ be the embedding. Then $S$ and $T$ are adjoint via $\phi \ulcorner A, B \urcorner = f|A$ (restriction of $f$). Again we have the commutative diagram

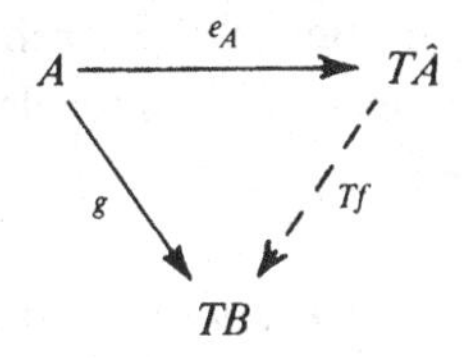


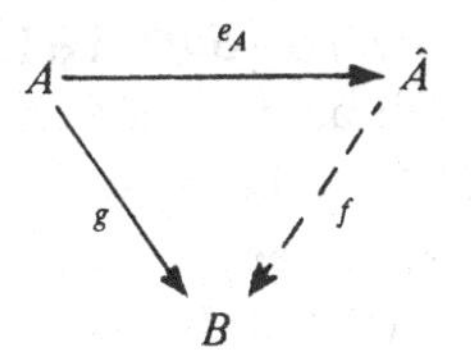


where $SA = \hat{A}$, the completion of $A$ in **B**.

In words: every isometry of $A \in \mathbf{A}$ into a complete metric space $B$ may be factored in a unique way through $e_A: A \to \hat{A}$, the embedding of $A$ into its completion. The first commutative diagram may be replaced by the second (just as in our previous example) due to the fact that $T$ is a full embedding of **B** as a reflective subcategory of **A**.

($g$) A *monoid* is a set with an associative multiplication for which there is a two-sided identity. Hence a group is a monoid for which every element has an inverse. A monoid is also called a *semigroup*. A *monoid-homomorphism* $f$ is a mapping with the property $f(ab) = f(a)f(b)$ that maps the identity onto the identity.

Let $\mathbf{B} = \mathbf{Rg}$ and let **A** be the category of monoids. For $A \in \mathbf{A}$ define $SA$ as the free abelian group with the elements of $A$ as basis and with multiplication defined by $(\sum_\alpha n_\alpha \alpha)(\sum_\beta m_\beta \beta) = \sum_{\alpha,\beta} m_\alpha n_\beta (\alpha\beta)$. The coefficients $n_\alpha$ and $m_\beta$ are in $\mathbb{Z}$ and $\alpha\beta$ denotes the product in $A$. We let only finitely many coefficients $n_\alpha$ and $m_\beta$ be different from null; clearly $SA$ is a ring with identity element $1 \,.\, 1$, the first 1 being in $\mathbb{Z}$, the second in $A$.

For $f: A \to A'$ in **A**, $Sf: SA \to SA'$ is defined by $Sf(\sum_\alpha n_\alpha \alpha) = \sum_\alpha n_\alpha (f\alpha)$. We define $T: \mathbf{B} \to \mathbf{A}$ as the embedding. Note that this is *not* a full embedding. Again, $S$ and $T$ are adjoint, as in the previous example, by $\phi \ulcorner A, B \urcorner: (SA, B) \to (A, TB)$ with $\phi \ulcorner A, B \urcorner (f) = f$. Again, **B** is a reflective subcategory of **A**. We may vary this example by taking for **A** the category **Gr** and then the adjoint situation expresses the universal property of the group ring.

($h$) $\mathbf{A} = \mathbf{Sets}_*$ and $\mathbf{B} = \mathbf{Gr}$. Let $T: \mathbf{Gr} \to \mathbf{Sets}_*$ be the forgetful functor where the neutral element of the group becomes the base-point of the underlying set. For $A \in \mathbf{A}$ we define $SA$ as the free group generated by the elements of $A$ different from $*$ (compare 1.4.8($g$)). We want to prove that $S$ and $T$ are adjoint using proposition 1.7.4.

For this purpose we define $\Phi$ and $\Psi$ in the following way. $\Phi: I_{\mathbf{Sets}_*} \to TS$; $\Phi(A)a = a$ for $a \in A$ and $a \neq *_A$. Check that $\Phi$ is a morphism of functors. $\Psi: ST \to I_{\mathbf{Gr}}$; $\Psi(B)$ assigns to each word in $STB$ the corresponding product in $B$. To the empty word is assigned the neutral element of $B$. Then $\Psi$ is a morphism of functors, $T\Psi \circ \Phi T = 1_T$

and $\Psi S \circ S\Phi = 1_S$. Here again, $T$ is not full but embeds **Gr** as a reflective subcategory of $\mathbf{Sets}_*$. The reader is invited to construct in a similar way left adjoints for the forgetful functors $\mathbf{Rg} \to \mathbf{Sets}_*$, $\mathbf{CRg} \to \mathbf{Sets}_*$ and $\mathbf{M}_R \to \mathbf{Sets}_*$.

(*i*) Let $\mathbf{B} = \mathbf{Ab}$. For $B \in \mathbf{B}$ let $TB$ be its *torsion subgroup*, i.e. the subgroup consisting of all elements of finite order. The embedding $S$ of the category of torsion groups into **Ab** is a left adjoint of $T$ making **Tors** a full coreflective subcategory of **Ab**. The category of torsionfree abelian groups is a full reflective subcategory of **Ab**.

(*j*) We now give an example of two contravariant functors which are adjoint on the right. Here **A** is the category of compact Hausdorff spaces with continuous mappings, **B** is the category of commutative complex Banach algebras with identity; its morphisms are the unitary algebra homomorphisms whose associated norms are $\leqslant 1$. The associated norm of $g: B \to B'$ is defined as $\sup \|g(b)\|$ for $b \in B$ and $\|b\| \leqslant 1$.

For $A \in \mathbf{A}$ let $SA$ be the space of all continuous functions on $A$ with complex values and with its norm the supremum norm defined by $\|f\| = \sup_{a \in A} |f(a)|$. For $f: A \to A'$ in $\mathbf{A}$ define $Sf: SA' \to SA$ by $(Sf)g = gf$ for $g \in SA'$. So $S$ may be considered as $h^{\mathbb{C}}$ with some extra structure on $(A, \mathbb{C})$. The map $Sf$ is a morphism from $SA'$ to $SA$ since it is an algebra-homomorphism and since $\|(Sf)g\| = \sup_{a \in A} |(Sf)g(a)| = \sup_{a \in A} |gf(a)| \leqslant \sup_{c \in C} |g(c)| = \|g\|$. Hence

$$\sup_{\substack{\|g\| \leqslant 1 \\ g \in SA'}} \|(Sf)g\| \leqslant \sup_{\substack{\|g\| \leqslant 1 \\ g \in SA'}} \|g\| \leqslant 1.$$

So $S$ is indeed a contravariant functor from **A** to **B**.

Now let $B \in \mathbf{B}$. For any $b \in B$ define

$$C_b = \{z \in \mathbb{C} \mid |z| \leqslant \|b\|\}.$$

Consider the topological product $X = \prod_{b \in B} C_b$. Then an element of this product is a complex function $f$ on $B$ with the special property that for all $b \in B$ one has $|f(b)| \leqslant \|b\|$. We now define $TB$ to be the subspace of $X$ consisting of all those functions on $B$ that are algebra-homomorphisms. In other words $TB$ is the subspace consisting of all those functions from $B$ to $\mathbb{C}$ that are morphisms in the category **B** (note that the associated norm of every function in $X$ is automatically $\leqslant 1$). We contend that $TB$ is closed in $X$.

To see this let $a, b \in B$. Define $h: X \to \mathbb{C}$ by $h(f) = f(a)f(b) - f(ab)$. This function is continuous. Then $h^{-1}(0)$ is closed and so $\{f \in X \mid f(ab) = f(a)f(b)\}$ is closed. So every relation to be satisfied by $f$ (in order to become an algebra-homomorphism) yields a closed subspace of $X$ and $TB$ is the intersection of all these closed subspaces. Hence $TB$ is closed. The space $X$ is compact and Hausdorff since it is the topological product of such spaces. So $TB$ is a compact Hausdorff space. For $g: B \to B'$ in $\mathbf{B}$ and $f \in TB'$ we define $(Tg)(f) = fg$. Then $Tg$ is a continuous map from $TB'$ to $TB$. This contravariant functor $T$ from $\mathbf{B}$ to $\mathbf{A}$ is called the *Gelfand functor*. It equals the contravariant functor $h^{\mathbb{C}}$ but has some more structure for each set $(B, \mathbb{C})_{\mathbf{B}}$. Next we define a morphism of functors $\Phi: I_{\mathbf{A}} \to TS$ by $\Phi(A)a = a^*$ and $a^*(b) = b(a)$ for $A \in \mathbf{A}$, $a \in A$ and $b \in SA$. Also $\Psi: ST \to I_{\mathbf{B}}$ by $\Psi(B)b = \hat{b}$ and $\hat{b}(f) = f(b)$ for $B \in \mathbf{B}$, $b \in B$, $f \in TB$.

The reader should check that $\Phi$ and $\Psi$ are indeed morphisms of functors. It follows that for $B \in \mathbf{B}$, $b \in B$, $f \in TB$ we have

$$T\Psi(B) \circ \Phi(TB)f(b) = (\Phi(TB)f \circ \Psi(B))(b) = f^*(\hat{b}) = \hat{b}(f) = f(b).$$

Hence $(T\Psi) \circ (\Phi T) = 1_T$.

Similarly it turns out that $(S\Phi) \circ (\Psi S) = 1_S$. From this it follows that $S$ and $T$ are right adjoint.

Remark. $\Phi$ is even an isomorphism of functors since for all $A \in \mathbf{A}$ the morphism $\Phi(A)$ is an isomorphism. This can be seen as follows. $A$ is normal, since it is a compact Hausdorff space. Using the lemma of Urysohn it follows that $\Phi(A)$ is injective. One may also prove that its image $\Phi(A)A$ is dense in $TSA$, for, if not, there must be $\rho \in TSA$ together with a neighbourhood $U$ such that $U \cap \Phi(A)A = \emptyset$. We may choose $U$ so that

$$U = \{\sigma \mid |\sigma(g_i) - (g_i)| < \varepsilon \text{ for } i = 1, \ldots, n\},$$

with $g_i \in SA$. Since $\rho$ is a homomorphism, we have $\rho(g^{-1}) = \rho(g)^{-1}$ for every $g \in SA$ that is invertible in $SA$. Choose

$$g = \sum_{i=1}^{n} (g_i - \rho(g_i))\overline{(g_i - \rho(g_i))} = \sum_{i=1}^{n} |g_i - \rho(g_i)|^2.$$

Then $\rho(g) = 0$ so that $g$ is not invertible in $SA$. Then there must be an element $a \in A$ with $|g(a)| < \varepsilon^2$. From

$$g(a) = \sum_{i=1}^{n} |g_i(a) - \rho(g_i)|^2$$

it follows that

$$|a^*(g_i) - \rho(g_i)| = |g_i(a) - \rho(g_i)| < \varepsilon$$

for $i = 1, \ldots, n$. Hence $a^* \in U$, contradicting the fact that $a^* \in \Phi(A)A$.

Thus we may say that $\Phi(A)$ is an injective map onto a dense subspace of $TSA$. Since $A$ is compact and $\Phi(A)$ continuous, $\Phi(A)A$ is compact. Hence $\Phi(A)A$ is closed, so that $\Phi(A)A = TSA$. This means that $\Phi(A)$ is a bijective continuous mapping of a compact space into a Hausdorff space so that it is a homeomorphism. One concludes that $\Phi(A)$: $A \rightrightarrows TSA$ in $\mathbf{A}$. It is customary in analysis to denote $\Phi$ by * and $\Psi$ by ^.

($k$) The previous example may be varied in the following way. A *Tychonoff space* is a completely regular $T_1$ space. A space $X$ is called *completely regular* provided that for every closed subset $A \subset X$ and for every point $a \notin A$ there exists a continuous function $f$ from $X$ to the closed segment $[0, 1]$ with $f(a) = 0$ and $f(A) = \{1\}$ [21, p. 57]. A compact Tychonoff space is a compact Hausdorff space. Conversely, a compact Hausdorff space is normal and so, by the Urysohn lemma, a compact Tychonoff space [21, p. 60].

Let $\mathbf{A}$ be the category of all Tychonoff spaces with continuous maps. Let $\mathbf{B}$ be as in the previous example. Let $\mathbf{C}$ be the category of compact Tychonoff spaces, which, then, is the same as the category of all compact Hausdorff spaces.

For $A \in \mathbf{A}$ let $SA$ be the space of all continuous bounded functions on $A$ with complex values, normed by the supremum norm. For $f$: $A \to A'$ in $\mathbf{A}$, $g \in SA'$, let $(Sf)g = gf$. Then we have a contravariant functor $S$: $\mathbf{A} \to \mathbf{B}$ while $T$: $\mathbf{B} \to \mathbf{C}$ is defined as above. Furthermore let $H$: $\mathbf{C} \to \mathbf{A}$ be the forgetful functor, which is full. Then the covariant functor $TS$ is a left adjoint of $H$ as we will now show. Define $\Phi'$: $I_{\mathbf{A}} \to HTS$ by $\Phi'(A)a = a^*$, $a^*(b) = b(a)$ for $A \in \mathbf{A}$, $a \in A$, $b \in SA$. We still have the isomorphism of functors $\Phi$: $I_{\mathbf{C}} \to TSH$, since $SH$ is our $S$ as above. Now $\Phi'$ and $\Phi$ coincide on $\mathbf{C}$ so that $\Phi' H = H\Phi$. Define $\Psi$: $TSH \to I_{\mathbf{C}}$ as the inverse of $\Phi$. For $A \in \mathbf{A}$ we then have the following commutative diagram:

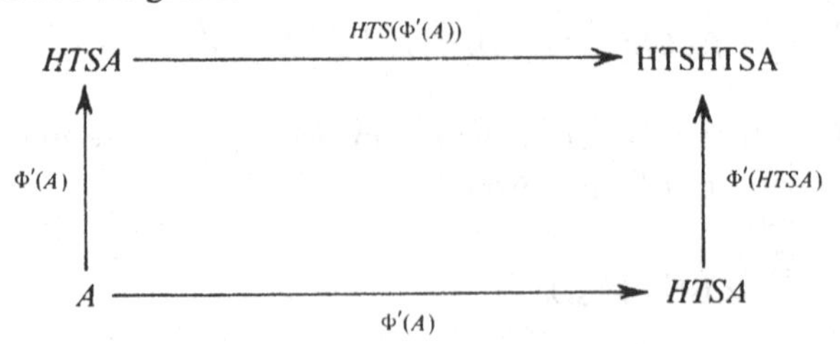


One proves just as before that $\Phi'(A)A$ is dense in $HTSA$. Since $HTS(\Phi'(A))$ and $\Phi'(HTSA)$ coincide on $\Phi'(A)A$ we have $HTS(\Phi'(A)) = \Phi'(HTSA)$. So $H\Psi(TSA) \circ HTS\Phi'(A) = H\Psi(TSA) \circ \Phi'(HTSA) = H\Psi(TSA) \circ H(\Phi TSA) = H1_{TSA}$. But $H$ is the forgetful functor, so

$\Psi TS \circ TS\Phi = 1_{TS}$. The reader may convince himself that $H\Psi \circ \Phi' H = 1_H$.

Another left adjoint of $H$ may be found as follows. Take $K = [0, 1]$. For $A \in \mathbf{A}$ let

$$X(A) = \prod_{f \in (A,K)_{\mathbf{A}}} K.$$

Denote the projections by $\rho_f$ and let $e(A)$: $A \to X(A)$ be the continuous mapping defined by $\rho_f(e(A)a) = f(a)$. Denote the closure of $e(A)A$ in $X$ by $FA$. For $g$: $A \to A'$ in $\mathbf{A}$, define $Xg$: $X(A) \to X(A')$ by $\rho_f((Xg)\vartheta) = \rho_{fg}(\vartheta)$ for $\vartheta \in X(A)$, $f \in (A', K)_{\mathbf{A}}$. Then $Xg$ is continuous and the following diagram commutes:

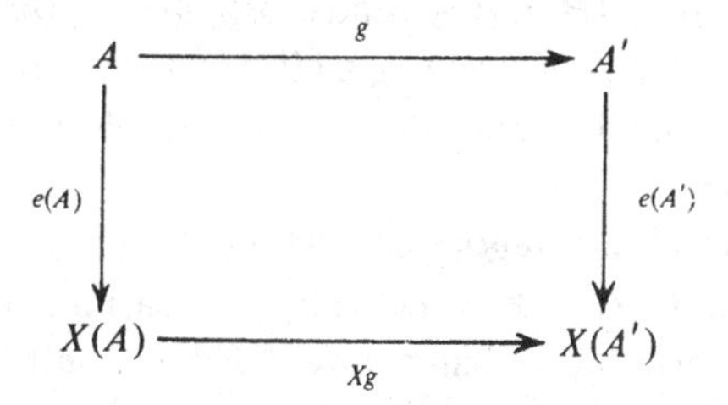


Hence the counter image of $FA'$ under $Xg$ contains $e(A)A$ and is closed; then it must contain $FA$. We define $Fg$: $FA \to FA'$ as the restriction of $Xg$ to $FA$. Then $F$ is a covariant functor from $\mathbf{A}$ to $\mathbf{C}$ and $e$: $I_{\mathbf{A}} \to HF$ a morphism of functors. The compact Hausdorff space $FA$ is called the Stone–Čech compactification of $A$.

The reader should check that the map $e(A)$ for $A \in \mathbf{A}$ yields a homeomorphism between $A$ and $e(A)A$. If $A$ is already compact, then $e(A)A = FA$. For $A \in \mathbf{A}$ and $C \in \mathbf{C}$ we have a bijection from $(A, HC)_{\mathbf{A}}$ to $(FA, C)_{\mathbf{C}}$ assigning $e(C)^{-1} \circ Ff$ to the element $f \in (A, HC)_{\mathbf{A}}$:

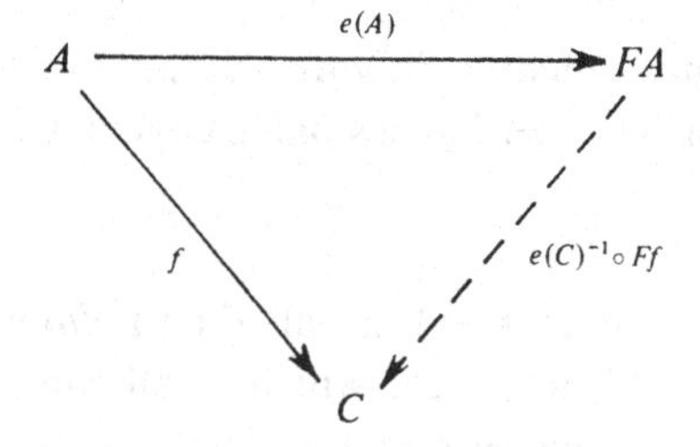


Thus both $F$ and $TS$ are left adjoints of $H$. Then they must be isomorphic. Apparently $FA$ and $TSA$ are isomorphic compactifications and $H$ fully embeds the category $\mathbf{C}$ of compact Hausdorff spaces reflectively into the category of Tychonoff spaces. The moral is that the usual

Stone–Čech compactification can also be obtained by utilizing the Gelfand functor [21, p. 105].

(*l*) Let $u: R \to S$ be a homomorphism of rings. Put $\mathbf{A} = \mathbf{M}_R$ and $\mathbf{B} = \mathbf{M}_S$. Write $G: \mathbf{B} \to \mathbf{A}$ for the functor which restricts scalars. This means $GB$ is $B$ considered as $R$-module by defining $b\rho = bu(\rho)$ for $b \in B \in \mathbf{B}$ and $\rho \in R$. Then $G$ has a left adjoint $F$ determined by $FA = A \otimes_R S$, $A \in \mathbf{A}$ ($F$ extends scalars). We shall content ourselves with giving $\phi: (FA, B)_\mathbf{B} \to (A, GB)_\mathbf{A}$ and its inverse $\psi$, leaving further verification to the reader. For $g: FA \to B$ put $\phi g(a) = g(a \otimes 1) \in GB$ and for $f: A \to GB$ take $\psi f(a \otimes \sigma) = f(a)\sigma$, $a \in A$, $\sigma \in S$.

The functor $G$ also has a right adjoint $H$ defined by $HA = (S, A)_\mathbf{A}$ where the right module structure is given by $h\sigma(\tau) = h(\sigma\tau)$, $h: S \to A$ in $\mathbf{A}$; $\sigma, \tau \in S$. Now $\phi: (GB, A)_\mathbf{A} \to (B, HA)_\mathbf{B}$ reads $\phi f(b)(\sigma) = f(b\sigma)$, $f \in (GB, A)$, $b \in B$, $\sigma \in S$ and $\psi g(b) = g(b)(1)$, $g \in (B, HA)_\mathbf{B}$. Again it is instructive to check the details. The functors in this example are often referred to as 'change of rings'.

(*m*) Let $\mathbf{C}$ be a category and $\mathbf{P}$ a category consisting of a single object and its identity morphism. If $F: \mathbf{C} \to \mathbf{P}$ is the unique covariant functor then $\mathbf{C}$ has an initial (terminal) object if and only if $F$ has a left (right) adjoint; cf. 1.4.8 (*i*).

### 1.8 Suprema and infima

In section 1.6 we defined limits, forming the category $\mathbf{I}$ associated with a directed preordered set $I$. It is desirable to consider a more general situation, assuming only that $\mathbf{I}$ is a small category. We develop a similar theory as before to obtain the concepts of supremum and infimum. These then appear as generalizations of the concepts of direct and inverse limits and even of sums and products (fibred or not) and are therefore sometimes also referred to as limits. We shall see that the theory of suprema and infima has much to do with adjoint functors as discussed in the previous section. Thus let $\mathbf{I}$ be a small category and $\mathbf{C}$ an arbitrary one.

**1.8.1 Definitions** A functor $D: \mathbf{I} \to \mathbf{C}$ is called an $\mathbf{I}$-*diagram in* $\mathbf{C}$, sometimes just a *diagram in* $\mathbf{C}$ if it is clear which small category is meant. We denote by $(\mathbf{I}, \mathbf{C})$ the category of all $\mathbf{I}$-diagrams in $\mathbf{C}$ (see 1.3.3). Let $M$ be the functor from $\mathbf{C}$ to $(\mathbf{I}, \mathbf{C})$ defined as follows. The object $MC$, for $C \in \mathbf{C}$, is the constant diagram $MCi = C$ for all objects $i \in \mathbf{I}$ and $MC\alpha = 1_C$ for $\alpha: i \to j$ in $\mathbf{I}$. For $u: C \to C'$ in $\mathbf{C}$ let $Mu$ be the morphism of functors defined by $Mu(i) = u$ for all $i \in \mathbf{I}$. Let $D$ be a fixed diagram in $(\mathbf{I}, \mathbf{C})$ and define the functor $F_D: \mathbf{C} \to \mathbf{Sets}$ by $F_D C = (D, MC)_{(\mathbf{I},\mathbf{C})}$ and for $u: C \to C'$, $F_D u: (D, MC)_{(\mathbf{I},\mathbf{C})} \to (D, MC')_{(\mathbf{I},\mathbf{C})}$ by $F_D u(c) = Mu \circ c$ where $c \in F_D C$.

All this may be visualized in the commutative triangles depicting an element $c \in F_D C$:

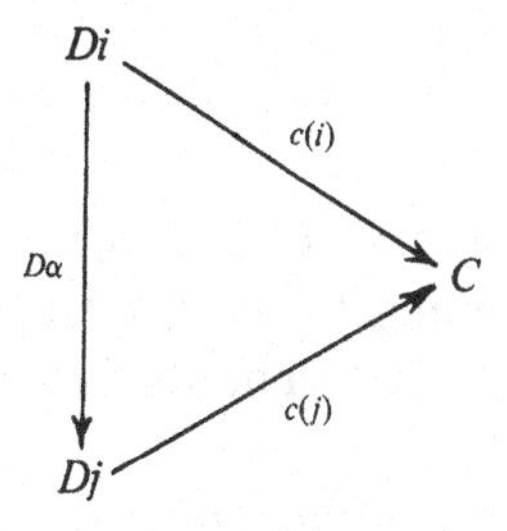


To obtain the element $F_D u(c)$ one only has to compose the above triangles with $C \xrightarrow{u} C'$, or, in other words, replace $C$ by $C'$ and all $c(i)$ by $u \circ c(i)$.

For any diagram $D$, for $C \in \mathbf{C}$ and $c \in F_D C$, the pair $\ulcorner C, c \urcorner$ is called an *upper bound* for the diagram $D$. If the functor $F_D$ is represented by $\ulcorner L, l \urcorner$, this last couple is called a *supremum* (*least upper bound*) for $D$.

If $F_D$ is represented by the couple $\ulcorner L, l \urcorner$, then $L \in \mathbf{C}$ and $l \in F_D L$; i.e. $l: D \to ML$ in $(\mathbf{I}, \mathbf{C})$. As we have seen before this means that for every object $C \in \mathbf{C}$ and for every element $c \in F_D C$ there is exactly one morphism $u: L \to C$ such that in the following picture all the triangles are commutative:

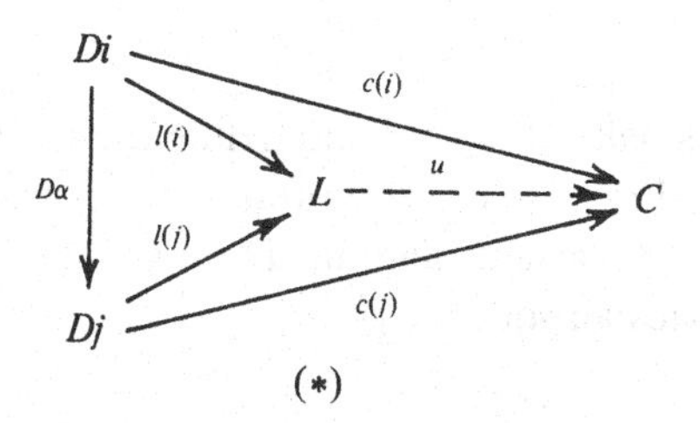


(*)

Suppose every $\mathbf{I}$-diagram $D$ is representable. Call the representing couple $\ulcorner LD, \Phi(D) \urcorner$. By 1.7.7 $L: (\mathbf{I}, \mathbf{C}) \to \mathbf{C}$ then is a functor which is a left adjoint of $M$. Therefore $(L\text{-}, \text{-})_{\mathbf{C}} \simeq (\text{-}, M\text{-})_{(\mathbf{I},\mathbf{C})}$ and in particular $(LD, C)_{\mathbf{C}} \simeq (D, MC)_{(\mathbf{I},\mathbf{C})}$.

Dually an infimum (greatest lower bound) is defined. For this purpose let $D$ now be a fixed contravariant diagram, i.e. $D \in (\mathbf{I}, \mathbf{C}^{\circ})$. The functor $M$ is defined as before, but now taken to be contravariant. The contravariant functor $F^D: \mathbf{C} \to \mathbf{Sets}$ is defined by $F^D C = (MC, D)_{(\mathbf{I},\mathbf{C}^{\circ})}$.

An element $c \in F^D C$ may then be depicted (for $i \leq j$) by

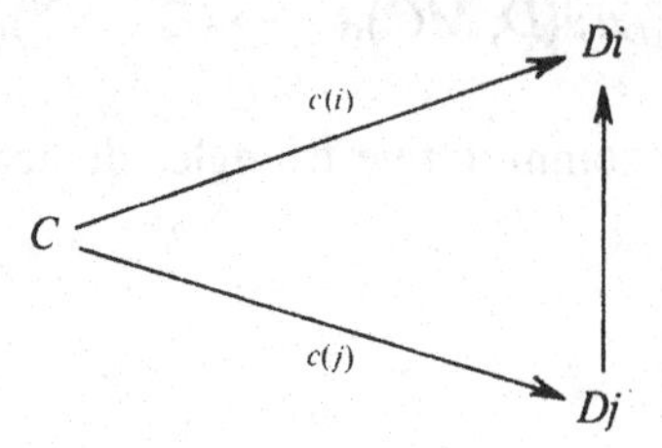


Such a couple $\ulcorner C, c \urcorner$ with $C \in \mathbf{C}$ and $c \in F^D C = (MC, D)_{(\mathbf{I}, \mathbf{C}^\circ)}$ is called a *lower bound* for $D$. A couple $\ulcorner L, l \urcorner$ representing $F^D$ is called a *greatest lower bound* or *infimum* for the diagram $D$. This amounts to reversing all the arrows in the diagram (*) above.

Now every $D$ has an infimum if and only if the functor $M: \mathbf{C} \to (\mathbf{I}, \mathbf{C}^\circ)$ has a right adjoint $L: (\mathbf{I}, \mathbf{C}^\circ) \to \mathbf{C}$. Thus $(\text{-}, L\text{-})_{\mathbf{C}} \simeq (M\text{-}, \text{-})_{(\mathbf{I}, \mathbf{C}^\circ)}$, in particular $(C, LD)_{\mathbf{C}} \simeq (MC, D)_{(\mathbf{I}, \mathbf{C}^\circ)}$.

At times, we also consider the infimum of a covariant functor $D: \mathbf{I} \to \mathbf{C}$. This is then the dual of a supremum in $\mathbf{C}^\circ$.

**1.8.2 Examples** (*a*) Consider the following objects and morphisms in $\mathbf{C}$:

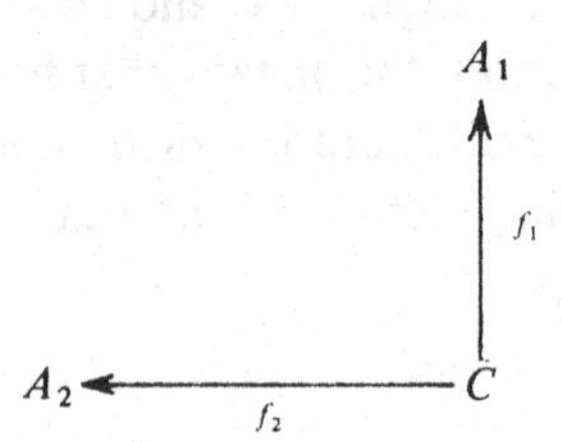


Make this configuration into a small category $\mathbf{I}$ and define the covariant functor $D: \mathbf{I} \to \mathbf{C}$ as the embedding. Then the supremum (if it exists in $\mathbf{C}$!) $\ulcorner A, q_1, q_2, q_1 f_1, q_2 f_2 \urcorner$ of the covariant diagram $D$ is just the fibred sum $A = A_1 \coprod_C A_2$ in the cocartesian square

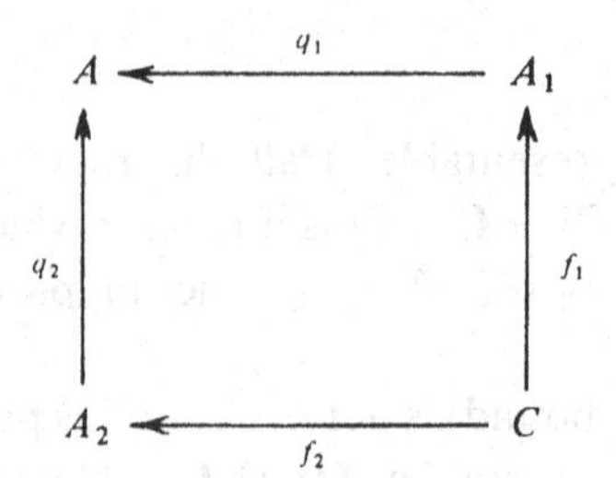


Dually one recognizes the fibred product $A = A_1 \prod_C A_2$ as the infimum of the diagram

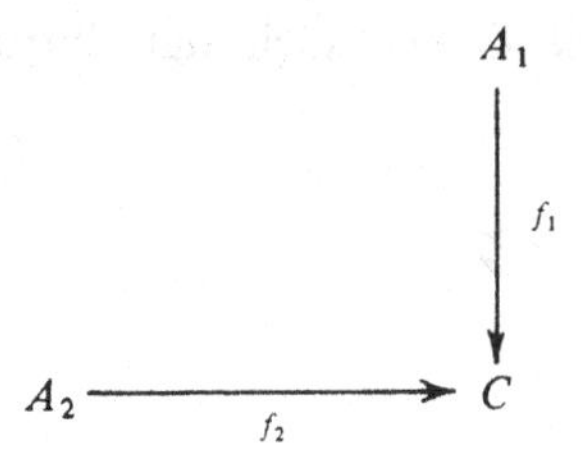


(*b*) Let $\{C_i \in \mathbf{C} | i \in I\}$ be a collection of objects in $\mathbf{C}$ indexed by a set $I$. Make $I$ into a category $\mathbf{I}$ with only identity morphisms. Suppose the product $P = \prod_{i \in I} C_i$ exists in $\mathbf{C}$.

Denote the projections by $p_i: P \to C_i$. Define $D: \mathbf{I} \to \mathbf{C}$ by $Di = C_i$ and $D1_i = 1_{C_i}$, for $i \in I$. Consider $D$ as a contravariant diagram. Then $\ulcorner P, \{p_i\} \urcorner$ is an infimum for $D$.

Take $D$ as a covariant diagram and assume $S = \coprod_{i \in I} C_i$ exists in $\mathbf{C}$; denote the injections by $q_i: C_i \to C$. Then $\ulcorner S, (q_i) \urcorner$ is a supremum for $D$.

(*c*) A rather different example of a supremum concerns contravariant functors from a small category $\mathbf{C}$ to **Sets**. To such a functor $F$ we associate a diagram in $(\mathbf{C}^\circ, \mathbf{Sets})$ by looking at all natural transformations $\alpha: h^C \to F, C \in \mathbf{C}$. More precisely, let $\mathbf{I}$ be the category consisting of objects $\ulcorner C, \alpha \urcorner$ with this property. As morphisms we take $v: \ulcorner C, \alpha \urcorner \to \ulcorner C', \alpha \urcorner$ with $v: C \to C'$ such that $\alpha' \circ h^v = \alpha$. This category is small since $\mathbf{C}$ is. A diagram $D: \mathbf{I} \to (\mathbf{C}^\circ, \mathbf{Sets})$ is determined by putting $D \ulcorner C, \alpha \urcorner = h^C$ and $Dv = h^v$. Then $F$ is the supremum of the diagram $D$. This is a consequence of the Yoneda lemma. Thus every functor $F: \mathbf{C}^\circ \to \mathbf{Sets}$ is in a canonical way a supremum of representable functors. Some authors say that the full subcategory of representable functors is dense in $(\mathbf{C}^\circ, \mathbf{Sets})$.

**1.8.3 Equalizers** Let $I = \{0, 1\}$ and let $\mathbf{I}$ be the category defined by

$$(0, 0) = \{1_0\}, (1, 1) = \{1_1\}, (0, 1) = \{\alpha | \alpha \in J\}, (1, 0) = \varnothing$$

with $J$ some nonvoid set. In other words all morphisms run from 0 to 1, or are identity morphisms. Let $\mathbf{C}$ be a category and $D: \mathbf{I} \to \mathbf{C}$ a diagram defined by $D0 = A$, $D1 = B$ and $D\alpha = d_\alpha: A \to B$. Suppose $\ulcorner L, l \urcorner$ is an infimum for $D$, i.e. $L \in \mathbf{C}$ and $l = \ulcorner l(0), l(1) \urcorner$ with $l(0): L \to A$ and $l(1): L \to B$.

The couple $\ulcorner L, l(0) \urcorner$ is called an *equalizer for the diagram D* or for the system $\{d_\alpha | \alpha \in J\}$.

This therefore means that for all $\alpha \in J$ the right hand triangle in the following diagram commutes, whereas if the big triangle commutes for all $\alpha \in J$ there is exactly one morphism $u: C \to L$ such that all triangles commute:

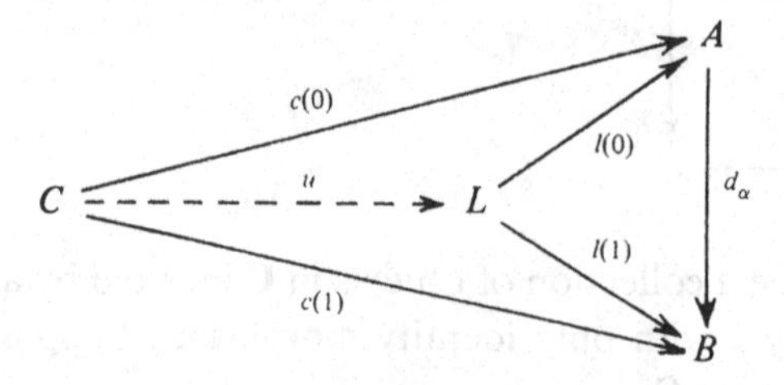


The dual notion is called a *coequalizer.*

If in the sequel the term equalizer (or coequalizer) is used without further reference, we always have in mind diagrams of type

$$A \overset{f}{\underset{g}{\rightrightarrows}} B$$

where the set $J$ has two elements.

**1.8.4** We say that **C** is a *category with sums* provided every collection of objects in **C** has a sum in **C**. If this is only true for every finite set of objects in **C** we say that **C** is a *category with finite sums.* A similar terminology is used for products.

Let **C** be such a category with sums and suppose we have a diagram $D: \mathbf{I} \to \mathbf{C}$ (with **I** a small category, as always). Then we can form the sum $\coprod_{i \in I} Di$ in **C**. This is, so to speak, the 'disjoint union' not taking into account the morphisms $D\alpha: Di \to Dj$ which might occur (due to $\alpha: i \to j$ in **I**). We wonder whether we can obtain another object in **C** that does take into account the identifications due to the above morphisms. What we thus are really looking for is a fibred sum, or supremum and not merely a sum: the sum is just an upper bound (obviously), but we want a least upper bound (supremum). A category **C** with the property that for every small category **I** every diagram $D; \mathbf{I} \to \mathbf{C}$ has a supremum is called *sup-complete* or *right complete* (similarly *inf-complete* or *left complete* is defined). *Complete* means left and right complete.

Saying that a category **C** is right complete is tantamount to saying that for every small category **I** the functor $M: \mathbf{C} \to (\mathbf{I}, \mathbf{C})$ has a left adjoint (see 1.8.1).

**Theorem** A category **C** with sums and with the property that every diagram $f, g: A \rightrightarrows B$ has a coequalizer is right complete.

*Proof.* Let $D: \mathbf{I} \to \mathbf{C}$ be a diagram. We divide the proof into two steps.

Step 1. Put $B = \coprod_{i \in I} Di$ with injections $q_i: Di \to B$. Attach to $D$ a new diagram

$$E: Di \overset{q_i}{\underset{q_j D\alpha}{\rightrightarrows}} B$$

where $i \in \mathbf{I}$ and $\alpha \in J(i)$, which is the set of all morphisms $\alpha: i \to j$ in $\mathbf{I}$ for $j \in \mathbf{I}$. Thus $J(i)$ is the set of morphisms originating in $i$, and $E$ consists of a set of pairs of morphisms into $B$, indexed by the set $\bigcup_{i \in \mathbf{I}} J(i)$. We first claim that if $h: B \to C$ is an upper bound for $E$, then $\ulcorner C, c \urcorner$, with $c(i) = hq_i$, is an upper bound for $D$. Indeed, $c(i) = hq_i = hq_j D\alpha = c(j)D\alpha$ for all $\alpha: i \to j$. On the other hand, if $\ulcorner C, c \urcorner$ is an upper bound of $D$, there is a unique morphism $h: B \to C$ such that $hq_i = c(i)$ for all $i \in \mathbf{I}$ and it follows that $h$ is an upper bound for the diagram $E$. Consequently it suffices to construct a supremum for $E$.

Step 2. Put $A = \coprod_\alpha D\ulcorner i, \alpha \urcorner$, where the sum is taken over $\alpha \in \bigcup_{i \in \mathbf{I}} J(i)$ and where $D\ulcorner i, \alpha \urcorner = Di$. Denote the injections by $q_{i\alpha}: D\ulcorner i, \alpha \urcorner \to A$. Let $f, g: A \rightrightarrows B$ be the unique morphisms such that $fq_{i\alpha} = q_i$ and $gq_{i\alpha} = q_j D\alpha$ for all $i \in \mathbf{I}$ and $\alpha \in J(i)$.

Thus for all these pairs $\ulcorner i, \alpha \urcorner$ the two triangles

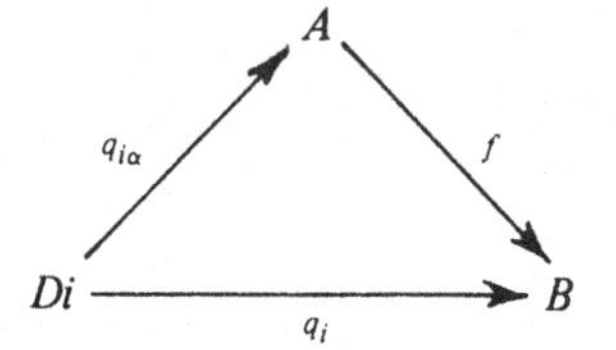


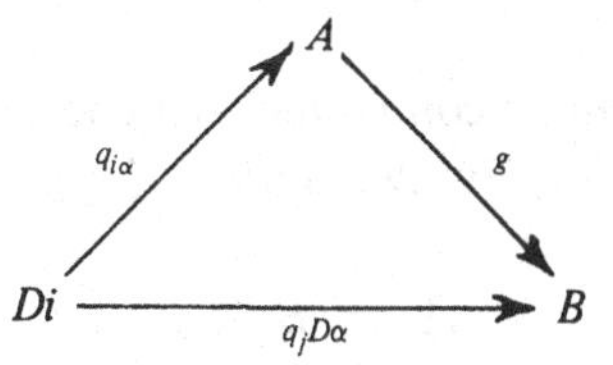


commute. Obviously if $c: B \to C$ is an upper bound of the pair $f, g: A \rightrightarrows B$, it is an upper bound of the diagram $E$. The converse follows because the morphisms $cf$ and $cg$ from the direct sum $A$ are uniquely determined. By assumption, there exists a coequalizer $l: B \to L$ for the pair of morphisms $f$ and $g$. It follows from step 1 that $\ulcorner L, l \urcorner$ with $l(i) = lq_i$ is a supremum for the diagram $D$.

Note. If $\mathbf{C}$ is a category with only finite sums one proves along the same lines as above that $\mathbf{C}$ is finitely right complete.

**Exercises** ($a$) Suppose we have the following diagram in a category with finite sums and coequalizers:

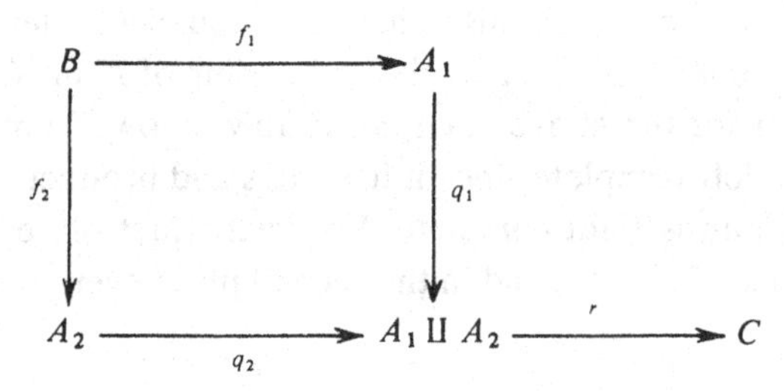

Prove that $r$ is a coequalizer for $q_1f_1$ and $q_2f_2$ if and only if the diagram

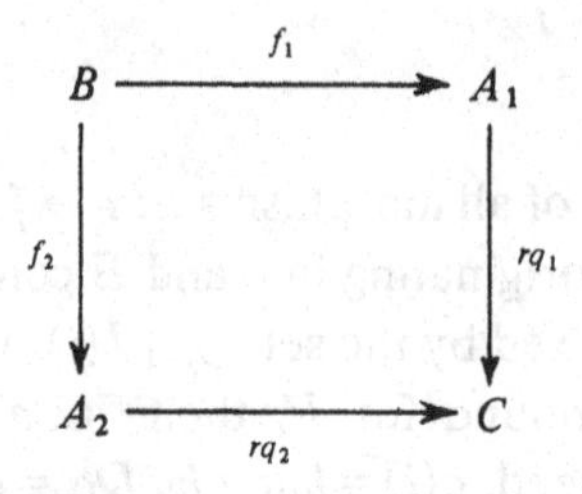


is cocartesian.

(*b*) Let **C** be a category with finite sums. Prove that the following assertions are equivalent.

(i) Every diagram of type

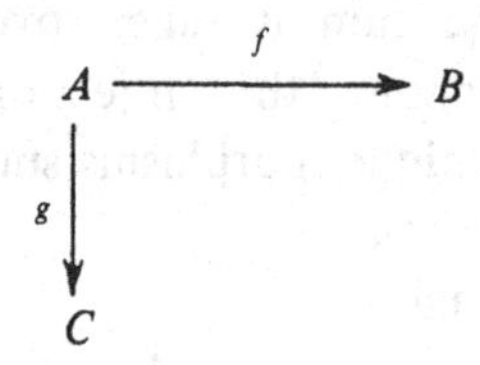


can be completed to a cocartesian square.

(ii) Every diagram of type

$$A \overset{f}{\underset{g}{\rightrightarrows}} B$$

has a coequalizer.

The dual statements of the results in this section are readily formulated.

**1.8.5 Examples** We will now use the theorem just proved as a means of recognizing whether a category is right or left complete.

(*a*) In the category **Sets** every diagram $f, g: X \rightrightarrows Y$ has a coequalizer. Consider to this end the equivalence relation $\equiv$ on $Y$ generated by the relation $f(x) \sim g(x)$. Let $Q$ be the set $Y/\equiv$ with the canonical map $q: Y \to Q$. Then $\ulcorner Q, q \urcorner$ is a coequalizer for the diagram $f, g: X \rightrightarrows Y$.

To prove that the diagram also has an equalizer, let $K = \{x \in X | f(x) = g(x)\}$ and let $I: K \to X$ be the embedding of $K$ in $X$. Then $\ulcorner K, i \urcorner$ is an equalizer for the above diagram. It now follows from 1.8.4 that **Sets** is right and left complete since it has sums and products.

(*b*) The category **Top** is right complete. For in the first place it is a category with sums (see 1.5.3(*b*)) and in the second place every diagram

$f, g: X \rightrightarrows Y$ has a coequalizer, constructed as above. The topology is introduced by defining $U \subset Q$ to be open if and only if $q^{-1}(U)$ is open in $Y$. Then it is easy to show that $\ulcorner Q, q \urcorner$ is a coequalizer of the diagram. It follows just as easily that **Top** is left complete.

($c$) The category **Gr** has sums (see 1.5.3($e$)). Consider a diagram $f, g: X \rightrightarrows Y$ in this category. Let $H$ be the normal subgroup of $Y$ generated by all elements $f(x)g(x^{-1})$ for $x \in X$. Set $Q = Y/H$ and let $q: Y \to Q$ be the canonical morphism. Then $\ulcorner Q, q \urcorner$ is a coequalizer for the diagram which shows that **Gr** is right complete. Verify that it is also left complete.

($d$) The category **Rg** of rings has products. For a diagram $f, g: X \rightrightarrows Y$ consider the set $K = \{x \in X | f(x) = g(x)\}$ as in example ($a$). It is a subring of $X$. The injection $i: K \to X$ makes $\ulcorner K, i \urcorner$ the equalizer for the above diagram. Thus **Rg** is left complete.

**1.8.6 Theorem (Kan extension)** Let $I: \mathbf{A} \to \mathbf{B}$ be a covariant functor between small categories. Let $\mathbf{C}$ be a right complete category. Then $T: (\mathbf{B}, \mathbf{C}) \to (\mathbf{A}, \mathbf{C})$, defined by $TG = GI$ for $G \in (\mathbf{B}, \mathbf{C})$ and $T\eta(A) = \eta(IA)$ for $\eta \in (G_1, G_2)_{(\mathbf{B},\mathbf{C})}$, $A \in \mathbf{A}$, has a left adjoint $S: (\mathbf{A}, \mathbf{C}) \to (\mathbf{B}, \mathbf{C})$. For $F \in (\mathbf{A}, \mathbf{C})$ we call $SF$ the *left Kan extension of F along I.*

Remark. For the construction to be defined below it is not necessary that $\mathbf{B}$ is small. We assume $\mathbf{B}$ is small in order to legitimately speak of $(\mathbf{B}, \mathbf{C})$ as a category.

*Proof.* We have to construct a functor $S: (\mathbf{A}, \mathbf{C}) \to (\mathbf{B}, \mathbf{C})$ in such a way that $(S\text{-}, \text{-}) \simeq (\text{-}, T\text{-})$.

For $B \in \mathbf{B}$ we consider the category $\mathbf{H}_B$ whose objects are $\ulcorner A, \alpha \urcorner$ with $A \in \mathbf{A}$ and $\alpha: IA \to B$ in $\mathbf{B}$ and whose morphisms from $\ulcorner A, \alpha \urcorner$ to $\ulcorner A', \alpha' \urcorner$ are those $v \in (A, A')_{\mathbf{A}}$ such that $\alpha' \circ Iv = \alpha$. For every $F \in (\mathbf{A}, \mathbf{C})$ we then obtain a functor $F_B: \mathbf{H}_B \to \mathbf{C}$ defined by $F_B \ulcorner A, \alpha \urcorner = FA$ and $F_B v = Fv$ for $v \in (\ulcorner A, \alpha \urcorner, \ulcorner A', \alpha' \urcorner)_{\mathbf{H}_B}$. Since $\mathbf{A}$ is small $\mathbf{H}_B$ is also small. Hence $F_B$ is a diagram in $\mathbf{C}$ which has a supremum since $\mathbf{C}$ is right complete. Denote this supremum by $\ulcorner L \ulcorner F, B \urcorner, l \ulcorner F, B \urcorner \urcorner$ and note that it is unique up to an isomorphism. We define $SF(B) = L \ulcorner F, B \urcorner$ for every $F \in (\mathbf{A}, \mathbf{C})$ and $B \in \mathbf{B}$. To each $f: B \to B'$ in $\mathbf{B}$ we want to attach $SF(f)$. For $v \in (\ulcorner A, \alpha \urcorner, \ulcorner A', \alpha' \urcorner)_{\mathbf{H}_B}$ we have the commutative diagram in $\mathbf{C}$ on page 50.

Since $\alpha' \circ Iv = \alpha$ we also have $f\alpha' \circ Iv = f\alpha$ and hence $v: \ulcorner A, f\alpha \urcorner \to \ulcorner A', f\alpha' \urcorner$ in $\mathbf{H}_{B'}$. Furthermore $F_B \ulcorner A, \alpha \urcorner = FA =$

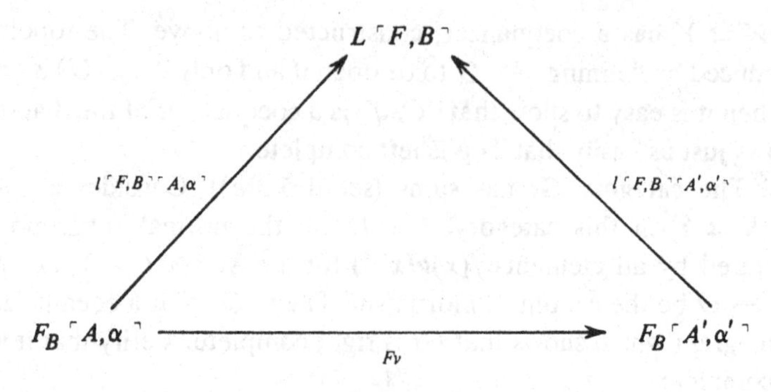


$F_{B'}\ulcorner A, f\alpha\urcorner$. This gives rise to the following diagram:

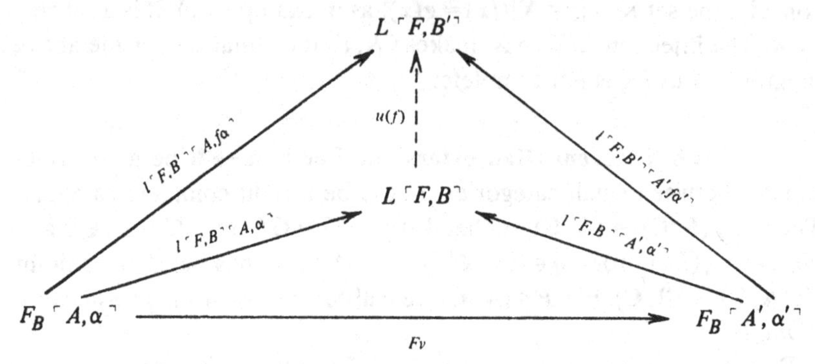


The dotted arrow stands for the unique morphism $u(f)\colon L\ulcorner F, B\urcorner \to L\ulcorner F, B'\urcorner$ with the property that

$$u(f)\circ l\ulcorner F, B\urcorner\ulcorner A, \alpha\urcorner = l\ulcorner F, B'\urcorner\ulcorner A, f\alpha\urcorner$$

for all $\ulcorner A, \alpha\urcorner \in \mathbf{H}_B$. We define $SF(f) = u(f)$. This completes the definition of $SF$ as a functor from $\mathbf{B}$ to $\mathbf{C}$.

Thus far we have obtained that $S$ assigns to every object of $(\mathbf{A}, \mathbf{C})$ an object of $(\mathbf{B}, \mathbf{C})$. For $F_1, F_2 \in (\mathbf{A}, \mathbf{C})$ and for $\eta \in (F_1, F_2)_{(\mathbf{A},\mathbf{C})}$ we now want to define $S\eta$. Let $B \in \mathbf{B}$ and consider the diagram on top of page 51.

The reader should convince himself that there is just one morphism

$$s\ulcorner \eta, B\urcorner\colon SF_1(B) = L\ulcorner F_1, B\urcorner \to L\ulcorner F_2, B\urcorner = SF_2(B)$$

such that

$$s\ulcorner \eta, B\urcorner \circ l\ulcorner F_1, B\urcorner\ulcorner A, \alpha\urcorner = l\ulcorner F_2, B\urcorner\ulcorner A, \alpha\urcorner \circ \eta(A)$$

for all $\ulcorner A, \alpha\urcorner \in \mathbf{H}_B$. Defining

$$S\eta(B) = s\ulcorner \eta, B\urcorner\colon SF_1(B) \to SF_2(B)$$

we obtain $S\eta\colon SF_1 \to SF_2$ as a morphism of functors. This at last completes the definition of the functor $S\colon (\mathbf{A}, \mathbf{C}) \to (\mathbf{B}, \mathbf{C})$.

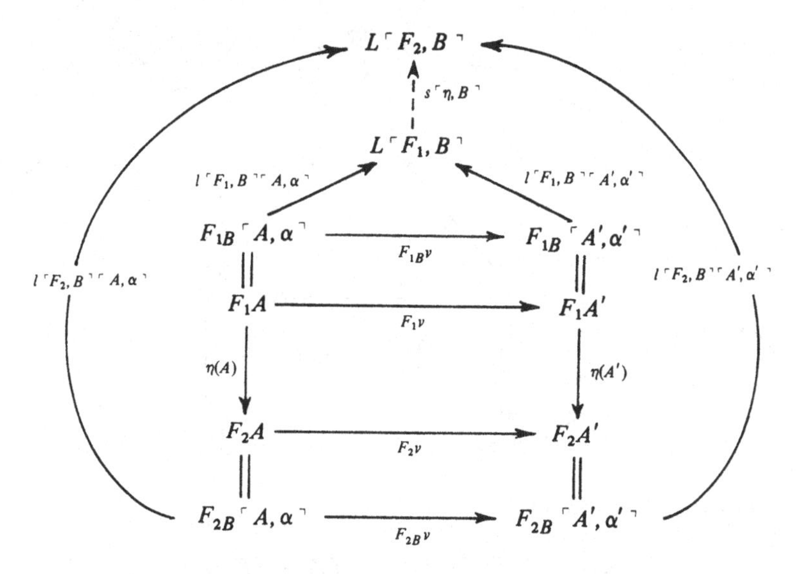


To show that $S$ is a left adjoint of $T$ we have to construct morphisms of functors $\Phi: I_{(\mathbf{A},\mathbf{C})} \to TS$ and $\Psi: ST \to I_{(\mathbf{B},\mathbf{C})}$. For $F \in (\mathbf{A}, \mathbf{C})$ we must define

$$\Phi(F): F \to T(SF) = (SF)I,$$

hence, for $A \in \mathbf{A}$,

$$\Phi(F)(A): FA \to SF(IA).$$

Since $SF(IA) = L\ulcorner F, IA\urcorner$ and since $FA = F_{IA}\ulcorner A, 1_{IA}\urcorner$, we may define $\Phi(F)(A) = l\ulcorner F, IA\urcorner\ulcorner A, 1_{IA}\urcorner$. The reader should check that $\Phi$ is indeed a morphism of functors.

Next we must define $\Psi(G)$ for $G \in (\mathbf{B}, \mathbf{C})$. That means we have to prescribe, for $B \in \mathbf{B}$,

$$\Psi(G)(B): (STG)B = S(GI)B \to GB.$$

Consider the diagram on top of page 52.

We define $\Psi(G)(B): S(GI)B \to GB$ as the unique morphism such that $\Psi(G)(B) \circ l\ulcorner GI, B\urcorner\ulcorner A, \alpha\urcorner = G\alpha$ for all $\ulcorner A, \alpha\urcorner \in \mathbf{H}_\mathbf{B}$. Then one should check that $\Psi: ST \to I_{(\mathbf{B},\mathbf{C})}$ thus defined is indeed a morphism of functors.

Finally we leave it to the reader to verify that $T\Psi \circ \Phi T = 1_T$ and $\Psi S \circ S\Phi = 1_S$, and thus to finish the proof that $S$ is a left adjoint of $T$.

Remark. In applications the functor $I: \mathbf{A} \to \mathbf{B}$ frequently is full and faithful. For a fixed $A \in \mathbf{A}$ to any morphism $\tilde{\alpha}': IA' \to IA$ in $\mathbf{B}$ there corresponds a unique $\alpha': A' \to A$ in $\mathbf{A}$ with $\tilde{\alpha}' = I\alpha'$.

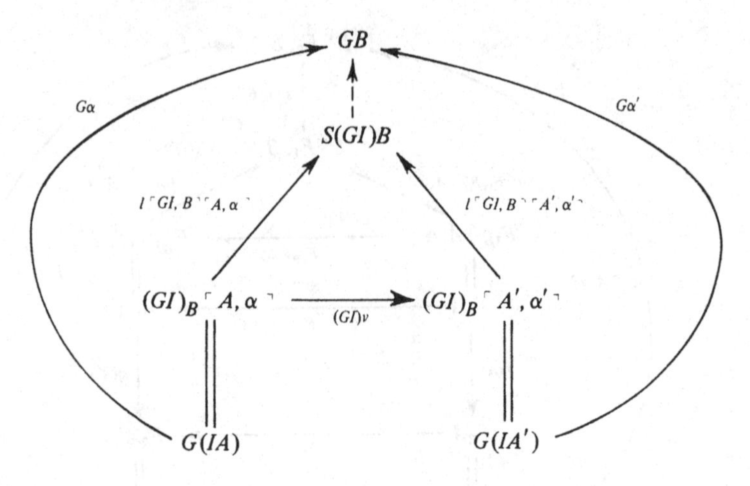


For $F \in (\mathbf{A}, \mathbf{C})$ we had

$$\Phi(F)(A) = l\ulcorner F, IA\urcorner\ulcorner A, 1_{IA}\urcorner : FA \to SF(IA).$$

Consider the following diagram

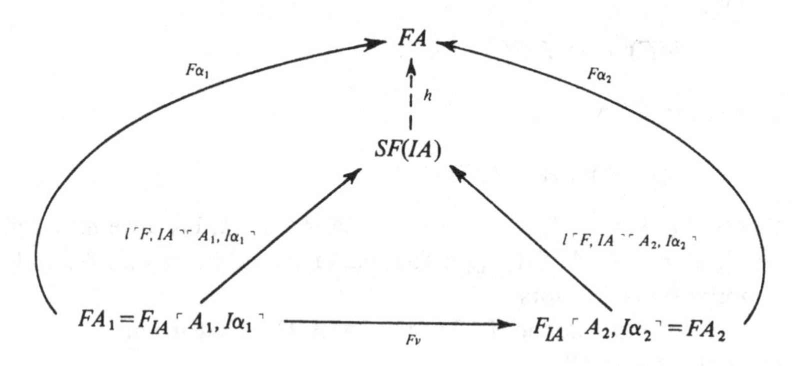


for $v \in (\ulcorner A_1, I\alpha_1\urcorner, \ulcorner A_2, I\alpha_2\urcorner)$.

Due to the fact that $SF(IA)$ is the supremum of the diagram $F_{IA}: \mathbf{H}_{IA} \to \mathbf{C}$ there exists a unique $h: SF(IA) \to FA$ such that $h \circ l\ulcorner F, IA\urcorner\ulcorner A', I\alpha'\urcorner = F\alpha'$ for all $\ulcorner A', I\alpha'\urcorner \in \mathbf{H}_{IA}$. Now the reader may convince himself that $SF(IA) \simeq FA$ by checking that $l\ulcorner F, IA\urcorner\ulcorner A, I1_A\urcorner \circ h = 1_{SF(IA)}$ and $h \circ l\ulcorner F, IA\urcorner\ulcorner A, I1_A\urcorner = 1_{FA}$. Assume now $I$ is an embedding, i.e. $IA = IA'$ implies $A = A'$. Since suprema are determined up to isomorphisms we may then take $L\ulcorner F, IA\urcorner = FA$ and $l\ulcorner F, IA\urcorner\ulcorner A', I\alpha'\urcorner = F\alpha'$ for $\ulcorner A', I\alpha'\urcorner \in \mathbf{H}_{IA}$. Furthermore

$$\Phi(F)(A) = l\ulcorner F, IA\urcorner\ulcorner A, 1_{IA}\urcorner = l\ulcorner F, IA\urcorner\ulcorner A, I1_A\urcorner = F1_A = 1_{FA}.$$

This means $TS = I_{(\mathbf{A},\mathbf{C})}$ and $\Phi = 1_{TS}$. Referring to 1.7.10° we see that $S$ embeds $(\mathbf{A}, \mathbf{C})$ as a full coreflective subcategory in $(\mathbf{B}, \mathbf{C})$. The formulas mentioned in 1.7.10° translated in terms of our situation here finally yield $\Psi S = 1_S$; $T\Psi = 1_T$; for $\eta: SF \to G$ we have $\phi \ulcorner F, G \urcorner \eta = T\eta$; for $\theta: F \to TG$ we have $T\psi \ulcorner F, G \urcorner \theta = \theta$. In this case $SF$ can be viewed as an extension of the functor $F$ from the subcategory $\mathbf{A}$ to $\mathbf{B}$, justifying the name 'Kan extension'.

We emphasize once again that $\mathbf{A}$ must be a small category for $F_B$ to be a diagram and hence have a supremum. For the construction of the Kan extension $SF$ it is unnecessary to assume $\mathbf{B}$ is small. The only advantage is that $(\mathbf{B}, \mathbf{C})$ is then a category and we could legitimately speak of an adjoint situation. The reader should set up the framework of the dual notion of a (right) Kan extension.

## 1.9 Continuous functors

Let $\mathbf{I}$ be a small category and $D: \mathbf{I} \to \mathbf{C}$ a diagram. Let $T: \mathbf{C} \to \mathbf{K}$ be a functor. Then $TD$ is a functor from $\mathbf{I}$ to $\mathbf{K}$, hence a diagram in $\mathbf{K}$. If, as before, $MC$ is the constant diagram in $\mathbf{C}$ defined by the object $C \in \mathbf{C}$, denote by $MTC$ the constant diagram in $\mathbf{K}$ defined by the object $TC \in \mathbf{K}$. Let $c: D \to MC$ be a morphism of functors (from $\mathbf{I}$ to $\mathbf{C}$). In other words, let $\ulcorner C, c \urcorner$ be an upper bound for the diagram $D$. The morphism of functors $Tc: TD \to MTC$ then gives rise to an upper bound $\ulcorner TC, Tc \urcorner$ in $\mathbf{K}$ for the diagram $TD$. If $\ulcorner L, l \urcorner$ is a supremum for $D$, $\ulcorner TL, Tl \urcorner$ is an upper bound for $TD$, but need not be a supremum of $TD$.

**1.9.1 Definitions** A functor $T: \mathbf{C} \to \mathbf{K}$ *preserves suprema* provided in all cases $\ulcorner TL, Tl \urcorner$ is a supremum of the diagram $TD$ whenever $\ulcorner L, l \urcorner$ is such for $D$. We usually call such a functor *right continuous*. In this section we shall explore the connection between continuous functors and adjoints. Dually one defines an *infimum preserving* or *left continuous functor*. A functor which is both left and right continuous is called *continuous*.

A functor preserving coequalizers (equalizers) is called *right exact* (*left exact*). A functor which is both right and left exact is called *exact*. A contravariant functor turning coequalizers into equalizers is called left exact, and so on.

**1.9.2 Theorem** For a covariant functor $T: \mathbf{C} \to \mathbf{K}$ the following assertions are equivalent if $\mathbf{C}$ has sums.

(i) $T$ is right continuous;

(ii) $T$ preserves sums and is right exact.

*Proof.* (i)⇒(ii) is trivial since sums and coequalizers are special kinds of suprema. Here the condition on **C** is not needed.

To prove (ii)⇒(i), let $D\colon \mathbf{I}\to\mathbf{C}$ be a diagram with supremum $\ulcorner L, l\urcorner$. As in 1.8.4 let $\ulcorner B, q_i\urcorner = \coprod_{i\in I} Di$. There is a unique $h\colon B\to L$ with $hq_i = l(i)$ for all $i\in I$. Since **C** has sums we can construct $A$ and $f, g\colon A \rightrightarrows B$ as considered there. The morphism $h$ then is a coequalizer of $f$ and $g$. The supremum of $D$ is characterized in terms of sums and coequalizers, which are preserved by $T$. It follows that $\ulcorner TB, Tl\urcorner$ is a supremum of the diagram $TD$ as in the proof of 1.8.4. The dual result is equally useful.

**1.9.3 Theorem** The functor $h_C\colon \mathbf{C}\to\mathbf{Sets}$ is left continuous for every $C\in\mathbf{C}$.

*Proof.* Let $D\colon \mathbf{I}\to\mathbf{C}$ be a contravariant diagram with infimum $\ulcorner L, l\urcorner$. Then $h_C\circ D\colon \mathbf{I}\to\mathbf{Sets}$ is also a contravariant diagram.

We have to prove that $\ulcorner h_C L, h_C l\urcorner$ is an infimum for $h_C\circ D$. Therefore, let $\ulcorner X, u\urcorner$ be a lower bound for $h_C\circ D$. This means that $h_C(D\alpha)\circ u(j) = u(i)$ whenever $\alpha\colon i\to j$ and hence $D\alpha\colon Dj\to Di$. We have to show that there is a unique $v\colon X\to h_C L$ satisfying, for all $i\in I$, $h_C l(i)\circ v = u(i)$:

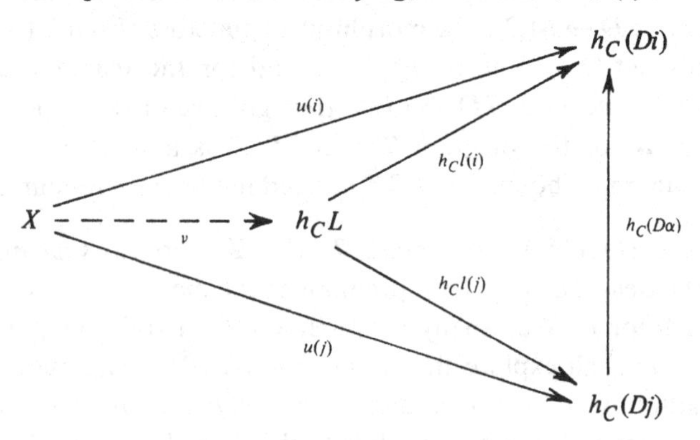


For $x\in X$, $u(j)(x)$ is a morphism from $C$ to $Dj$. Denote this morphism by $u_x(j)$. Then we have

$$\begin{aligned}(D\alpha)\circ u_x(j) &= h_C(D\alpha)(u_x(j))\\ &= h_C(D\alpha)(u(j)(x))\\ &= u(i)(x)\\ &= u_x(i).\end{aligned}$$

This means that $\ulcorner C, u_x\urcorner$ is a lower bound for $D$. But then there is a unique $v(x)\colon C\to L$ making the following diagram commute:

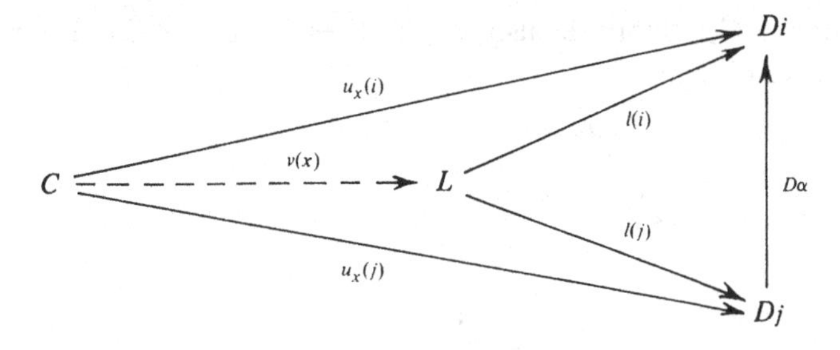


Consider the mapping $x \mapsto v(x)$ of $X$ into $h_C L$. Since for all $x$ we have $(h_C l(i))v(x) = l(i) \circ v(x) = u_x(i) = u(i)(x)$, we proved that $h_C l(i) \circ v = u(i)$. This completes the demonstration.

Applying the above result to the functor $h_C$: $\mathbf{C}^\circ \to \mathbf{Sets}$ one shows that the contravariant functor $h^C$: $\mathbf{C} \to \mathbf{Sets}$ carries suprema to infima.

A converse to 1.9.3 holds if $\mathbf{C}$ is a small left complete category.

**1.9.4 Theorem** Let $\mathbf{C}$ be a small left complete category. Then every left continuous functor $T$: $\mathbf{C} \to \mathbf{Sets}$ is representable.

*Proof.* Consider the category $\mathbf{C}_T$ (see 1.4.8($j$)). We have to prove that $\mathbf{C}_T$ has an initial object. Define the covariant functor $D$: $\mathbf{C}_T \to \mathbf{C}$ by $D\ulcorner C, c\urcorner = C$ and $D\alpha = \alpha$ for $\alpha: \ulcorner B, b\urcorner \to \ulcorner C, c\urcorner$ in $\mathbf{C}_T$. Because $\mathbf{C}$ is small $\mathbf{C}_T$ is also small and hence $D$ is a diagram which must have an infimum $\ulcorner L, u\urcorner$, where $L \in \mathbf{C}$ and $u$: $ML \to D$. Define

$$X = \prod_{\ulcorner C, c\urcorner \in \mathbf{C}_T} TD\ulcorner C, c\urcorner$$

and denote the projections by $p\ulcorner C, c\urcorner$: $X \to TD\ulcorner C, c\urcorner$. Note that $X$ is just a product in the category **Sets**. Now choose $x \in X$ in such a way that for all $\ulcorner C, c\urcorner \in \mathbf{C}_T$ we have $p\ulcorner C, c\urcorner(x) = c$. Denote the embedding of the element $x$ in $X$ by $\Theta$: $\{x\} \to X$. Then we have for $\alpha: \ulcorner B, b\urcorner \to \ulcorner C, c\urcorner$

$$TD\alpha(p\ulcorner B, b\urcorner \Theta(x)) = TD\alpha(b) = c = p\ulcorner C, c\urcorner \Theta(x).$$

This means $\ulcorner\{x\}, p \circ \Theta\urcorner$ is a lower bound for $TD$. The functor $T$ being left continuous, the couple $\ulcorner Tl, Tu\urcorner$ must be an infimum for $TD$. Thus there must be a unique $\theta: \{x\} \to TL$ such that $Tu\ulcorner C, c\urcorner \circ \theta = p\ulcorner C, c\urcorner \circ \Theta$ for every $\ulcorner C, c\urcorner \in \mathbf{C}_T$. Set $\theta(x) = l \in TL$. Now we assert that $\ulcorner L, l\urcorner$ represents the functor $T$, or, in other words, $\ulcorner L, l\urcorner$ is an initial object in $\mathbf{C}_T$. Suppose, therefore, $\ulcorner C, c\urcorner \in \mathbf{C}_T$. Since $u\ulcorner C, c\urcorner$: $L \to C$ in $\mathbf{C}$, we have $Tu\ulcorner C, c\urcorner$: $TL \to TC$. For $l \in TL$ this yields $Tu\ulcorner C, c\urcorner(l) = Tu\ulcorner C, c\urcorner(\theta(x)) = p\ulcorner C, c\urcorner \Theta(x) = c$. Thus $u\ulcorner C, c\urcorner$: $\ulcorner L, l\urcorner \to \ulcorner C, c\urcorner$ is a morphism in $\mathbf{C}_T$. Suppose also $u': \ulcorner L, l \urcorner \to \ulcorner C, c \urcorner$ in $\mathbf{C}_T$. From the commutative diagram

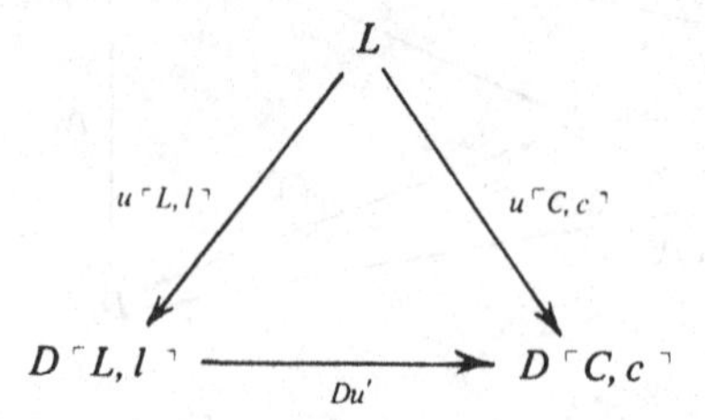


we see that $u \ulcorner C, c \urcorner = Du' \circ u \ulcorner L, l \urcorner$. We shall show that $u \ulcorner L, l \urcorner = 1_L$. Since $Du' = u'$, we then have proved that $u' = u \ulcorner C, c \urcorner$, settling uniqueness. Since $\ulcorner L, u \urcorner$ is an infimum of $D$, for every morphism $X \to L$ in $\mathbf{C}$ for which the following diagrams are commutative (for all $\alpha: \ulcorner A, a \urcorner \to \ulcorner B, b \urcorner$) there is just the single morphism indicated by the dotted arrow making the diagrams commute:

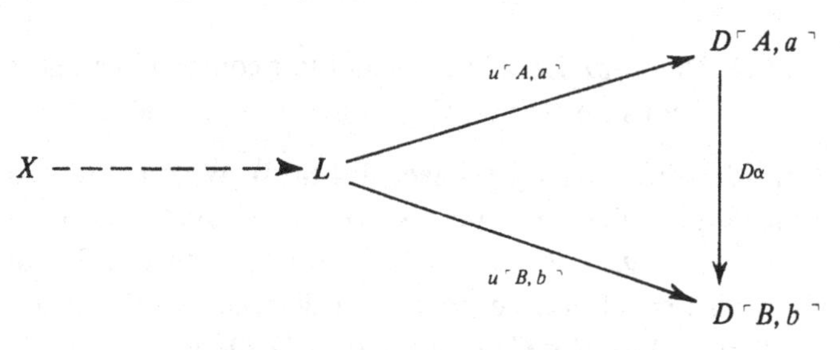


Realizing that $L = D \ulcorner L, l \urcorner$ and taking for $X$ the object $L$ one sees that for the dotted arrow both $1_L$ and $u \ulcorner L, l \urcorner$ may be taken, the diagrams being commutative in both cases. This implies $u \ulcorner L, l \urcorner = 1_L$, finishing the proof.

Remark. The assumption that $\mathbf{C}$ should be left complete may be weakened. We only need that $D$ has an infimum.

**1.9.5 Theorem** Let $S: \mathbf{A} \to \mathbf{B}$ be a left adjoint of $T: \mathbf{B} \to \mathbf{A}$. Then $S$ is right continuous and $T$ is left continuous.

*Proof.* We will show that $S: \mathbf{A} \to \mathbf{B}$ preserves suprema. The other assertion follows from duality arguments. Let $\mathbf{I}$ be a small category and $D: \mathbf{I} \to \mathbf{A}$ a diagram, with supremum $\ulcorner L, l \urcorner$. Then we already know that $\ulcorner SL, Sl \urcorner$ is an upper bound for $SD$. We have to show that it is a supremum. According to the second definition of 1.8.1 this means that the functor $F_{SD}$ is represented by $\ulcorner SL, Sl \urcorner$. We recall (see the first definition of 1.8.1) that $F_{SD}: \mathbf{B} \to \mathbf{Sets}$ and $Sl \in F_{SD}(SL) = (SD, MSL)_{(\mathbf{I}, \mathbf{B})}$. Referring to the Yoneda lemma (1.4.2) we thus have to show that $\Theta(Sl)\colon h_{SL} \rightrightarrows F_{SD}$; i.e. we have to show that for all $B \in \mathbf{B}$ $\Theta(Sl)(B)\colon (SL, B)_{\mathbf{B}} \rightrightarrows F_{SD}(B) = (SD, MB)_{(\mathbf{I},\mathbf{B})}$. For this purpose we are going to look for an inverse mapping $(SD, MB) \to (SL, B)$. Since $S$ and $T$ are adjoint we have inverse isomorphisms $\phi\colon (S\text{-}, \text{-})_{\mathbf{B}} \rightrightarrows (\text{-}, T\text{-})_{\mathbf{A}}$ and $\psi\colon (\text{-}, T\text{-})_{\mathbf{A}} \rightrightarrows (S\text{-}, \text{-})_{\mathbf{B}}$. Suppose $\eta \in (SD, MB)_{(\mathbf{I},\mathbf{B})}$. This means $\eta\colon SD \to MB$ is a functor morphism making

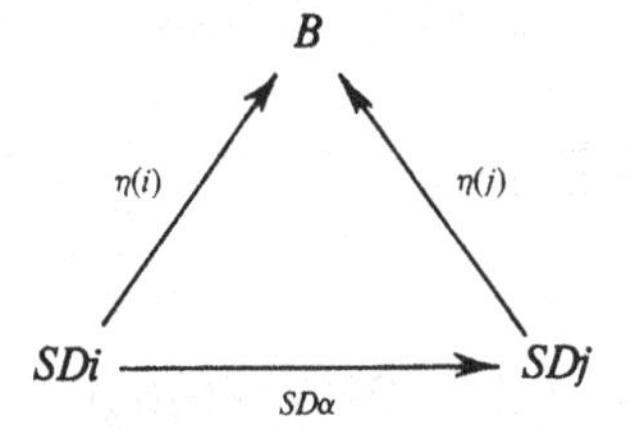


commute for $\alpha\colon i \to j$ in $\mathbf{I}$. Applying $\phi$ we obtain

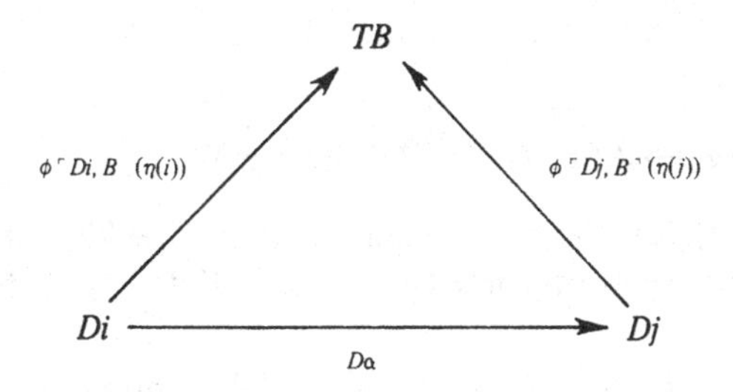


Since $\ulcorner L, l \urcorner$ is a supremum for $D$ there is a unique morphism $u(\eta)\colon L \to TB$, such that $u(\eta) \circ l(i) = \phi \ulcorner Di, B \urcorner (\eta(i))$ for all $i \in \mathbf{I}$. In turn applying $\psi$ we obtain $\psi \ulcorner L, B \urcorner (u(\eta))\colon SL \to B$ and thus we have constructed a mapping $(SD, MB) \to (SL, B)$. It remains to be proved that

$$\Theta(Sl)(B)\colon (SL, B) \to (SD, MB) \quad \text{and}$$

$$\psi \ulcorner L, B \urcorner u\colon (SD, MB) \to (SL, B)$$

are inverse to each other.

We first consider the composite mapping $(SD, MB) \to (SL, B) \to (SD, MB)$ and show that it is equal to $1_{(SD,MB)}$.

For $\eta \in (SD, MB)$ we have

$$\eta \mapsto \psi \ulcorner L, B \urcorner (u(\eta)) \mapsto \Theta(Sl)(B)(\psi \ulcorner L, B \urcorner (u(\eta)).$$

According to the definition of $\Theta$ in the Yoneda lemma the last item can be rewritten as

$$F_{SD}\psi \ulcorner L, B \urcorner (u(\eta))(Sl) = M(\psi \ulcorner L, B \urcorner (u(\eta)) \circ Sl.$$

Applied to $i$ this yields $\psi\ulcorner L, B\urcorner(u(\eta))\circ Sl(i)$. According to the commutative diagram

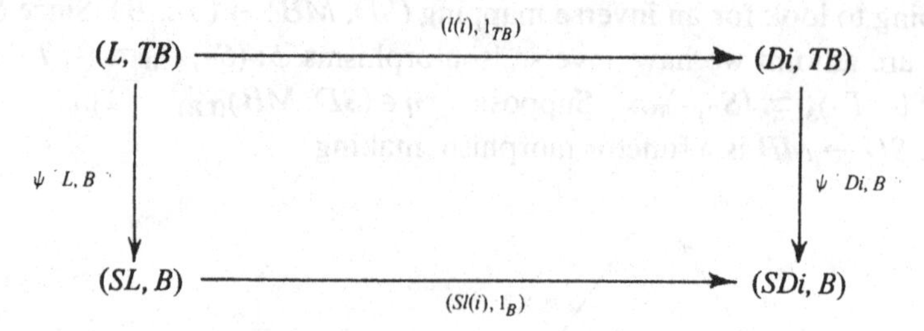


the last expression may be written as

$$\psi\ulcorner Di, B\urcorner(u(\eta))\circ l(i))=\psi\ulcorner Di, B\urcorner\phi\ulcorner Di, B\urcorner\circ(\eta(i))=\eta(i),$$

which implies the desired result.

Next we look at the composite mapping $(SL, B)\to(SD, MB)\to(SL, B)$ and show that this equals $1_{(SL,B)}$.

For $g\in(SL, B)$ we have

$$g\mapsto\Theta(Sl)(B)(g)\mapsto\psi\ulcorner L, B\urcorner(u(\Theta(Sl)(B)(g))).$$

Writing $\eta=\Theta(Sl)(B)(g)$ for short we examine $u(\eta): L\to TB$, defined as the unique morphism such that $u(\eta)\circ l(i)=\phi\ulcorner Di, B\urcorner(\eta(i))$ for all $i\in\mathbf{I}$.

Again, due to the definition of $\Theta$ in Yoneda's lemma we have $\Theta(Sl)(B)(g)=F_{SD}g(Sl)=Mg\circ Sl$; thus $\eta(i)=g\circ Sl(i)$. But $\phi\ulcorner Di, B\urcorner(g\circ Sl(i))=\phi\ulcorner L, B\urcorner(g)\circ l(i)$ we see from the following commutative diagram:

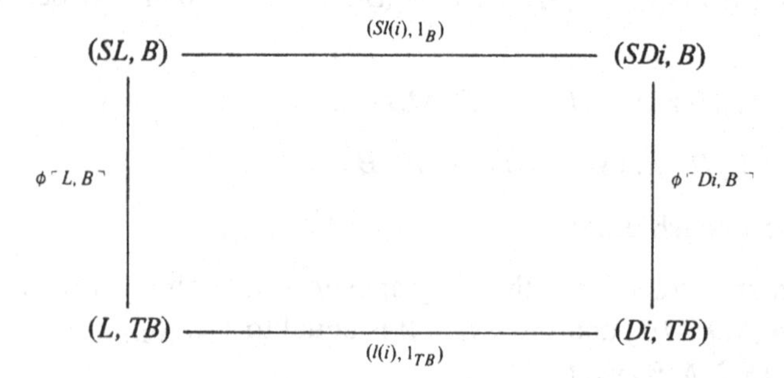


Hence $\phi\ulcorner Di, B\urcorner(\eta(i))=\phi\ulcorner L, B\urcorner(g)\circ l(i)$ for all $i\in\mathbf{I}$. From the uniqueness of $u$ we conclude $u(\eta)=\phi\ulcorner L, B\urcorner(g)$. This yields $\psi\ulcorner L, B\urcorner(u(\eta))=\psi\ulcorner L, B\urcorner\phi\ulcorner L, B\urcorner(g)=g$ as we were hoping for. The tedious proof is now complete. A more intuitive argument is easy to give if one is willing to take functoriality for granted.

A kind of converse is expressed in the next theorem.

**1.9.6 Theorem** Let $T: \mathbf{B} \to \mathbf{A}$ be a fully faithful functor with left adjoint $S$. Let $D: \mathbf{I}^{\circ} \to \mathbf{B}$ be a diagram in $\mathbf{B}$. If the diagram $TD$ has an infimum $\ulcorner L, l \urcorner$ in $\mathbf{A}$, then $D$ has an infimum $\ulcorner \bar{L}, \bar{l} \urcorner$ in $\mathbf{B}$. Note that, according to 1.9.5, the couple $\ulcorner T\bar{L}, T\bar{l} \urcorner$ is an infimum of $TD$. In particular the result holds if $\mathbf{B}$ is a full reflective subcategory of $\mathbf{A}$. Then $\bar{L}$ is a reflection of $L$ in the subcategory $\mathbf{B}$.

*Proof.* Invoking 1.7.7 we may assert that $F_L = h_L \circ T$ is representable. A representing couple $\ulcorner \bar{L}, e \urcorner$ is characterized by the property that every $f: L \to TB$ can be factored uniquely in $\mathbf{A}$ through $e$ (see 1.7.9):

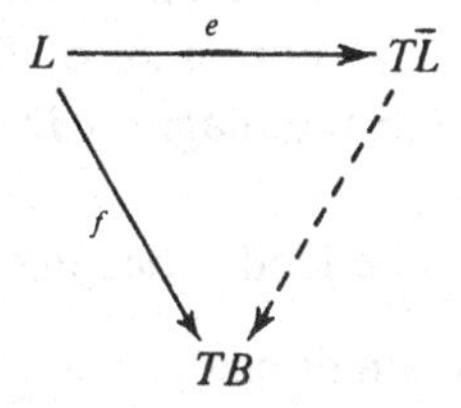


For $\alpha: i \to j$ in $\mathbf{I}$ we then have the following situation

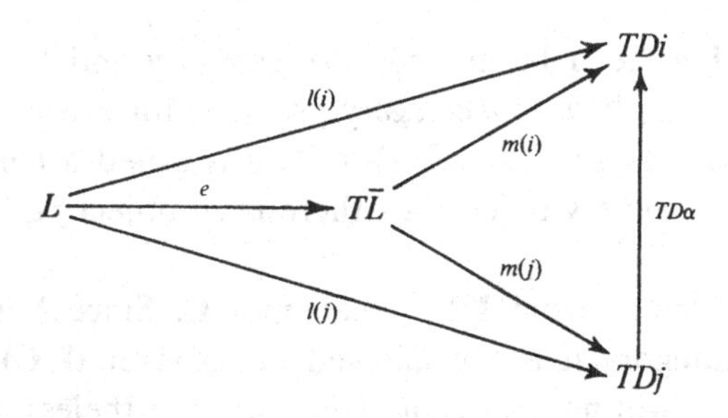


in which $m(i) \circ e = l(i)$ for every $i \in \mathbf{I}$. Since $TD\alpha \circ m(j) \circ e = m(i) \circ e$ the unicity of the factorization through $e$ implies $TD\alpha \circ m(j) = m(i)$. Hence $\ulcorner T\bar{L}, m \urcorner$ is a lower bound for the diagram $TD$. Due to the morphism $e$ it is in fact an infimum of $TD$. Since $T$ is full, the triangles

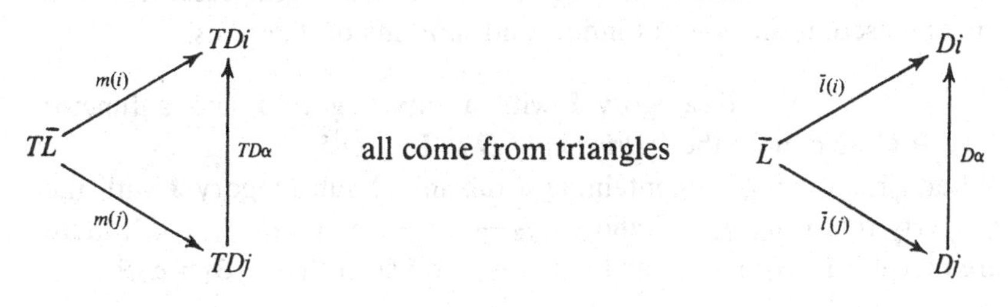

in **B**, where $T\bar{l}(i) = m(i)$. These triangles are uniquely determined and commute because $T$ is faithful. Hence $\ulcorner \bar{L}, \bar{l} \urcorner$ is a lower bound of the diagram $D: \mathbf{I} \to \mathbf{B}$. The proof that $\ulcorner \bar{L}, \bar{l} \urcorner$ is actually an infimum of $D$ is now easily completed, again using the fact that $T$ is full and faithful. The case of a reflection is clear. The dual result should also be noted.

**1.9.7 Theorem** Let $T: \mathbf{B} \to \mathbf{A}$ be a fully faithful functor with left adjoint $S$. Then:

(i) if **A** is (finitely) right complete, so is **B**;

(ii) if **A** is (finitely) left complete, so is **B**.

*Proof.* $\Psi: ST \to I_{\mathbf{B}}$ must be an isomorphism due to 1.7.8. Let $D: \mathbf{I} \to \mathbf{B}$ be a diagram in **B** and let $\ulcorner L, l \urcorner$ be a supremum of $TD$ in **A**. Then $\ulcorner SL, Sl \urcorner$ is a supremum of $STD$ in **B**, due to 1.9.5. But since $\Psi D: STD \rightrightarrows D$, the couple $\ulcorner SL, Sl \circ \Psi^{-1}D \urcorner$ is a supremum of $D$. This proves (i). If $TD$ has an infimum in **A**, by 1.9.6 the diagram $D$ has an infimum in **B**. This proves (ii).

In particular, again, this applies to a fully embedded subcategory.

So far, we have considered diagrams from small categories only. In the following sections we shall show how to relax this restriction somewhat.

**1.9.8 Definition** Let **I** be an arbitrary category and **J** a subcategory of **I**. **J** is called an *initial subcategory* provided for every object $i \in \mathbf{I}$ there is an object $j \in \mathbf{J}$ such that $(j, i)_{\mathbf{I}} \neq \varnothing$. **J** is called a *terminal subcategory* provided for every object $i \in \mathbf{I}$ there is an object $j \in \mathbf{J}$ such that $(i, j)_{\mathbf{I}} \neq \varnothing$.

Now let $D$ be a functor from **I** to a category **C**. Since **I** is not necessarily a small category it is not allowed to consider $(\mathbf{I}, \mathbf{C})$ as a category. Therefore we will not call $D$ a diagram. Nevertheless a pair $\ulcorner C, c \urcorner$ with $C \in \mathbf{C}$ and $c: MC \to D$ a morphism of functors may be called a lower bound for $D$. Then a pair $\ulcorner L, l \urcorner$ will be called an infimum provided for every lower bound $\ulcorner C, c \urcorner$ there is exactly one morphism $u: C \to L$ such that $l(i) \circ u = c(i)$ for all $i \in \mathbf{I}$.

Remark. The terminology 'left complete' and 'right complete' will only be used in the cases of infima and suprema of diagrams.

**1.9.9** For a category **I** with a subcategory **J** and a functor $D: \mathbf{I} \to \mathbf{C}$ we denote the *restriction of D to* **J** by $D|\mathbf{J}$.

Let **I** be a category containing a full initial subcategory **J** with the property that if $\alpha_1: j_1 \to i$ and $\alpha_2: j_2 \to i$ in **I** for certain $j_1, j_2 \in \mathbf{J}$, there are morphisms $\beta_1: i' \to j_1$ and $\beta_2: i' \to j_2$ in **I** such that $\alpha_1\beta_1 = \alpha_2\beta_2$.

**Lemma** Let $D: \mathbf{I} \to \mathbf{C}$ be a functor for which $D|\mathbf{J}: \mathbf{J} \to \mathbf{C}$ has an infimum $\ulcorner L, l \urcorner$. Then $\ulcorner L, l \urcorner$ is also an infimum for $D: \mathbf{I} \to \mathbf{C}$ (after extension of $l$).

*Proof.* We first extend $l$. Let $i \in \mathbf{I}$. Then there is an object $j_1 \in \mathbf{J}$ with $(j_1, i) \neq \varnothing$. Thus there exists $\alpha_1: j_1 \to i$. Define $l(i) = D(\alpha_1) \circ l(j_1)$. The reader should verify, making use of the special property of the subcategory $\mathbf{J}$, that this definition is independent of the choice of $j_1$. Then one easily checks that $l$, thus constructed, is a morphism of functors from $ML \to D$, which implies that $\ulcorner L, l \urcorner$ is a lower bound for $D: \mathbf{I} \to \mathbf{C}$.

Suppose $\ulcorner C, c \urcorner$ is also a lower bound for $D$. Then it may be considered as a lower bound for $D|\mathbf{J}$ (by taking $c: MC \to D|\mathbf{J}$). Thus there is exactly one morphism $g: C \to L$ such that $l(j) \circ g = c(j)$ for all $j \in \mathbf{J}$. But then also $l(i) \circ g = D\alpha \circ l(j) \circ g = D\alpha \circ c(j) = c(i)$ for all $i \in \mathbf{I}$, since there exists an $\alpha: j \to i$ with $j$ in the initial subcategory $\mathbf{J}$. Since $g$ was already unique this proves that $\ulcorner L, l \urcorner$ is an infimum for $D$.

**1.9.10 Definition** A functor $T: \mathbf{C} \to \mathbf{Sets}$ is called *proper* provided there is a full, small subcategory $\mathbf{B}$ of $\mathbf{C}$ so that $\mathbf{B}_T$ is an initial subcategory of $\mathbf{C}_T$ (see 1.4.8($j$)).

Remark that for a small category properness is automatic.

**Example** Suppose $T$ is represented by $\ulcorner B, b \urcorner$, $B \in \mathbf{C}$ and $b \in TB$. Take for $\mathbf{B}$ the category with only one object $B$ and with $(B, B)_{\mathbf{B}} = (B, B)_{\mathbf{C}}$. For $\ulcorner C, c \urcorner \in \mathbf{C}_T$ there is a (unique) morphism $\ulcorner B, b \urcorner \to \ulcorner C, c \urcorner$ in $\mathbf{C}_T$, hence $\mathbf{B}_T$ is an initial subcategory in $\mathbf{C}_T$.

As a consequence, representable functors are proper.

Let $T: \mathbf{C} \to \mathbf{Sets}$ be a proper functor. Let $\mathbf{B}$ be as above. If we know $TB$ for all $B \in \mathbf{B}$ and $Tf(b)$ for all morphisms $f: B \to C$ and $b \in TB$ then we know the functor $T$ completely. For $C \in \mathbf{C}$ and $c \in TC$ there are $B \in \mathbf{B}$, $b \in TB$ and $f: B \to C$ with $Tf(b) = c$. Thus $TC = \bigcup Tf(TB)$, where the union is taken over all $B \in \mathbf{B}$ and over all $f \in (B, C)_{\mathbf{C}}$. If moreover $g: C \to C'$ in $\mathbf{C}$, then $Tg(c) = Tg(Tf(b)) = T(gf)(b)$ which element is known.

Let $T$ and $T'$ be two functors from $\mathbf{C}$ to $\mathbf{Sets}$ with $T$ proper and $\mathbf{B}$ as above. Suppose $\theta: T \to T'$ is a morphism of functors. Then $\theta$ is completely determined by all $\theta(B)$ with $B \in \mathbf{B}$. For if $C \in \mathbf{C}$ and thus $\theta(C): TC \to TC'$, we want to know $\theta(C)(c)$ for $c \in TC$. But there is $\ulcorner B, b \urcorner \in \mathbf{B}_T$ and $f: B \to C$ with $Tf(b) = c$. Since $\theta$ is a morphism of functors we have $\theta(C)(c) = \theta(C)(Tf(b)) = T'f(\theta(B)(b))$. We write $\theta|\mathbf{B}$ for the restriction of $\theta$ to $\mathbf{B}$ so that $\theta|\mathbf{B}: T|\mathbf{B} \to T'|\mathbf{B}$ is a morphism of functors from $\mathbf{B}$ to $\mathbf{Sets}$.

In the sequel we will denote by $(T, T')_{(\mathbf{C},\mathbf{Sets})}$ the restrictions $\Theta|\mathbf{B}$ of all $\Theta: T \to T'$. Since **B** is a small category this is a set. In the sequel we will also mean by (**C**, **Sets**) the category of all proper functors from **C** to **Sets**. The next result generalizes 1.9.4.

**1.9.11 Theorem** Let $T: \mathbf{C} \to \mathbf{Sets}$ be a proper functor. Suppose that **C** is left complete and suppose that $T$ is left continuous. Then $T$ is representable.

*Proof.* Let **B** be a full, small subcategory of **C** such that $\mathbf{B}_T$ is an initial subcategory of $\mathbf{C}_T$. Let $D: \mathbf{C}_T \to \mathbf{C}$ be as in 1.9.4. Here $D$ is not necessarily a diagram but $D|\mathbf{B}_T: \mathbf{B}_T \to \mathbf{C}$ is; $\mathbf{B}_T$ is a full subcategory of $\mathbf{C}_T$. We will show that $\mathbf{B}_T$ satisfies the requirements of lemma 1.9.9.

Let $\alpha_1: \ulcorner B_1, b_1 \urcorner \to \ulcorner C, c \urcorner$ and $\alpha_2: \ulcorner B_2, b_2 \urcorner \to \ulcorner C, c \urcorner$ in $\mathbf{C}_T$ with $\ulcorner B_i, b_i \urcorner \in \mathbf{B}_T$ $(i = 1, 2)$. We have to find $\ulcorner C', c' \urcorner \in \mathbf{C}_T$ with $\beta_i: \ulcorner C', c' \urcorner \to \ulcorner B_i, b_i \urcorner$ $(i = 1, 2)$ so that $\alpha_1\beta_1 = \alpha_2\beta_2$. Since **C** is left complete there are $\beta_i: C' \to B_i$ $(i = 1, 2)$ making the diagram below-left cartesian. The fact that $T$ preserves infima makes the diagram below-right cartesian.

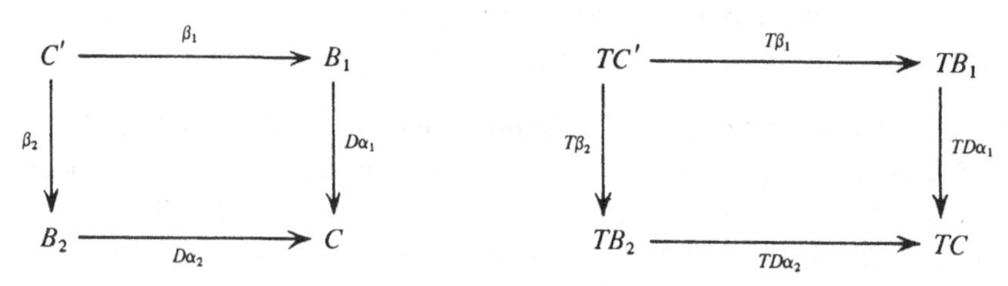


From the construction of the fibred product in **Sets** (see 1.5.3) we know that $TC'$ contains exactly one element $c'$ such that $T\beta_i(c') = b_i$ $(i = 1, 2)$, since $TD\alpha_i(b_i) = c$. Abusing language, we call by the same name $\beta_i$ the corresponding morphisms $\ulcorner C', c' \urcorner \to \ulcorner B_i, b_i \urcorner$ in $\mathbf{C}_T$. These clearly satisfy $\alpha_1\beta_1 = \alpha_2\beta_2$, so $\mathbf{B}_T$ has the right properties. Now $D|\mathbf{B}_T \to \mathbf{C}$ is a diagram and thus has an infimum, say $\ulcorner L, u \urcorner$. But then lemma 1.9.9 implies that this infimum is also an infimum for $D$, after $u$ has been extended. Just as in the proof of 1.9.4 it is shown now that this gives rise to a representing couple $\ulcorner L, l \urcorner$ for $T$.

The next result, originally due to Freyd [13, ch. 3], subsumes theorems 1.7.7 and 1.9.5. Our version is lifted from Lambek [23, p. 38].

**1.9.12 Adjoint functor theorem** Let **B** be a left complete category and $T: \mathbf{B} \to \mathbf{A}$ a covariant functor. Then the following assertions are equivalent.

(i) $T$ has a left adjoint.

(ii) $T$ is left continuous and the functor $F_A: \mathbf{B} \to \mathbf{Sets}$ defined by $F_A B = (A, TB)_\mathbf{A}$ is proper for every $A \in \mathbf{A}$.

*Proof.* (i)$\Rightarrow$(ii). Since $T$ has a left adjoint the functor $F_A$ is representable (for all $A \in \mathbf{A}$) according to theorem 1.7.7. Then $F_A$ is proper (see 1.9.10, clear from example). Theorem 1.9.5 says that $T$ is left continuous.

(ii)$\Rightarrow$(i). Since $h_A$ and $T$ are both left continuous, so is their composition $F_A = h_A \circ T$. From 1.9.11 it follows that $F_A$ is representable (for all $A \in \mathbf{A}$), so $T$ has a left adjoint (1.7.7).

The force of this theorem is that it allows one to conclude that many functors have adjoints. Using 1.8.4° and 1.9.2° left completeness and left continuity are often easy to check, but verifying properness for $F_A$ can give more trouble. One may impose conditions on $\mathbf{B}$ for this to be realized, leading to various forms of the 'special' adjoint functor theorem; cf. [13, p. 89], [23, p. 42].

Dually to 1.9.12 one has

**1.9.12° Theorem** Let $\mathbf{A}$ be a right complete category and $S: \mathbf{A} \to \mathbf{B}$ a covariant functor. Then the following assertions are equivalent.

(i) $S$ has a right adjoint.

(ii) $S$ is right continuous and the contravariant functor $G^B: \mathbf{A} \to \mathbf{Sets}$ defined by $G^B A = (SA, B)_\mathbf{B}$ is proper for every $B \in \mathbf{B}$.

An amusing consequence of the adjoint functor theorem 1.9.12 is that a small left complete category is also right complete. This is a special case of the following.

**1.9.13 Corollary** A left complete category $\mathbf{C}$ is right complete if and only if for every diagram $D: \mathbf{I} \to \mathbf{C}$ from some small category $\mathbf{I}$ the functor $h_D \circ M: \mathbf{C} \to \mathbf{Sets}$ is proper.

*Proof.* The functor $M: \mathbf{C} \to (\mathbf{I}, \mathbf{C})$ of the first definition of 1.8.1 is trivially left continuous. By 1.9.12 properness of $h_D \circ M$ for every $D$ is equivalent to $M$ having a left adjoint $L: (\mathbf{I}, \mathbf{C}) \to \mathbf{C}$. But this is the condition for $\mathbf{C}$ to be right complete discussed in 1.8.1 and 1.8.4.

# 2

# Internal structure of categories

## 2.1 Epimorphisms and monomorphisms

**2.1.1 Definition** A morphism $u: A \to B$ in category $\mathbf{C}$ is called an *epimorphism* ($u$ is called *epic* for short) provided $fu = gu$, with $f$ and $g \in (B, C)_{\mathbf{C}}$, implies $f = g$. Notation: $u$: $A \twoheadrightarrow B$. The following assertions hold.

(i) $f: A \twoheadrightarrow B$ and $g: B \twoheadrightarrow C$ imply $gf: A \twoheadrightarrow C$.

(ii) If, for $u: A \to B$ and $v: B \to C$, $vu: A \twoheadrightarrow C$, then $v$ is epic.

(iii) For all $C \in \mathbf{C}$ the morphism $1_C$ is an epimorphism.

(iv) A coequalizer is always epic.

(v) $u: A \to B$ is an epimorphism if and only if the following square is cocartesian:

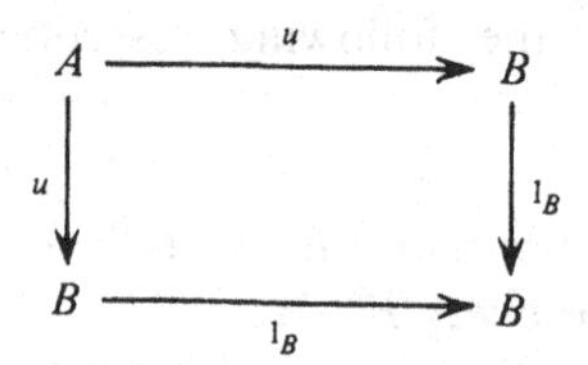

As a corollary of (v) one obtains: a right continuous functor certainly preserves cocartesian squares and hence also epimorphisms.

**2.1.2 Proposition** If the square

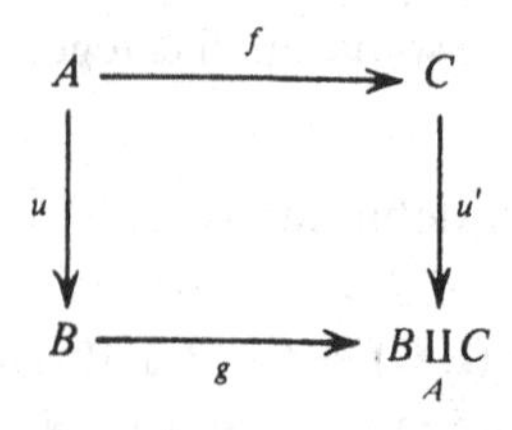

in which $u$ is epic, is cocartesian, $u'$ is also epic.

*Proof.* Let $h_1$ and $h_2 \in (B \coprod_A C, D)_{\mathbf{C}}$ with $h_1 u' = h_2 u'$. Since $(h_1 g)u = h_1(gu) = h_1(u'f) = (h_2 u')f$ there is exactly one morphism $h: B \coprod_A C \to D$ such that $hg = h_1 g$ and $hu' = h_2 u'$. Since both $h_1$ and $h_2$ satisfy the requirements for $h$ we must have $h = h_1 = h_2$.

## *2.1 Epimorphisms and monomorphisms*

**2.1.1° Definition** A morphism $U: A \to B$ in a category $\mathbf{C}$ is called a *monomorphism* (*monic* for short) provided $uf = ug$, with $f$ and $g \in (C, A)_{\mathbf{C}}$, implies $f = g$. In other words provided $u^\circ: B \to A$ is an epimorphism in $\mathbf{C}^\circ$. Notation: $u: A \rightarrowtail B$.

Dualizing the assertions of 2.1.1 one obtains:

(i)° $f: A \rightarrowtail B$ and $g: B \rightarrowtail C$ imply $gf: A \rightarrowtail C$;

(ii)° if, for $u: A \to B$ and $v: B \to C$, $vu: A \rightarrowtail C$, then $u$ is monic;

(iii)° for all $C \in \mathbf{C}$ the morphism $1_C$ is a monomorphism;

(iv)° an equalizer is always monic;

(v)° $u: A \to B$ is a monomorphism if and only if the following square is cartesian.

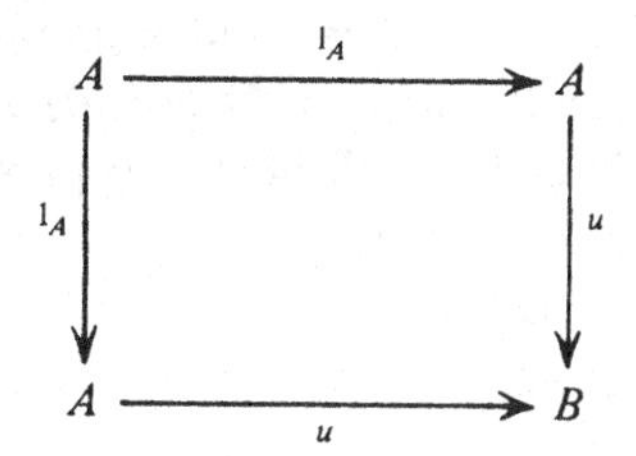


Corollary of (v)°: a left continuous functor will also preserve monomorphisms.

**2.1.2° Proposition** If the square

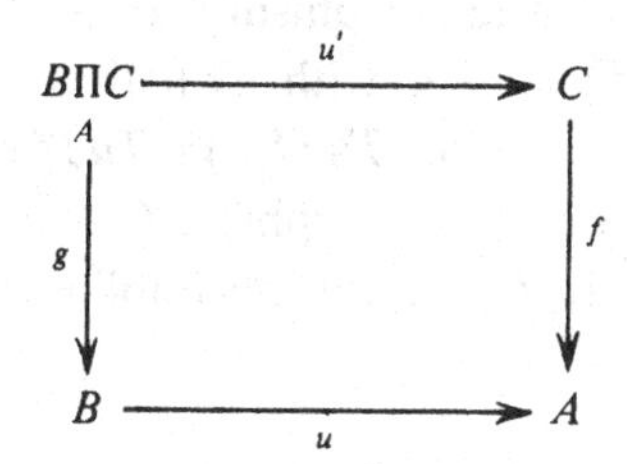


where $u$ is monic, is cartesian, $u'$ is also monic.

Remarks. If

$$\mathbf{A} \underset{T}{\overset{S}{\rightleftarrows}} \mathbf{B}$$

are adjoint functors, $S$ preserves epimorphisms and $T$ preserves monomorphisms (see 1.9.5). Moreover theorem 1.7.8 may be refined as follows: $\Psi$ is epic if and only if $T$ is faithful and $\Psi$ is monic if $T$ is full. Similarly in 1.7.8° the morphism of functors $\Phi$ is monic if and only if $S$ is faithful and is epic if $S$ is full. The proofs are straightforward.

**2.1.3 Definition** A morphism $u: A \to B$ which is monic as well as epic is called a *bimorphism*. Notation: $u: A \rightarrowtail\!\!\!\twoheadrightarrow B$. (Note that an isomorphism is always a bimorphism.)

**2.1.4 Examples** (*a*) For **Sets** monomorphisms are injections, epimorphisms are surjections and bimorphisms are bijections.

(*b*) Let **C** be a concrete category with $T: \mathbf{C} \to \mathbf{Sets}$ a faithful forgetful functor (see 1.2.3). If $u: A \to B$ is a morphism in **C** such that $Tu$ is injective (surjective) then $u$ is a monomorphism (epimorphism).

(*c*) Consider the embedding $i: \mathbb{Z} \to \mathbb{Q}$ in the category **CRg**. Then $i$ is obviously a monomorphism. We will show that $i$ is also an epimorphism. For this purpose let $R$ be any object in **CRg** and suppose that $u, v \in (\mathbb{Q}, R)_{\mathbf{CRg}}$ with $ui = vi$. Since the only ideals of $\mathbb{Q}$ are $(0)$ and $\mathbb{Q}$ itself we see that either $u$ and $v$ are both the null-homomorphism or $\ker u = \ker v = (0)$. But then, for $x = m/n \in \mathbb{Q}$, we have $u(x) = u(im/in) = ui(m)/ui(n) = vi(m)/vi(n) = v(x)$.

As a partial converse of 2.1.4(*b*) we have

**2.1.5 Proposition** Let **C** be a concrete category with $T: \mathbf{C} \to \mathbf{Sets}$ a faithful forgetful functor. If $T$ is representable $Tu$ is injective for each monomorphism $u$.

*Proof.* Let $u: A \rightarrowtail B$ in **C**. Then $Tu: TA \to TB$ in **Sets**. Suppose $\ulcorner C, c \urcorner$ represents $T$. If $a, a' \in TA$ there is exactly one morphism $f: C \to A$ with $Tf(c) = a$ and exactly one morphism $f': C \to A$ with $Tf'(c) = a'$. Suppose furthermore $Tu(a) = Tu(a')$. That means $TuTf(c) = TuTf'(c)$ or $T(uf)(c) = T(uf')(c)$. But there can only be one morphism $C \to B$ with that property so $uf = uf'$. Since $u$ is monic, $f = f'$. But this implies $a = a'$. Hence $Tu$ is injective.

**Corollary** In the categories **Gr**, **Ab**, $\mathbf{M}_R$, **Rg**, **CRg**, **Top**, **TopGr**, **Hausd**, monomorphisms are injective (by which we mean that $Tu$ is injective if $u$ is monic). For these categories $T$ stands for the forgetful functor and this functor is representable in all these cases: for $\mathbf{M}_R$ by $\ulcorner R, 1 \urcorner$, $R$ considered as right $R$-module; for **Rg**, **CRg** by $\ulcorner \mathbb{Z}[X], X \urcorner$, $\mathbb{Z}$ considered as the commutative ring of integers; for **Gr**, **Ab** by $\ulcorner \mathbb{Z}, 1 \urcorner$, $\mathbb{Z}$ considered as the infinite cyclic group with generator 1; for **Top**, **Hausd** by $\ulcorner \{a\}, a \urcorner$; for **TopGr** by $\ulcorner \mathbb{Z}, 1 \urcorner$, $\mathbb{Z}$ considered as the infinite cyclic group with generator 1 and with discrete topology.

*Remark.* For the category of finite groups the forgetful functor is not representable. Nevertheless in this category also every monomorphism is injective. For let $f: G \to H$ be a monomorphism. Then $\{x \in G | f(x) = 0\}$ is a subgroup $K$ in $G$ and thus a finite group. Let $i$ be the embedding of $K$ in $G$ and $j$ the homomorphism that maps every element of $K$ onto the neutral element of $G$. Then $fi = fj$ and, since $f$ is monic, $i = j$. Thus $K$ only contains the neutral element and $f$ is injective.

If **C** is a concrete category with faithful functor $T: \mathbf{C} \to \mathbf{Sets}$, then $Tf$ need not be surjective if $f$ is epic. See 2.1.4($c$). In some concrete categories, however, epimorphisms are surjective. This is the case for **Ab**, $\mathbf{M}_R$, **Gr** and the category of finite groups. In the first two cases this is not difficult to show. We shall prove it for **Gr**. Let $u: K \twoheadrightarrow G$ in **Gr**. We have to show that $u(K)$ is a subgroup $H$ of $G$ that equals $G$. Suppose there are at least two right cosets different from $H$ itself. Denote these by $Ha$ and $Hb$. Let $P$ be the group of all bijections of $G$ onto itself, i.e. the permutation group of $G$. Define the involution $\sigma \in P$ as follows. For $x = ha$, $\sigma(x) = hb$; for $x = hb$, $\sigma(x) = ha$; and for $x \notin Ha \cup Hb$, $\sigma(x) = x$. Further we define $f: G \to P$ by $f(y)x = yx$, $y \in G$; $x \in G$; and $g: G \to P$ by $g(y) = \sigma f(y)\sigma$. Then $fu = gu$, which, since $f$ and $g$ are group-homomorphisms, implies $f = g$. This means that $f(x) = \sigma f(x)\sigma$ for all $x \in G$. Taking $x = b^{-1}$ this yields, operating on $a$, $b^{-1}a = ((\sigma f b^{-1})\sigma)(a) = e$. Thus $a = b$, contradicting the assumption $Ha \neq Hb$. Hence $H$ has at most one right coset besides itself. That makes $H$ a normal subgroup. Let $k$ be the canonical homomorphism $k: G \to G/H$ and let $l$ be the homomorphism that maps all of $G$ onto the neutral element of $G/H$. Then $ku = lu$, so that $k = l$. This means $H = G$. The same proof works in the category of finite groups.

This and the following examples illustrate the fact that to prove the expected characterization of epimorphisms a little ingenuity is sometimes helpful; cf. [8], [37].

**2.1.6 Examples and exercises** ($a$) In **Hausd** $f: A \to B$ is epic if and only if $f(A)$ is dense in $B$.

($b$) Let $\mathbf{K}_p$ be the category of fields of characteristic $p > 0$. For $k \in \mathbf{K}_p$, let $k^p$ be the subfield consisting of all elements $x^p$ for $x \in k$. Then the embedding $i: k^p \to k$ is epic.

($c$) Let $A$ be a commutative ring and $S \subset A$ a multiplicative system. Let $S^{-1}A$ be the corresponding ring of fractions. The canonical ring-homomorphism $f: A \to S^{-1}A$ is epic in **CRg**. Example 2.1.4($c$) is a special case.

($d$) Let $A$ and $B$ be objects in **CRg** and let $f: A \to B$ be such that $B$ is a finitely generated $A$-module. Then $f$ is surjective if and only if it is an epimorphism. See [32, p. 165].

**2.1.7** Let $C$ be a fixed object of an arbitrary category **C**. Consider the collection of all monomorphisms into the object $C$: $\{h_i\colon C_i \rightarrowtail C\}_{i\in I}$. The collection $I$ may be preordered by the stipulation that $i \leqslant j$ provided there is a morphism $h_{ij}\colon C_i \to C_j$ such that $h_j h_{ij} = h_i$(then $h_{ij}$ is necessarily monic). We will call such a morphism $h_{ij}$ a *morphism over C*. When both relations $i \leqslant j$ and $j \leqslant i$ hold one proves easily that the objects $C_i$ and $C_j$ are isomorphically embedded in the object $C$:

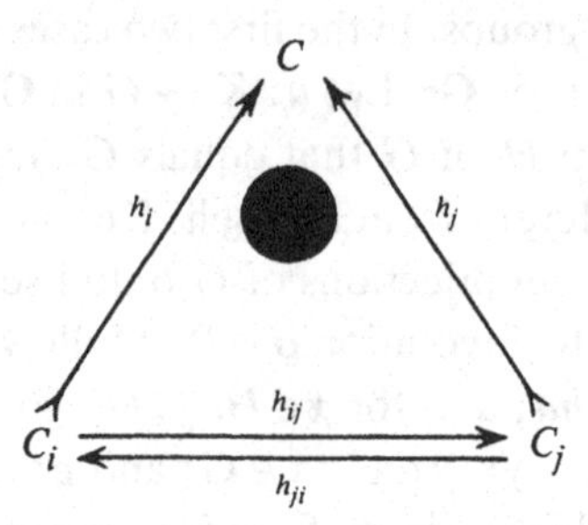


The symbol ● denotes that the triangle is commutative. Defining $i \sim j$ if and only if $i \leqslant j$ and $j \leqslant i$, one obtains an equivalence relation on the collection $I$. Going over to equivalence classes one may partially order these in the obvious way (using representatives). Carrying over this division into equivalence classes to the collection of monomorphisms $\{h_i\colon C_i \rightarrowtail C\}_{i\in I}$ and inducing the partial order from $I$ one obtains a partially ordered collection of equivalence classes of monomorphisms. Such an equivalence class is called a *subobject* of $C$. To justify this terminology, consider **Ab**. Then a subobject of an abelian group $A$ is an object $A_i$ equipped with a particular monomorphism $h_i\colon A_i \rightarrowtail A$. This is what we commonly call a subgroup. We emphasize the fact that a subobject of an object $C$ is usually confounded with a representative $h_i\colon C_i \rightarrowtail C$.

Dually, *quotient objects* are defined.

## 2.2 Punctured categories

**2.2.1 Definition** A *null object* (also called a *point object* or a *zero object*) in a category is an initial object which is also a terminal object (cf. 1.4.8($i$)). Such an object will be denoted by $*$ or 0.

Examples are **Sets**$_*$ with $*$, **Ab** with (0), $\mathbf{M}_R$ with (0), **Gr** with (1) and **Top**$_*$ with $*$.

For a category with null object there is, for any two objects $A$ and $B$, a distinguished morphism from $A$ to $B$ called the *null morphism* which is defined as the composition of the morphisms $A \to *$ and $* \to B$. This morphism is denoted by $*_{AB}$. Two immediate properties are:

$$A \xrightarrow[*_{AB}]{} B \xrightarrow[f]{} X = A \xrightarrow[*_{AX}]{} X$$

$$Y \xrightarrow[g]{} A \xrightarrow[*_{AB}]{} B = Y \xrightarrow[*_{YB}]{} B$$

Let $\mathbf{C}$ be a category and $\{*_{AB}\}_{A,B \in \mathbf{C}}$ a family of morphisms in $\mathbf{C}$. This family is called a *null family* provided:

(i) for every pair of objects $A$ and $B$ in $\mathbf{C}$ there is exactly one morphism $*_{AB}: A \to B$ in the family;

(ii) if $f: B \to X$ then $f*_{AB} = *_{AX}$ and if $g: Y \to A$ then $*_{AB}g = *_{YB}$.

A category with a null family is called *punctured.*

Remark. The family of null morphisms in a category with a null object is an example of a null family. Therefore a category with a null object (a *pointed category*) is a punctured category.

**Exercises** (*a*) A category has at most one null family.

(*b*) Let $\mathbf{C}$ be a punctured category. We can construct a category $\mathbf{D}$ with null object such that $\mathbf{C}$ is a full subcategory of $\mathbf{D}$ by adding a null object $*$ to $\mathbf{C}$. The reader should check which morphisms should be added and how composition of morphisms in $\mathbf{D}$ should be defined.

(*c*) Let $\mathbf{C}$ be a category with null object and let $\{*_{AB}\}_{A,B \in \mathbf{C}}$ be the associated null family. Prove that an object $C \in \mathbf{C}$ is a null object if and only if $1_C = *_{CC}$.

(*d*) If $\mathbf{C}$ is a punctured category so is $\mathbf{C}^\circ$.

Remark. It follows from exercise (*b*) that punctured categories are very close to categories with null object. When convenient we shall henceforth assume that a punctured category is equipped with a null object.

**2.2.2 Proposition** Let $\mathbf{C}$ be a punctured category. Suppose that the objects $A$ and $B$ in $\mathbf{C}$ have a product $A \prod B$ with canonical projections $p_A$ and $p_B$. Then $p_A$ and $p_B$ are epic.

*Proof.* One defines $i_A$ by

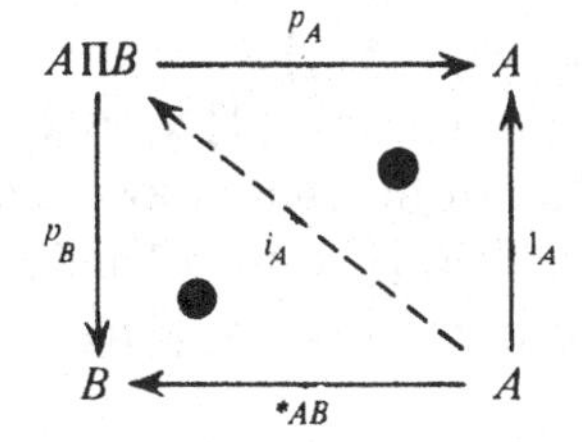

Since $1_A = p_A i_A$ is epic so is $p_A$. Similarly one proves that $p_B$ is epic.

The dual statement is: for $A$ and $B$ in **C** with sum $A \coprod B$ and with canonical injections $q_A$ and $q_B$, these injections are monic. More generally we have: if $\{A_i\}_{i \in I}$ has a product $\prod_i A_i$ in a punctured category **C**, the canonical projections $p_i$ are epic, and if there is a sum $\coprod_i A_i$ the canonical injections $q_i$ are monic.

Remark. The reason that we have labelled the canonical morphisms associated with a product (a sum) projections (injections) is that for punctured categories these morphisms are epic (monic).

In the following we use a notation introduced in 1.5.2.

**2.2.3** Let **C** be a category with null family and with finite sums and products. Then for $A_1, A_2 \in \mathbf{C}$ we write

$$(\delta_{ij}^A) = \begin{pmatrix} 1 & * \\ * & 1 \end{pmatrix}^A : A_1 \coprod A_2 \to A_1 \prod A_2 \ (i, j = 1, 2)$$

as an abbreviation for the morphism

$$\begin{pmatrix} 1_{A_1} & *_{A_2 A_1} \\ *_{A_1 A_2} & 1_{A_2} \end{pmatrix}.$$

Suppose furthermore that $u_i: A_i \to B_i (i = 1, 2)$. Then we have $u_1 \coprod u_2: A_1 \coprod A_2 \to B_1 \coprod B_2$ and $u_1 \prod u_2: A_1 \prod A_2 \to B_1 \prod B_2$ (see 1.5.1).

**Proposition** The following diagram commutes:

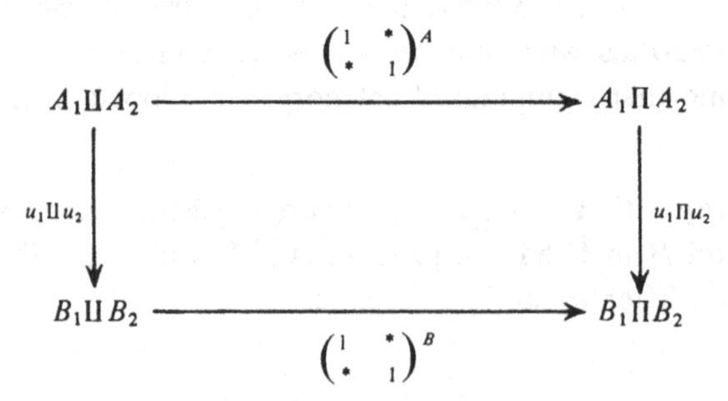


*Proof.* Denoting the injections associated with $A_1 \coprod A_2$ and $B_1 \coprod B_2$ respectively by $q_i^A$ and $q_i^B (i = 1, 2)$ the morphism $u_1 \coprod u_2$ is characterized by the property $(u_1 \coprod u_2) q_i^A = q_i^B u_i$. In the same way, denoting the projections associated with $A_1 \prod A_2$ and $B_1 \prod B_2$ by $p_i^A$ and $p_i^B (i = 1, 2)$ the morphism $u_1 \prod u_2$ is characterized by $p_i^B (u_1 \prod u_2) = u_i p_i^A$. Consider the following diagram

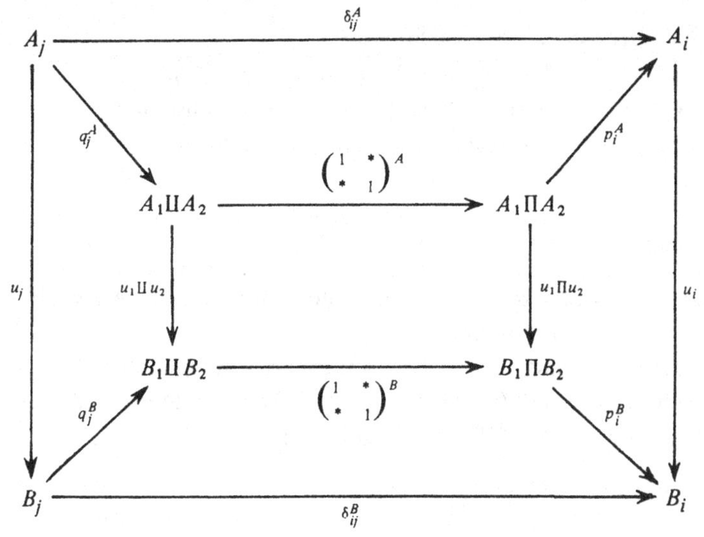


where

$$p_i^B\begin{pmatrix}1 & * \\ * & 1\end{pmatrix}^B(u_1 \coprod u_2)q_j^A = \delta_{ij}^B u_j$$

and similarly

$$p_i^B(u_1 \prod u_2)\begin{pmatrix}1 & * \\ * & 1\end{pmatrix}^A q_j^A = u_i \delta_{ij}^A.$$

But for all $i$ and $j$ $\delta_{ij}^B u_j = u_i \delta_{ij}^A$ which is seen as follows. For $i \neq j$, $\delta_{ij}^B u_j = *_{A_j B_i} = u_i \delta_{ij}^A$. For $i = j$, $\delta_{ij}^B u_j = u_j = u_i = u_i \delta_{ij}^A$. In view of 1.5.2 this implies the assertion.

**Corollary** For a punctured category with finite sums and products, the morphism $\begin{pmatrix}1 & * \\ * & 1\end{pmatrix}$ is a morphism of bifunctors:

$$\begin{pmatrix}1 & * \\ * & 1\end{pmatrix}: \text{-}\coprod\text{-} \rightarrow \text{-}\prod\text{-}.$$

A similar statement holds for $n \times n$ matrices.

**2.2.4 Definitions** Consider a diagram

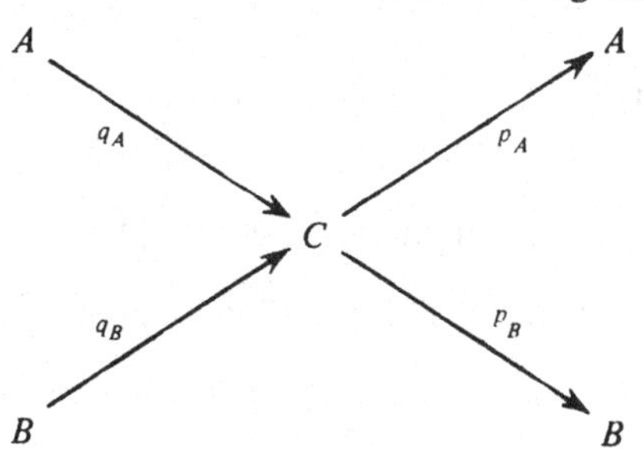

in a punctured category **C** such that:

(i) $C = A \coprod B$ with canonical injections $q_A$ and $q_B$;

(ii) $C = A \prod B$ with canonical projections $p_A$ and $p_B$;

(iii) $p_A q_A = 1_A$ and $p_B q_B = 1_B$, $p_A q_B = *$ and $p_B q_A = *$, i.e.

$$q_A = \begin{pmatrix} 1 \\ * \end{pmatrix}, q_B = \begin{pmatrix} * \\ 1 \end{pmatrix} \quad \text{and} \quad p_A = (1*), p_B = (*1).$$

The system $\ulcorner C, p_A, q_A, p_B, q_B \urcorner$ is called a *biproduct of A and B* and is denoted by $A \oplus B$ (compare 1.5.3($d$)).

Let $f: A \to B$ and $g: C \to D$ be morphisms in a punctured category **C** in which $A \oplus C$ and $B \oplus D$ both exist. Then we define the *direct sum* $f \oplus g$ in one of the following equivalent ways: (i) $f \oplus g = f \coprod g$; (ii) $f \oplus g = f \prod g$; (iii) $f \oplus g = \begin{pmatrix} f & * \\ * & g \end{pmatrix}$.

**2.2.5 Lemma** Let **C** be punctured. Suppose the morphism $\begin{pmatrix} 1 & * \\ * & 1 \end{pmatrix}$ defined by

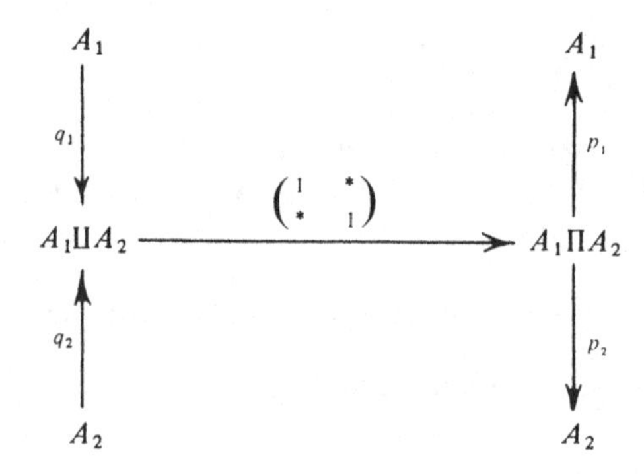


is an isomorphism.

With $q_i' = \begin{pmatrix} 1 & * \\ * & 1 \end{pmatrix} q_i$ the system $\ulcorner p_1, q_1', p_2, q_2' \urcorner$ then makes $A_1 \prod A_2$ into a biproduct $A_1 \oplus A_2$.

*Proof.* We first prove that $A_1 \prod A_2$ is a sum with canonical injections $q_i'$. Suppose, therefore, $f_i: A_i \to B$. Then $(f_1\ f_2): A_1 \coprod A_2 \to B$. Let $\zeta$ denote the inverse of $\begin{pmatrix} 1 & * \\ * & 1 \end{pmatrix}$; then $(f_1\ f_2)\zeta: A_1 \prod A_2 \to B$ and $(f_1\ f_2)\zeta q_i' = (f_1\ f_2) q_i = f_i$. If also $g q_i' = f_i$ for a certain $g: A_1 \prod A_2 \to B$ we must have $g \begin{pmatrix} 1 & * \\ * & 1 \end{pmatrix} = (f_1\ f_2)$, or $g = (f_1\ f_2)\zeta$. This proves that

$A_1 \prod A_2$ is a sum with canonical injections $q'_i$. Finally we have to show $p_i q'_j = \delta_{ij}$. However $p_i \begin{pmatrix} 1 & * \\ * & 1 \end{pmatrix} q_j = \delta_{ij}$.

Remark. One easily generalizes subsections 2.2.2 to 2.2.5 to more than two factors for which the notation $\oplus A_i$ will be used. When, in the sequel, we refer to a biproduct without further specification we mean a finite biproduct (usually consisting of only two factors).

## 2.3 Additive categories

**2.3.1 Definition** Let **C** be any category. Let $S$: **Ab** → **Sets** be the forgetful functor. **C** is called *additive* provided the bifunctor $(-,-)_{\mathbf{C}}$ can be factored through $S$. This means that there is a bifunctor $(-,-)'_{\mathbf{C}}: \mathbf{C}^\circ \times \mathbf{C} \to \mathbf{Ab}$ such that the diagram

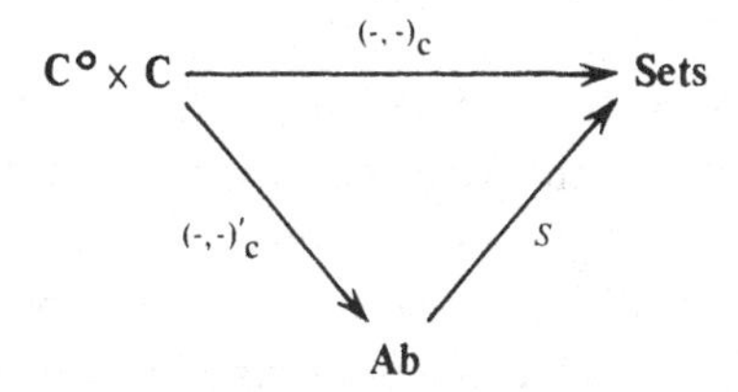


commutes.

As before (see 1.2.5) we use the notation $(-,-)_{\mathbf{C}}\ulcorner A, B\urcorner = (A, B)_{\mathbf{C}}$ and $(-,-)_{\mathbf{C}}\ulcorner f, g\urcorner = (f, g)_{\mathbf{C}}$ where $f: A \to A_1$ and $g: B \to B_1$ are in **C** (so that $f^\circ: A_1 \to A$ in $\mathbf{C}^\circ$). Thus for $u \in (A, B)_{\mathbf{C}}$ we have $(f, g)_{\mathbf{C}} u = g \circ u \circ f^\circ \in (A_1, B_1)_{\mathbf{C}}$.

Remark. For $A$ and $B$ in **C** we have $S(A, B)'_{\mathbf{C}} = (A, B)_{\mathbf{C}}$. This means that the set $(A, B)_{\mathbf{C}}$ is equipped with an abelian group structure and is then called $(A, B)'_{\mathbf{C}}$. Hence for $u: A \to B$ and $v: A \to B$ in **C** there is a morphism $u + v: A \to B$. For $f: A \to A_1$ and $g: B \to B_1$ in **C** we have $S(f, g)'_{\mathbf{C}} = (f, g)_{\mathbf{C}}$. This implies that the mapping $(f, g)_{\mathbf{C}}$ may be considered as a group-homomorphism: $(f, g)_{\mathbf{C}}(u + v) = (f, g)_{\mathbf{C}} u + (f, g)_{\mathbf{C}} v$ or $g \circ (u + v) \circ f^\circ = (g \circ u \circ f^\circ) + (g \circ v \circ f^\circ)$. In the sequel we will frequently not distinguish between $(A, B)_{\mathbf{C}}$ and $(A, B)'_{\mathbf{C}}$ or between $(f, g)_{\mathbf{C}}$ and $(f, g)'_{\mathbf{C}}$.

**Exercise** Let **C** be any category. Then the following two assertions are equivalent.

(i) **C** is additive.

(ii) On each of the $(X, Y)_{\mathbf{C}}$ there is the structure of an abelian group so that $(A, B)_{\mathbf{C}} \times (C, A)_{\mathbf{C}} \to (C, B)_{\mathbf{C}}$ is always biadditive.

If we denote the null elements of the abelian groups $(A, B)'_{\mathbf{C}}$ by $0_{AB}$ it follows that these elements $\{0_{AB}\}_{A,B \in \mathbf{C}}$ form a null family in the category **C**. Since a category can have at most one null family we see that if **C** was a punctured category with null family $\{*_{AB}\}_{A,B \in \mathbf{C}}$ this family would coincide with the null family $\{0_{AB}\}_{A,B \in \mathbf{C}}$. We may therefore always assume that the additive category has a null object 0.

**2.3.2 Definition** An *abelian semigroup* or an *abelian monoid* is a set with a composition (which mostly is denoted additively) that is commutative and associative with a neutral element (null element) denoted by 0.

Let **AbMon** denote the category of abelian monoids with as morphisms the semigroup-homomorphisms $f(a+b)=f(a)+f(b)$ and $f(0)=0$. The latter has to be stipulated because there need not be (additive) inverses in a monoid.

Replacing **Ab** in 2.3.1 by this category **AbMon** we might define a *semi-additive category* **C**. Just as in the case of **Ab** we may try to construct a null family by taking the null elements of the abelian monoids. However, in this case, condition (ii) of the definition of a null family (2.2.1) is not automatically satisfied since we may not be able to subtract. Therefore we require in addition for semi-additivity that the neutral elements $0_{AB}$ in the monoids $(A, B)'_{\mathbf{C}}$ form a null family $\{0_{AB}\}$ in **C**. From the definitions it is clear that a semi-additive category is additive if and only if every identity morphism $1_A$ has an additive inverse $-1_A$ in $(A, A)$ such that $1_A+(-1_A)=0_A$.

In an additive category the morphism $u: A \to B$ is an epimorphism (monomorphism) if and only if $fu=0_{AC}$ for $f: B \to C$ implies $f=0_{BC}$ ($ug=0_{CB}$ for $g: C \to A$ implies $g=0_{CA}$).

**2.3.3 Proposition** Let **C** be semi-additive. Consider

$$\coprod_{k=1}^{l} A_k \xrightarrow[(f_{jk})]{} \bigoplus_{j=1}^{m} B_j \xrightarrow[(g_{ij})]{} \prod_{i=1}^{n} C_i$$

in **C**. Then $(f_{jk})$ and $(g_{ij})$ can be composed in the following way: $(h_{ik})=(g_{ij})(f_{jk})$ with the matrix multiplication rule $h_{ik}=\sum_{j=1}^{m} g_{ij}f_{jk}$.

*Proof.* Choose $f_j=(f_{j1}\ f_{j2} \ldots f_{jl})$ and choose for $g_j$ the column with elements $g_{1j} \ldots g_{nj}$. In other words

$$f_j: \coprod_{k=1}^{l} A_k \to B_j \quad \text{and} \quad g_j: B_j \to \prod_{i=1}^{n} C_i.$$

We assert that

$$(f_{jk}) = \sum_{t=1}^{m} q_t^B f_t.$$

This follows from

$$p_j^B\left(\sum_{t=1}^{m} q_t^B f_t\right) q_k^A = \sum_{t=1}^{m} \delta_{jt}^B f_{tk} = f_{jk}.$$

Similarly one finds

$$(g_{ij}) = \sum_{s=1}^{m} g_s p_s^B.$$

Therefore

$$(g_{ij})(f_{jk}) = \left(\sum_{s=1}^{m} g_s p_s^B\right)\left(\sum_{t=1}^{m} q_t^B f_t\right) = \sum_{s,t=1}^{m} g_s \delta_{st}^B f_t = \sum_{t=1}^{m} g_t f_t.$$

This means

$$h_{ik} = p_i^C\left(\sum_{t=1}^{m} g_t f_t\right) q_k^A = \sum_{t=1}^{m} (p_i^C g_t)(f_t q_k^A) = \sum_{t=1}^{m} g_{it} f_{tk}.$$

Remark. This proof goes through even if the index sets for the sum $\coprod A_k$ and for the product $\prod C_i$ are arbitrary (not necessarily finite).

**2.3.4 Proposition** Let **C** be a semi-additive category with finite sums and products. Then

$$\begin{pmatrix} 1 & 0 \\ 0 & 1 \end{pmatrix} : -\coprod- \rightrightarrows -\prod-$$

is an isomorphism of bifunctors and so **C** has biproducts.

*Proof.* We shall establish that for all $A_1$ and $A_2$ in **C**

$$\begin{pmatrix} 1 & 0 \\ 0 & 1 \end{pmatrix}^A : A_1 \coprod A_2 \to A_1 \prod A_2$$

is an isomorphism. According to the corollary in 2.2.3 $\begin{pmatrix} 1 & 0 \\ 0 & 1 \end{pmatrix}$ then is an isomorphism of bifunctors and hence, according to lemma 2.2.5, **C** has biproducts.

To prove the isomorphism consider

$$A_1 \prod A_2 \xrightarrow[p_i]{} A_i \xrightarrow[q_i]{} A_1 \coprod A_2$$

and so

$$q_1p_1+q_2p_2\colon A_1\prod A_2 \to A_1\coprod A_2$$

(**C** is semi-additive). Now we have

$$p_1\begin{pmatrix}1 & 0\\ 0 & 1\end{pmatrix}^A(q_1p_1+q_2p_2)=p_1\begin{pmatrix}1 & 0\\ 0 & 1\end{pmatrix}^A q_1p_1+p_1\begin{pmatrix}1 & 0\\ 0 & 1\end{pmatrix}^A q_2p_2=p_1$$

and just so

$$p_2\begin{pmatrix}1 & 0\\ 0 & 1\end{pmatrix}^A(q_1p_1+q_2p_2)=p_2.$$

But this means

$$\begin{pmatrix}1 & 0\\ 0 & 1\end{pmatrix}^A(q_1p_1+q_2p_2)=1_{A_1\prod A_2}.$$

Dually $(q_1p_1+q_2p_2)\begin{pmatrix}1 & 0\\ 0 & 1\end{pmatrix}^A=1_{A_1\coprod A_2}$. Therefore $\begin{pmatrix}1 & 0\\ 0 & 1\end{pmatrix}^A$ is an isomorphism.

**Exercise** The following somewhat sharper statement is true in a semi-additive category. If $\ulcorner\coprod A_i, q_i\urcorner$ is a finite sum, define $p_j\colon \coprod A_i \to A_j$ by putting $p_jq_i=\delta_{ji}$. Then $p_jq_j=1_{A_j}$ for every $j$, while $\sum q_jp_j=1_{\coprod A_i}$ and $\ulcorner\coprod A_i, p_i\urcorner$ is a product of the $A_i$s.

**2.3.5** Suppose **C** is semi-additive with a biproduct $A\oplus A$ for every $A\in\mathbf{C}$. Furthermore let $f_1, f_2\colon A \rightrightarrows B$ in **C**. Then

$$\begin{pmatrix}f_1\\ f_2\end{pmatrix}\colon A \to B\oplus B \quad\text{and}\quad p_1+p_2\colon B\oplus B \to B.$$

Since $(p_1+p_2)q_i=1 (i=1, 2)$ we have $p_1+p_2=(1\ \ 1)$.

We also have

$$f_1+f_2=(1\ \ 1)\begin{pmatrix}f_1\\ f_2\end{pmatrix}$$

(matrix multiplications, see 2.3.3). Now the morphism $(1\ \ 1)\begin{pmatrix}f_1\\ f_2\end{pmatrix}$ is defined with the aid of the canonical injections and projections of the biproduct $B\oplus B$. Thus this expression may even be defined in a punctured category without semi-additive structure. Moreover $f_1+f_2$ does not depend on the choice of the bimorphism $(\text{-}, \text{-})'_{\mathbf{C}}$ in 2.3.2. This means that the bimorphism $(\text{-}, \text{-})'_{\mathbf{C}}$ is determined by the addition of morphisms (compare the exercise in 2.3.1). Summarizing we have

**Lemma** In a semi-additive category **C**, having a biproduct $A \oplus A$ for each object $A \in \mathbf{C}$, addition is uniquely determined. There is just one bimorphism $(-,-)'_{\mathbf{C}}$ making the diagram in 2.3.1, with **Ab** changed into **AbMon**, commutative.

Remark. According to 2.3.4 the lemma holds for a semi-additive category with sums and products of the type $A \coprod A$, $A \prod A$. In view of the exercise there, we need only assume the existence of either of these.

**2.3.6 Theorem** Let **C** be a punctured category having a biproduct $A \oplus A$ for every $A \in \mathbf{C}$. Then **C** is semi-additive.

*Proof.* Consider, for $A$, $B$ and $C \in \mathbf{C}$, the diagram

$$A \underset{f_2}{\overset{f_1}{\rightrightarrows}} B \underset{g}{\rightarrow} C.$$

We want to show that the following diagram commutes.

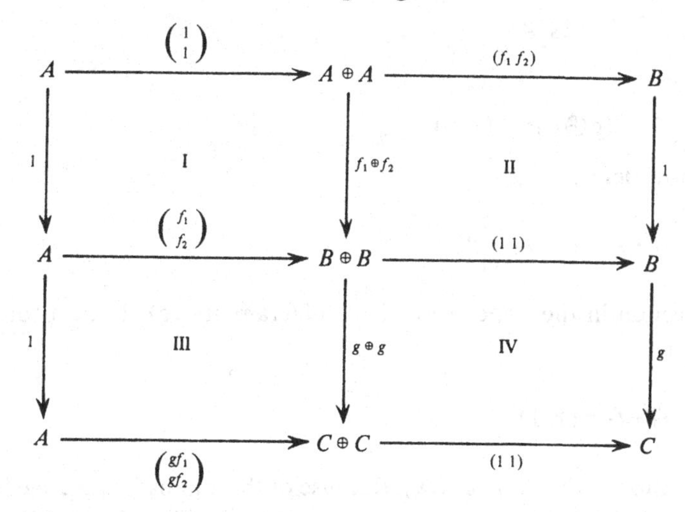


For I we have

$$p_i^B(f_1 \oplus f_2)\binom{1}{1} = f_i p_i^A \binom{1}{1} = f_i \qquad (i = 1, 2)$$

and hence

$$(f_1 \oplus f_2)\binom{1}{1} = \binom{f_1}{f_2}.$$

Dually for II. For III:

$$p_i^C(g \oplus g)\binom{f_1}{f_2} = g p_i^B \binom{f_1}{f_2} = g f_i \qquad (i = 1, 2),$$

so that

$$(g \oplus g)\binom{f_1}{f_2} = \binom{gf_1}{gf_2}.$$

For IV:

$$(1\ \ 1)(g \oplus g)q_i^B = (1\ \ 1)q_i^C g = g \qquad (i = 1, 2),$$

so that

$$(1\ \ 1)(g \oplus g) = (g\ \ g).$$

Since also

$$g(1\ \ 1)q_i^B = g \qquad (i = 1, 2),$$

we have

$$g(1\ \ 1) = (g\ \ g).$$

Together:

$$(1\ \ 1)(g \oplus g) = g(1\ \ 1).$$

We now define

$$f_1 + f_2 = (f_1\ \ f_2)\binom{1}{1}$$

(composition in the upper row of the diagram above). Then it follows that also

$$f_1 + f_2 = (1\ \ 1)\binom{f_1}{f_2}$$

(composition in the middle row). Because of the lemma in 2.3.5 we have no other choice for $f_1 + f_2$ than the one made here. Since for $A \in \mathbf{C}$ there may be more than just one biproduct we make a fixed choice for $A \oplus A$ throughout.

We shall now verify that this definition makes $\mathbf{C}$ into a semi-additive category by showing that the addition satisfies requirement (ii) of the exercise in 2.3.1 for semi-additivity. In the first place, the addition turns the sets $(A, B)_\mathbf{C}$ into abelian monoids with the null morphisms $0_{AB}$ as neutral elements as we will show below. In the second place, if

$$A \underset{f_2}{\overset{f_1}{\rightrightarrows}} B \underset{g}{\rightarrow} C,$$

the equivalence relation on $B$ defined by $x \sim y$ if $x, y \in A$ or $x = y$. Take $C$ equal to $B/\sim$ and $*_C$ the class containing $*_B$. For $c: B \to C$ one takes the canonical mapping.

(*b*) For **Top**$_*$ kernels and cokernels are constructed as in (*a*). For $K$ the topology induced by the topology of $A$ is used. Then $k$ is continuous as is the map $s$ in 2.4.1. The topology for $C$ is again defined (see 1.4.8(*d*)) by the requirement that $U$ is open in $C$ if and only if $c^{-1}(U)$ is open in $B$. Then $c$ (and also $t$ in 2.4.1°) are both continuous.

(*c*) For **Gr** the kernel of $f: G \to H$ is just the embedding of

$$K = \{x \in G \mid fx = e\}$$

in $G$. The cokernel is the canonical mapping $c: H \to H/N$, where $N$ equals the intersection of all normal subgroups of $H$ containing the set-theoretic image of $f$.

(*d*) For **Ab** we have the same situation as in (*c*), only now $N$ equals the set-theoretic image of $f$.

(*e*) For $\mathbf{M}_R$ the constructions of kernels and cokernels are as in (*d*).

**2.4.3** In an additive category **A** the converse of 2.4.1(*b*) holds. For if $0_{0A}$ is the kernel of $f: A \to B$ and $g, h: C \rightrightarrows A$ with $fg = fh$ then $f(g-h) = 0_{CB}$ implies there is a unique $s: C \to 0$ such that $g - h = 0_{0A}s = 0_{CA}$. Hence $g = h$. The reader should formulate the dual statement.

## 2.5 Exact and abelian categories

In this section all categories are punctured, unless explicitly stated otherwise. We list a number of axioms for future reference.

N. Every monomorphism is the kernel of a morphism.

N°. Every epimorphism is the cokernel of a morphism.

K. Every morphism has a kernel (kernels exist).

K°. Every morphism has a cokernel (cokernels exist).

P. Finite products exist.

P°. Finite sums exist.

Z. Every morphism may be factored through an epimorphism and a monomorphism:

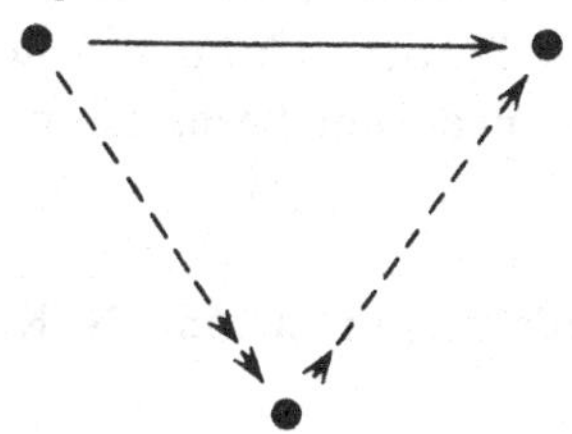

AB2. Putting $k = \ker f$, $c = \operatorname{coker} f$, $s = \operatorname{coker} k$, $t = \ker c$, there is a unique morphism $\tilde{f}$ such that $f = t\tilde{f}s$, with $\tilde{f}$ an isomorphism:

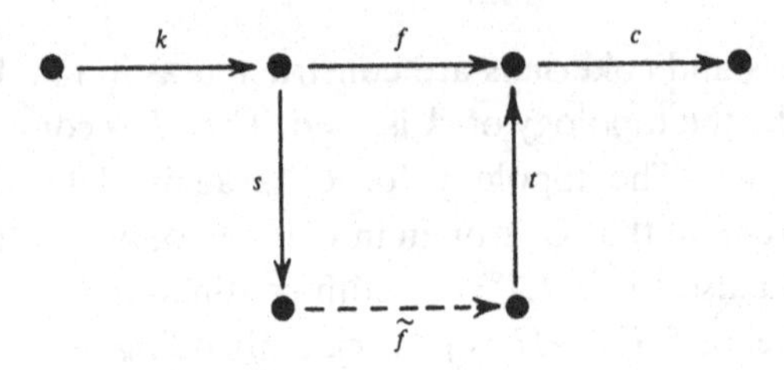


Notice that the axioms Z and AB2 are self dual (i.e. they have the same formulation in the dual category).

**2.5.1** Suppose a punctured category satisfies axiom N. If a monomorphism $f$ has a cokernel, then $f$ is a kernel of this cokernel: put $c = \operatorname{coker} f$ and let $s$ be the morphism such that $f = \ker s$. According to 2.4.2 we have $f = \ker \operatorname{coker} \ker s$, or $f = \ker c$, or $f = \ker \operatorname{coker} f$.

Dually: a category satisfying axiom N° enjoys the property that every epimorphism $f$ with a kernel equals the cokernel of its kernel: $f = \operatorname{coker} \ker f$.

**2.5.2 Lemma** In a category satisfying axiom N every bimorphism is an isomorphism.

*Proof.* Let $f: A \to B$ be a bimorphism. Consider $A \overset{f}{\twoheadrightarrow} B \overset{*}{\to} *$. Then $*$ is a cokernel of $f$. Since $f$ is a monomorphism, 2.5.1 yields $f = \ker *$. Also $1_B = \ker *$. Since kernels are isomorphic it follows that $f$ is an isomorphism (see 2.4.1).

Dually: for a category satisfying axiom N° every bimorphism is an isomorphism.

**2.5.3 Definition** A category in which every bimorphism is an isomorphism is called *balanced*. Balanced categories are, for example, **Sets**, **Gr**, **Ab.** An example of a category that is not balanced is **CRg**: for $\mathbb{Z} \rightarrowtail\!\!\!\!\twoheadrightarrow \mathbb{Q}$ is a bimorphism which is not an isomorphism; see 2.1.4($c$). Another example is **Hausd**: for $f: X \rightarrowtail\!\!\!\!\twoheadrightarrow Y$ with $f(X)$ dense in $Y$ the monomorphism $f$ is a bimorphism that need not be an isomorphism (take $\mathbb{Q} \rightarrowtail\!\!\!\!\twoheadrightarrow \mathbb{R}$); see 2.1.6($a$).

**2.5.4 Lemma** In a category satisfying the axioms N, K and K° every diagram

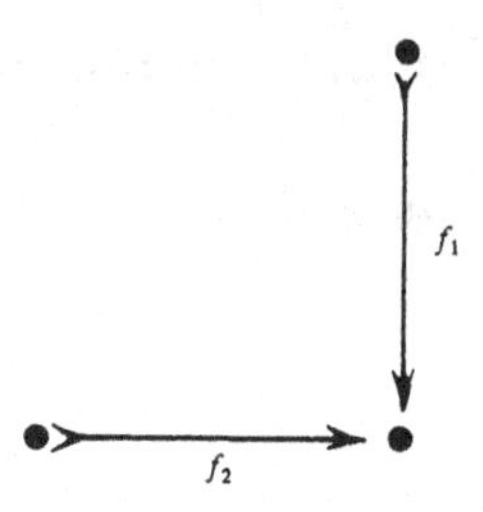


may be completed to a cartesian square.

*Proof.* Let $c = \text{coker } f_1$ and $k_2 = \ker cf_2$. Then $f_1 = \ker c$ (2.5.1). Since $cf_2k_2 = *$, there is a unique $k_1$ such that $f_1k_1 = f_2k_2$. Let $f_1s_1 = f_2s_2$:

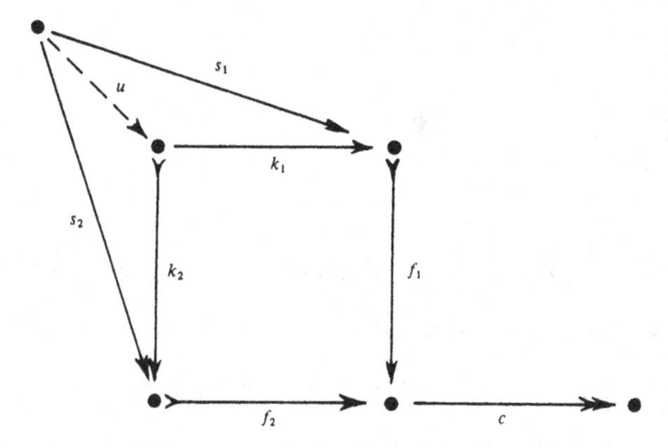


Since $cf_1 = *$, assuredly $cf_1s_1 = *$, and therefore also $cf_2s_2 = *$. Because $k_2 = \ker cf_2$ there is a unique $u$ such that $k_2u = s_2$. We also have $f_1k_1u = f_2k_2u = f_2s_2 = f_1s_1$. Since $f_i$ is monic, it follows that $k_iu = s_i$ $(i = 1, 2)$.

Dually: in a category satisfying the axioms N°, K and K° every diagram

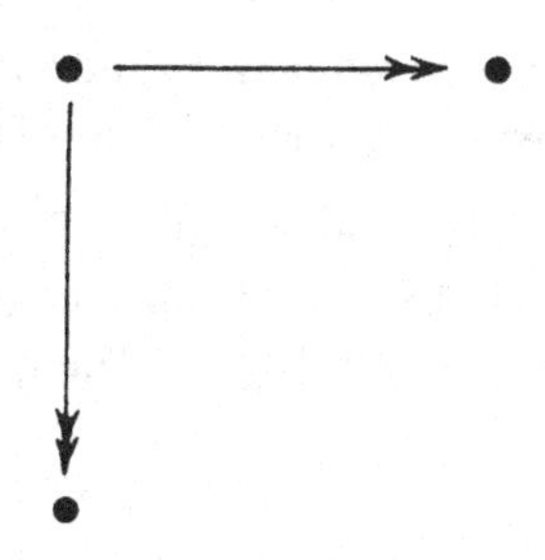

may be completed into a cocartesian square.

**2.5.5 Lemma** A category satisfying the axioms N, K, K° and P has equalizers.

*Proof.* Consider $f_1, f_2 \colon A \rightrightarrows B$. Then $\begin{pmatrix}1\\f_i\end{pmatrix}$ $(i=1,2)$ are monomorphisms from $A$ to $A \prod B$, since $p_1\begin{pmatrix}1\\f_i\end{pmatrix} = 1_A$. According to lemma 2.5.4 the diagram

$$\begin{pmatrix}1\\f_1\end{pmatrix}, \begin{pmatrix}1\\f_2\end{pmatrix} \colon A \rightrightarrows A \prod B$$

may be completed into a cartesian square:

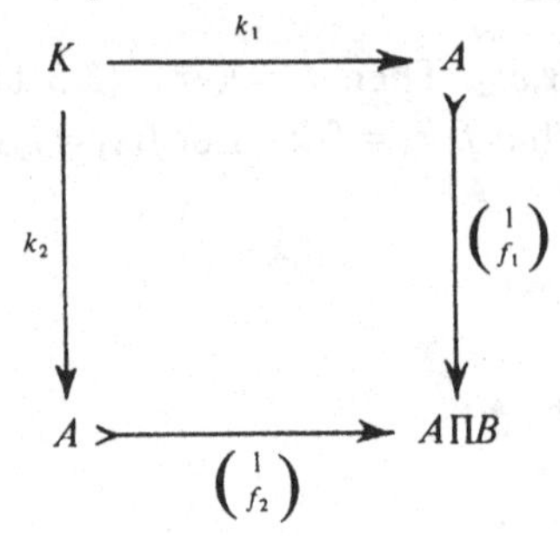

We then have $\begin{pmatrix}1\\f_1\end{pmatrix} k_1 = \begin{pmatrix}1\\f_2\end{pmatrix} k_2$ and hence $p_1\begin{pmatrix}1\\f_1\end{pmatrix} k_1 = p_1\begin{pmatrix}1\\f_2\end{pmatrix} k_2$ so that $k_1 = k_2$ $(=k$, say). We assert that $k \colon K \to A$ is an equalizer for $f_1, f_2 \colon A \rightrightarrows B$. For in the first place it follows from $p_2\begin{pmatrix}1\\f_1\end{pmatrix} k = p_2\begin{pmatrix}1\\f_2\end{pmatrix} k$ that $f_1 k = f_2 k$. The universal property of $k$ comes from the fact that the square is cartesian.

Dually: a category satisfying N°, K, K° and P° has coequalizers.

Remark. Now it follows from the exercise in 1.8.4 that a category satisfying N°, K, K° and P° has cocartesian squares. Dually a category satisfying N, K°, K and P has cartesian squares.

**2.5.6 Lemma** Let **C** be a category with equalizers and suppose **C** satisfies axiom N. Let $f$ be a morphism with cokernel $*$. Then $f$ is epic.

*Proof.* Suppose $u_1 f = u_2 f$. Let $k$ be an equalizer for $u_1$ and $u_2$. Then $k$ is monic. Furthermore there is a unique $v$ such that $kv = f$. Since coker $f =$ $*$ (assumption) we also have coker $k = *$ (see 2.4.1°($f$)°). Then, using 2.5.1, we find $k = \ker \operatorname{coker} k = \ker *$:

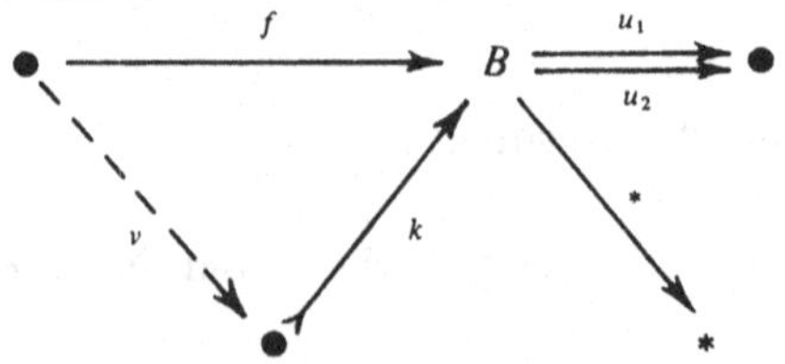

Since $B \xrightarrow{1_B} B \to * = B \to *$, there is a unique morphism $w$ such that $B \xrightarrow{w} \bullet \xrightarrow{k} B = B \xrightarrow{1_B} B$. Then the fact that $kw = 1_B$ yields that $k$ is epic, so that $u_1 = u_2$.

Dually: let $C$ be a category with coequalizers and suppose **C** satisfies axiom N°. Let $f$ be a morphism with kernel $*$. Then $f$ is monic.

**2.5.7 Proposition** If **C** satisfies axioms N, K, K° and P, then **C** also satisfies axiom Z.

*Proof.* We have to show that every morphism $\bullet \xrightarrow{f} \bullet$ may be factored in the following way:

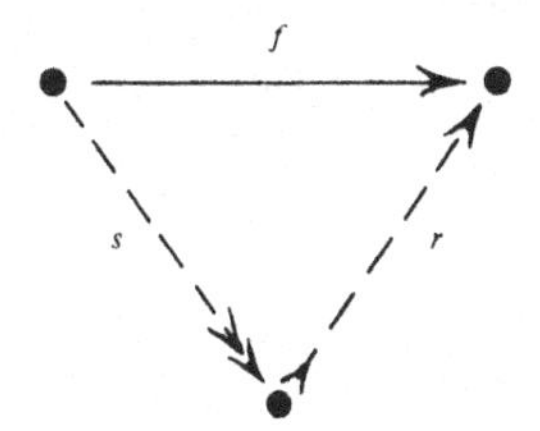


Let $c = \operatorname{coker} f$ and $r = \ker c$. Then $r$ is monic and $c$ is epic:

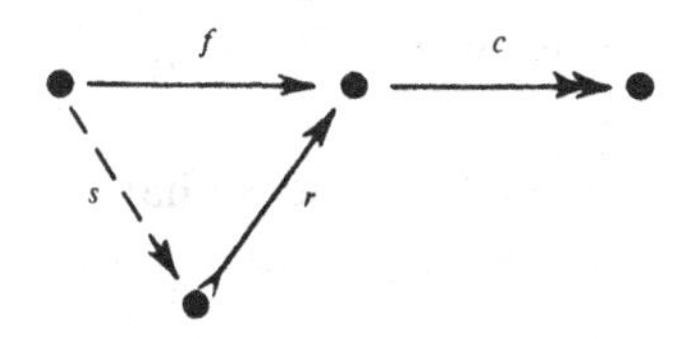


Now there is a unique $s$ such that $rs = f$. We will show that $s$ is epic. Let $g = \operatorname{coker} s$ and $u = \ker g$. Then there is a unique $v$ such that $uv = s$:

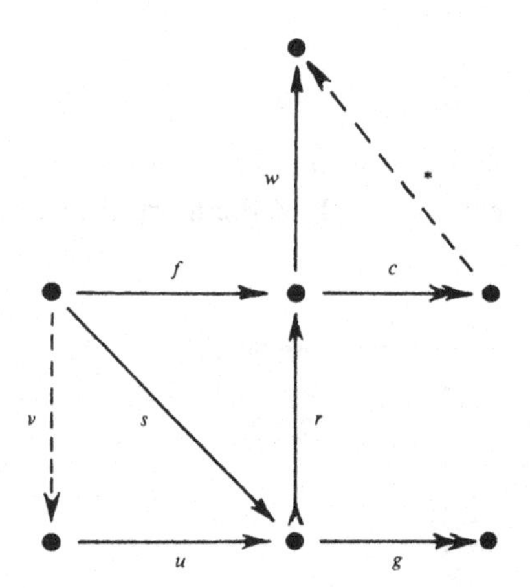

Since $r = \ker c$, we have $cr = *$ and so $cru = *$. We contend that $c = \operatorname{coker} ru$. For, if we have $w$ with $wru = *$, then $* = wruv = wrs = wf$ so that there is a unique $x$ with $xc = w$ (due to the fact that $c = \operatorname{coker} f$).

As both $r$ and $u$ are monic, so is $ru$. According to 2.5.1 we may conclude that $ru = \ker \operatorname{coker} ru = \ker c$. Since also $r = \ker c$, $u$ must be an isomorphism. From $gu = *$ ($u = \ker g$) we then find $g = *$. Finally $* = g = \operatorname{coker} s$ implies $s$ is epic (by 2.5.6).

**2.5.8 Lemma** The (punctured!) category **C** satisfies axioms K, K°, AB2 if and only if it satisfies axioms N, N° and Z.

*Proof.* Assuming N, N° and Z we first prove the validity of axiom K. Let $f$ be a morphism, factored according to axiom Z:

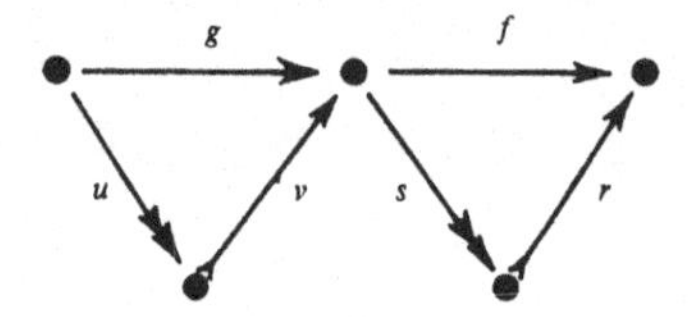


In virtue of N° there is a morphism $g$ with $s = \operatorname{coker} g$. We factor $g$ again as we did $f$. Since $u$ is epic we have $s = \operatorname{coker} g = \operatorname{coker} vu = \operatorname{coker} v$ (2.4.1°(g)°). Now $v$ is monic, hence (2.5.1) $v = \ker s$. Furthermore $\ker f = \ker rs = \ker s$ (2.4.1(g)) since $r$ is monic. Thus K is proved. Dually one proves K°.

Now we show that axiom AB2 is satisfied. Suppose we have

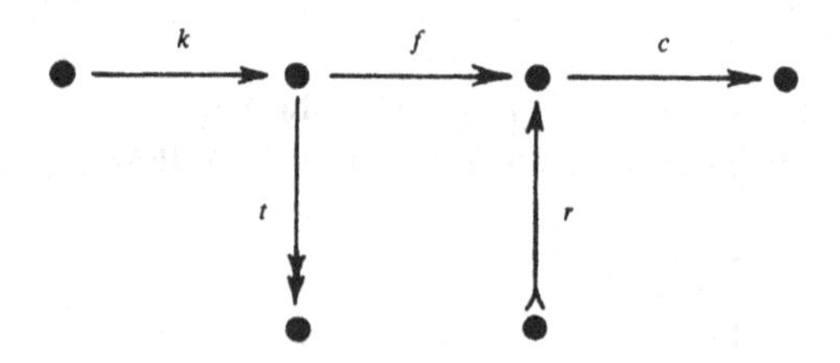


with $k = \ker f$, $c = \operatorname{coker} f$, $t = \operatorname{coker} k$, $r = \ker c$. From $cf = *$, $r = \ker c$ it follows that there is a unique $s$ with $rs = f$. By Z the morphism $s$ may be factored:

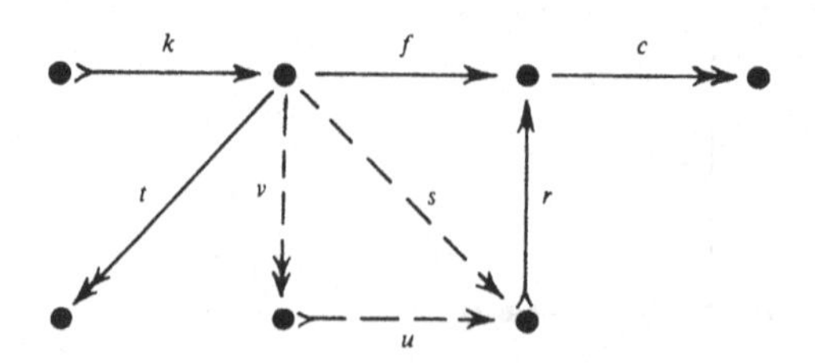

Just as in 2.5.7 it turns out that $u$ is an isomorphism. This makes $s$ an epimorphism (since $v$ is one). Since $k = \ker f$, we have $* = fk = rsk$. As $r$ is monic it then follows that $sk = *$. Since $t = \operatorname{coker} k$ there is a unique $\tilde{f}$ with $\tilde{f}t = s$. The morphism $s$ is epic so $\tilde{f}$ is. Now we have $r\tilde{f}t = rs = f$. Suppose also $r\bar{f}t = f$. But $r$ is monic and $t$ is epic, so this would imply $\bar{f} = \tilde{f}$. Hence $\tilde{f}$ is unique.

The above construction with $s$, $u$ and $v$ may be repeated in a dual fashion: from $fk = *$, $t = \operatorname{coker} k$, it follows that there is a unique $s'$ with $s't = f$, etc. We will then find a monomorphism $\hat{f}$ with $r\hat{f}t = f$. Then uniqueness implies $\hat{f} = \tilde{f}$, proving that $f$ is a bimorphism. Since the category **C** is balanced (it satisfies axiom N, see 2.5.2) $\tilde{f}$ must be an isomorphism, which finishes the first half of the proof.

Going the other way, we first prove the validity of axiom Z. Let $f$ be a morphism and $k = \ker f$, $c = \operatorname{coker} f$, $t = \operatorname{coker} k$ and $r = \ker c$. Then there is an isomorphism $\tilde{f}$ with $r\tilde{f}t = f$, guaranteed by AB2. One factorization of $f$ through an epimorphism and a monomorphism is given by

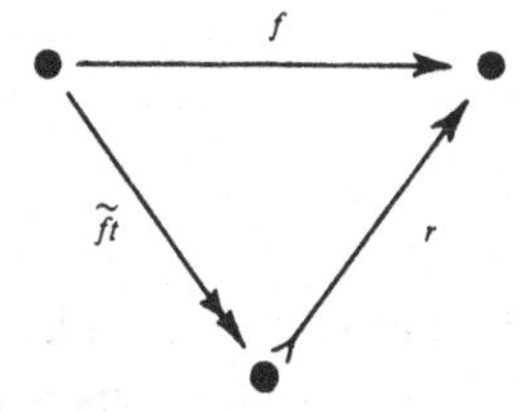


This proves Z.

To prove N, let $f$ be a monomorphism. Then $\ker f = *$ and $1_A = \operatorname{coker} *$. Let $c = \operatorname{coker} f$, $r = \ker c$. Then AB2 gives the following diagram

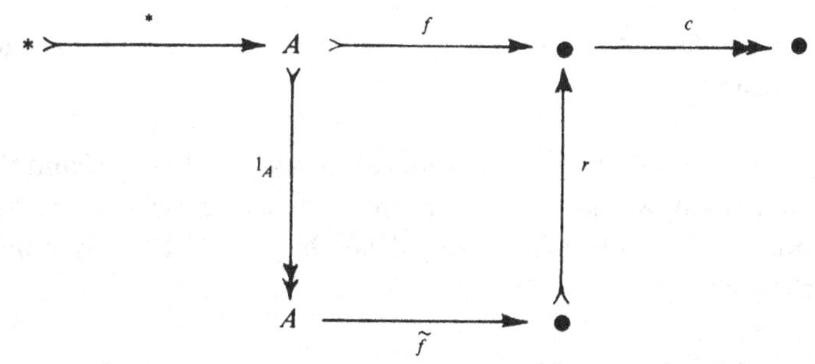


with $\tilde{f}$ an isomorphism. Since $1_A$ is an isomorphism, so is $\tilde{f}1_A$. But then $r = \ker c$ implies $f = \ker c$. This proves N. Dually one proves N°.

**2.5.9 Definition** A punctured category is called *exact* provided it satisfies K, K° and AB2 or, equivalently, N, N° and Z.

Remark. A punctured category satisfying N, N°, K, K° and P or P° is exact according to proposition 2.5.7.

**2.5.10 Lemma** In an exact category the factoring of a morphism through an epimorphism and a monomorphism (*epic–monic factorization* for short) is natural.

*Proof.* Consider the following commutative diagram:

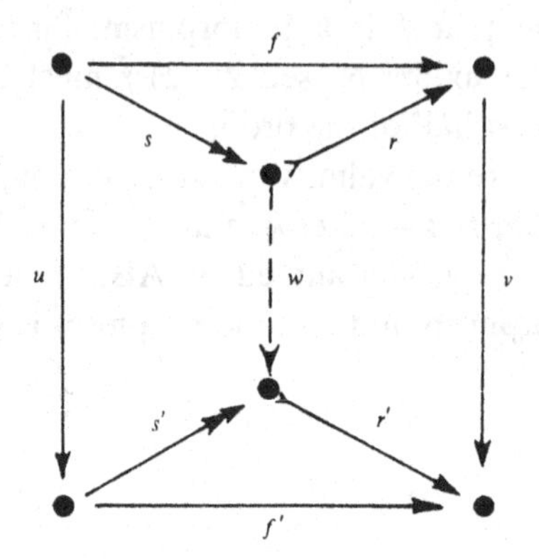


Put $k = \ker f$ and $k' = \ker f'$. Then $\ker f = \ker s$, hence $\operatorname{coker} k = s$ (2.4.1(*g*) and 2.5.1°). Similarly $s' = \operatorname{coker} k'$. Since $r's'uk = f'uk = vfk = *$ and $r'$ is monic we have $s'uk = *$ and there is a unique $w$ with $ws = s'u$ since $s = \operatorname{coker} k$. Now $r'ws = r's'u = f'u = vf = vrs$. Since $s$ is epic it follows that $r'w = vr$.

**Corollary** An epic–monic factorization is unique up to isomorphism.

*Proof.* Taking $u$ and $v$ as identity morphisms we find $ws = s'$ and $r'w = r$. In the same way we have a morphism $\tilde{w}$ satisfying $\tilde{w}s' = s$ and $r\tilde{w} = r'$. From these we infer $w\tilde{w}s' = ws = s'$ so that $w\tilde{w} = 1$ ($s'$ is epic) and similarly $\tilde{w}w = 1$.

**2.5.11 Lemma** For a category satisfying P° the morphism $(1 \ \ *): A \coprod B \to A$ is a cokernel for $q_2: B \to A \coprod B$. If the category also satisfies N then $q_2 = \ker (1 \ \ *)$.

*Proof.* Firstly, we have $(1 \ \ *)q_2 = *$ (see diagram on top of page 91). Suppose further that $(f \ \ g): A \coprod B \to C$ has the property that $(f \ \ g)q_2 = *$.

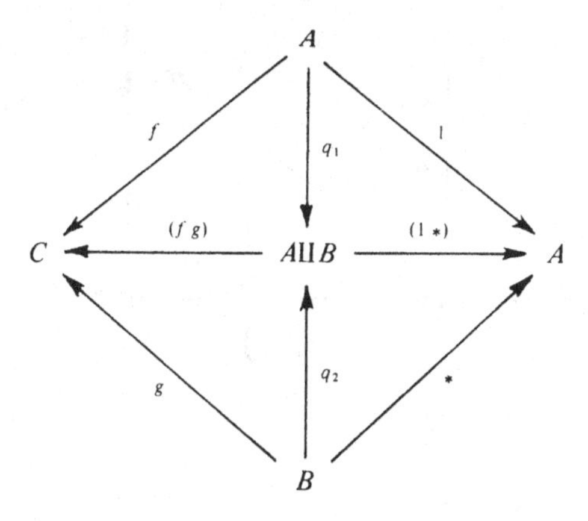


Then $g=*$ and $(f\ *)\colon A \coprod B \to C$ can be factored in a unique way through $A$:

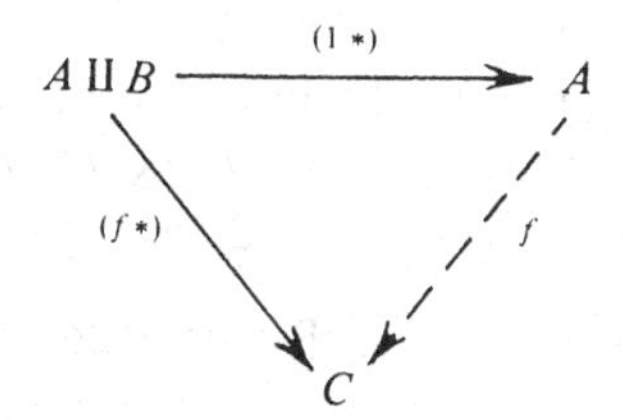


This triangle commutes since $(f\ *)q_1 = f = f(1\ *)q_1$ and $(f\ *)q_2 = * = f(1\ *)q_2$.

It should be clear that $f$ is unique. This proves that $(1\ *) = \operatorname{coker} q_2$. According to proposition 2.2.2° $q_2$ is monic so that if the category satisfies axiom N lemma 2.5.1 implies that $q_2 = \ker(1\ *)$. Similar statements of course apply to the morphism $(*\ 1)\colon A \coprod B \to B$.

**2.5.11° Lemma** For a category satisfying P the morphism $\begin{pmatrix}1\\ *\end{pmatrix}\colon A \to A \prod B$ is a kernel of $p_2\colon A \prod B \to B$ and $\begin{pmatrix}*\\ 1\end{pmatrix}\colon B \to A \prod B$ is a kernel of $p_1\colon A \prod B \to A$. If the category also satisfies N° then $p_2 = \operatorname{coker}\begin{pmatrix}1\\ *\end{pmatrix}$ and $p_1 = \operatorname{coker}\begin{pmatrix}*\\ 1\end{pmatrix}$.

**2.5.12 Theorem** A category satisfying N, N°, K, K°, P and P° is additive.

*Proof.* In the diagram

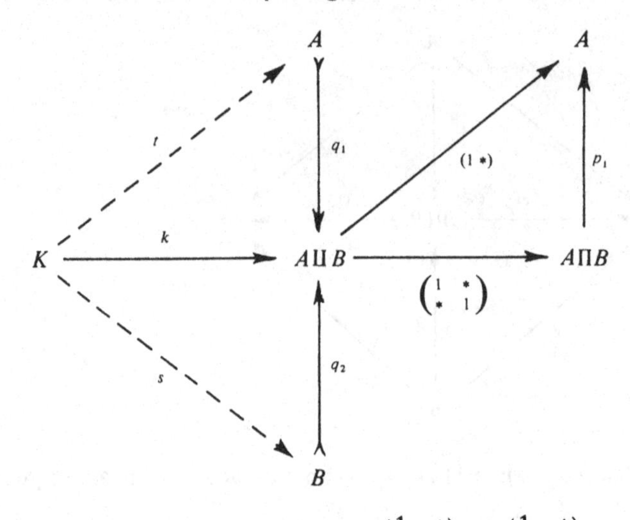


$q_1$ and $q_2$ are monomorphisms; $k = \ker\begin{pmatrix}1 & * \\ * & 1\end{pmatrix}$; $p_1\begin{pmatrix}1 & * \\ * & 1\end{pmatrix} = (1\ \ *)$. Since $q_2 = \ker(1\ \ *)$ in virtue of lemma 2.5.11, there is a unique $s$ with $q_2 s = k$.

In the same way one proves that there is a unique $t$ with $q_1 t = k$. Now we have $* = (1\ \ *)q_2 s = (1\ \ *)k = (1\ \ *)q_1 t = 1t = t$. But then $k = *$, which implies (lemma 2.5.6°) that $\begin{pmatrix}1 & * \\ * & 1\end{pmatrix}$ is monic. By considering $c = \operatorname{coker}\begin{pmatrix}1 & * \\ * & 1\end{pmatrix}$ one shows in a similar way that $\begin{pmatrix}1 & * \\ * & 1\end{pmatrix}$ is epic. Since the category is balanced this proves that $\begin{pmatrix}1 & * \\ * & 1\end{pmatrix}$ is an isomorphism. According to lemma 2.2.5 and theorem 2.3.6 the category is semi-additive, so we write 0 for $*$. To prove additivity it is sufficient to show that for every object $A$ the morphism $1_A$ has an inverse $-1_A$ (see 2.3.2).

We have already noted that lemma 2.2.5 tells us that $A \coprod A = A \prod A$ is a biproduct $A \oplus A$. Then consider $\begin{pmatrix}1 & 1 \\ 0 & 1\end{pmatrix}: A \oplus A \to A \oplus A$.

Let $k = \ker\begin{pmatrix}1 & 1 \\ 0 & 1\end{pmatrix}$, $k = \begin{pmatrix}f \\ g\end{pmatrix}$:

$$K \xrightarrow{k} A \oplus A \xrightarrow{\begin{pmatrix}1 & 1 \\ 0 & 1\end{pmatrix}} A \oplus A.$$

From $\begin{pmatrix}1 & 1 \\ 0 & 1\end{pmatrix} k = 0$ it follows by matrix multiplication that $0 = \begin{pmatrix}f+g \\ g\end{pmatrix}$. Thus $g = 0$ and $f + g = 0$ so that $k = 0$. This means that $\begin{pmatrix}1 & 1 \\ 0 & 1\end{pmatrix}$ is monic

(2.5.6°). In a similar fashion, taking $c = \text{coker}\begin{pmatrix}1 & 1\\0 & 1\end{pmatrix}$ one proves that $\begin{pmatrix}1 & 1\\0 & 1\end{pmatrix}$ is epic. So it must be an isomorphism. Let $\chi: A\oplus A \to A\oplus A$ be the inverse of $\begin{pmatrix}1 & 1\\0 & 1\end{pmatrix}$, say $\chi = \begin{pmatrix}u & v\\w & x\end{pmatrix}$.

Then a short computation shows

$$1 = \begin{pmatrix}1 & 0\\0 & 1\end{pmatrix} = \begin{pmatrix}1 & 1\\0 & 1\end{pmatrix}\begin{pmatrix}u & v\\w & x\end{pmatrix} = \begin{pmatrix}u+w & v+x\\w & x\end{pmatrix}.$$

Thus $x = 1_A$ and $v + x = 0$ so that $v = -1_A$.

**2.5.13 Definition** A punctured category satisfying N, N°, K, K°, P and P° is called *abelian.* As we have seen above, an abelian category is both exact and additive. But also P, P° and exactness imply that a category is abelian. From the remark after 2.5.5° it follows that an abelian category has both cartesian and cocartesian squares.

Remark. It was shown by D. Puppe [34], whose presentation we have mainly followed, that in the presence of exactness we actually only need P or P°, but this requires a bit more work.

**2.5.14. Examples** Denote by $\mathbf{V}_{1k}$ the category whose objects are vector spaces over the field $k$ of dimension at most one and whose morphisms are the linear maps. Then we have

| | satisfies | | | | | | |
|---|---|---|---|---|---|---|---|
| $\mathbf{V}_{1k}$ | N | N° | K | K° | | | Z |
| **Gr** | | N° | K | K° | P | P° | Z |
| **Sets**$_*$ | N | | K | K° | P | P° | Z |
| **Ab** | N | N° | K | K° | P | P° | Z |
| $\mathbf{M}_R$ | N | N° | K | K° | P | P° | Z |

Therefore **Ab** and $\mathbf{M}_R$ are abelian; $\mathbf{V}_{1k}$ is exact and additive; **Gr** and **Sets**$_*$ are neither exact, nor additive.

We end this section with a useful property of (co)cartesian squares in an arbitrary category.

**2.5.15 Definition** An epimorphism $u$ is called a *universal epimorphism* provided that for every cartesian square in which $u$ is epic $u'$ is also epic. (See diagram on top of page 94.)

Compare with 2.1.2. A *universal monomorphism* is defined dually.

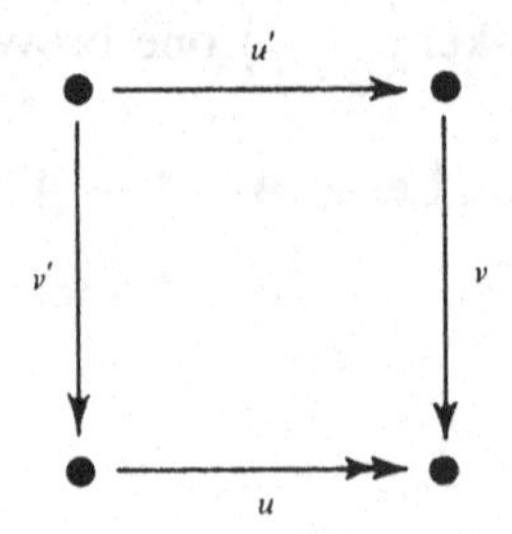


**2.5.16 Lemma** In an abelian category the square

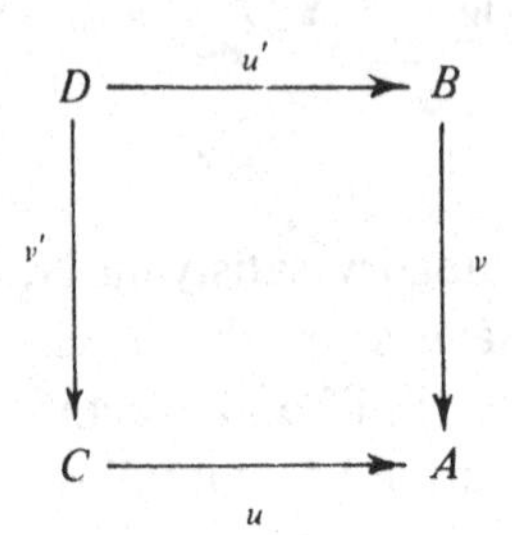


is cartesian if and only if in the sequence

$$D \xrightarrow{\binom{u'}{v'}} B \oplus C \xrightarrow{(-v\ \ u)} A,$$

$\begin{pmatrix} u' \\ v' \end{pmatrix} = \ker(-v\ \ u)$.

*Proof.* Suppose the square is cartesian. Then $(-v\ \ u)\begin{pmatrix} u' \\ v' \end{pmatrix} = -vu' + uv' =$ $0$. If $(-v\ \ u)\begin{pmatrix} f \\ g \end{pmatrix} = 0$ it follows in the first place that $-vf + ug = 0$, or $vf = ug$. Since our square is cartesian there exists a unique $h$ with $u'h = f$ and $v'h = g$; in other terms, $\begin{pmatrix} u' \\ v' \end{pmatrix} h = \begin{pmatrix} f \\ g \end{pmatrix}$. Thus $\begin{pmatrix} u' \\ v' \end{pmatrix} = \ker(-v\ \ u)$.

The converse is readily verified. Also formulate the dual statement.

**2.5.17 Lemma** In an abelian category every epimorphism (monomorphism) is a universal epimorphism (monomorphism).

*Proof.* Consider the above cartesian square in which $u$ is now supposed to be epic. But then it follows from the previous lemma that $\begin{pmatrix} u' \\ v' \end{pmatrix} =$ $\ker(-v\ \ u)$ and so $(-v\ \ u) = \operatorname{coker}\begin{pmatrix} u' \\ v' \end{pmatrix}$. We will prove that $u'$ is epic. Let

$s: B \to X$ be such that $su' = 0$. Then $(s \;\; 0)\binom{u'}{v'} = su' + 0 = 0$. Since $(-v \;\; u) = \operatorname{coker}\binom{u'}{v'}$ there is a unique $k: A \to X$ with $k(-v \;\; u) = (s \;\; 0)$. This implies $(-kv \;\; ku) = k(-v \;\; u) = (s \;\; 0)$. Hence $ku = 0$ and $-kv = s$. As $u$ is epic the morphism $k$ must equal 0 and therefore $s = 0$. It follows that $u'$ is epic.

Dually one proves that every monomorphism is a universal monomorphism.

**Example** Let $N$ be the following normal subgroup of the group $A_4$ of all even permutations of four elements: {(1), (12)(34), (13)(24), (14)(23)}. Consider the following square:

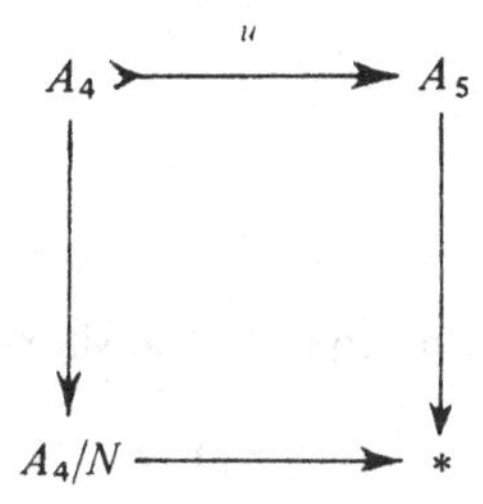


Check that the square is cocartesian using the fact that $A_5$ has no nontrivial normal subgroups. Thus we see that in **Gr** not every monomorphism is a universal monomorphism.

**2.5.18 Exercises** (*a*) Prove that in **Gr** every epimorphism is a universal epimorphism. Recall that epimorphisms are just surjections (2.1.5).

(*b*) In an abelian category every cartesian square of the following type (with $u$ epic) is also cocartesian, hence bicartesian.

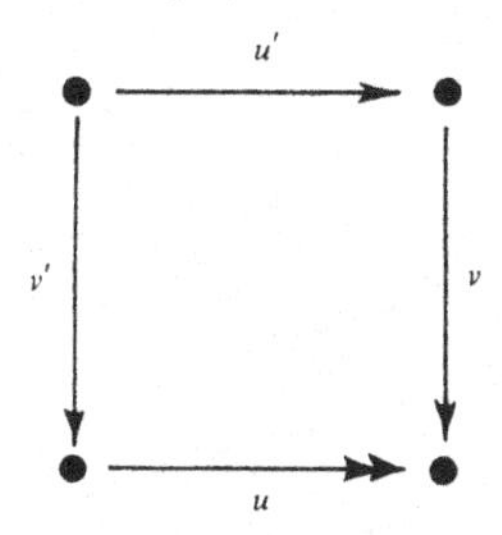


Hint: use lemma 2.5.16 and its dual (and the fact that $u$ is epic).

(*c*) In an additive category **C** with K and K°, P and P°, axiom N is equivalent with **C** being balanced and every monomorphism being universal; see [31].

## 2.6 Exact sequences

In this section we gather together some facts concerning exactness in exact and abelian categories. The reader is advised to read 2.6.1 to 2.6.5 inclusive and then 2.6.11. The rest will be used in chapter 3 and is best referred to when required.

**2.6.1 Definition** A sequence $A \xrightarrow{f} B \xrightarrow{g} C$ in an exact category $\mathbf{C}$ is called a *short exact sequence* provided $f = \ker g$ and $g = \operatorname{coker} f$.

**Exercise** In an abelian category the square of lemma 2.5.16 is bicartesian if and only if the sequence

$$D \xrightarrow{\binom{u'}{v'}} B \oplus C \xrightarrow{(-v\ u)} A$$

is short exact.

It follows immediately from the definition that for a short exact sequence $A \xrightarrow{f} B \xrightarrow{g} C$ the kernel of $f$ is given by $* \to A$ since $f = \ker g$ implies that $\ker f = *$. Dually one sees that the cokernel of $g$ is given by $C \to *$ $(\operatorname{coker} g = *)$.

**Examples** The sequences $* \to A \xrightarrow{1} A$ and $A \xrightarrow{1} A \to *$ are short exact. In an abelian category the sequence $A \xrightarrow{q_A} A \oplus B \xrightarrow{p_B} B$ is a short exact sequence (2.5.11).

If $N$ is a submodule of a right $R$-module $\mathbf{M}$, then the sequence $N \to M \to M/N$ is short exact in $\mathbf{M}_R$.

**2.6.2 Definition** In a situation

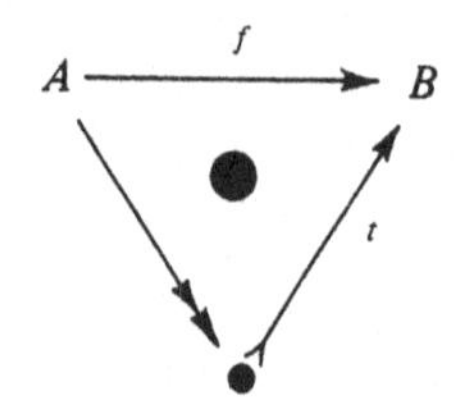


$t$ is called an *image* of $f$ and we denote $t$ by $\operatorname{im} f$ ($t$ is unique up to isomorphism by 2.5.10).

Example. In $\mathbf{Ab}$ we have, for $f: A \to B$, that $\operatorname{im} f$ equals the embedding of $\{f(x) | x \in A\}$ in $B$.

**2.6.3 Definition** A sequence

$$\dots \to C_{i-1} \xrightarrow{d_{i-1}} C_i \xrightarrow{d_i} C_{i+1} \to \dots$$

in an exact category **C** is called *exact at* $C_i$ provided in the commutative diagram

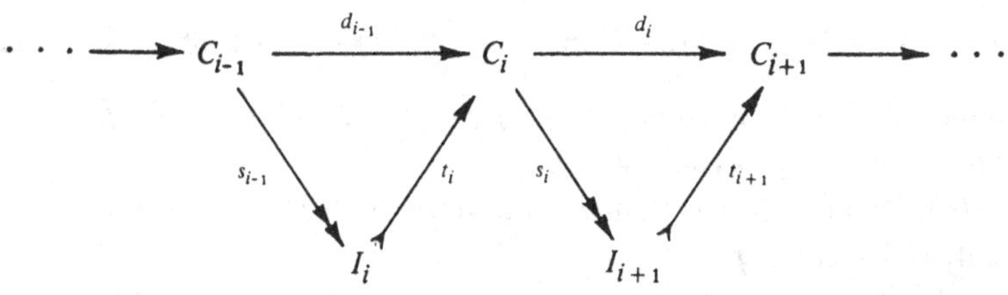


the sequence $I_i \xrightarrow{t_i} C_i \xrightarrow{s_i} I_{i+1}$ is short exact. The sequence is called *exact* provided it is exact at all objects $C_j$, except for the extremities if they exist (the sequence may extend *ad infinitum* at either or both ends). Since epic–monic factorizations are unique up to isomorphism, exactness does not depend on the chosen factorizations.

**2.6.4** Let **C** be an exact category. Then we have the following properties.

(*a*) A sequence

$$\dots \to C_{i-1} \xrightarrow{d_{i-1}} C_i \xrightarrow{d_i} C_{i+1} \to \dots$$

is exact at $C_i$ if and only if im $d_{i-1} = \ker d_i$.

*Proof.* The exactness at $C_i$ means (by definition) (i) $t_i = \ker s_i$, and (ii) $s_i = \operatorname{coker} t_i$. But by 2.5.1° (ii) is implied by (i). Since $t_{i+1}$ is monic $\ker d_i = \ker s_i$. So exactness at $C_i$ is equivalent to $t_i = \ker d_i$ or im $d_{i-1} = \ker d_i$.

As an immediate consequence we point out that $* \to A \to B$ is exact if and only if $A \to B$ is monic and dually $A \to B \to *$ is exact if and only if $A \to B$ is epic.

(*b*) The sequence $* \to A \xrightarrow{f} B \xrightarrow{g} C$ is exact if and only if $f = \ker g$.

*Proof.* From $f = \ker g$ it follows that $f$ is monic; cf. 2.4.1(*d*). Therefore $* \to A$ equals $\ker f$; cf. 2.4.1(*b*). Thus we may factor

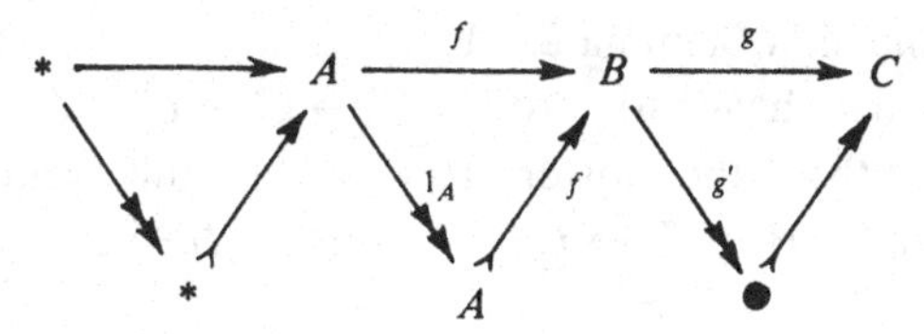

Since $\ker g = \ker g'$ (cf. 2.4.1(g)) we have $f = \ker g'$. But $g'$ is epic so $g' = \operatorname{coker} \ker g'$; cf. 2.5.1. This means $\operatorname{coker} f = g'$. Hence the sequence is exact at $B$. Since there is also exactness at $A$ the sequence is exact.

Suppose, conversely, that $* \to A \xrightarrow{f} B \xrightarrow{g} C$ is exact. Factoring $f$ as $A \overset{f'}{\twoheadrightarrow} A' \overset{f''}{\rightarrowtail} B$ it follows from $f' = \operatorname{coker}(* \to A)$ that $f'$ is monic and hence, since $f''$ is monic, that $f$ is monic. Thus $\operatorname{im} f = f$. Since also $\operatorname{im} f = \ker g$ we have $f = \ker g$.

(*c*) Dually one has that the sequence $A \to B \to C \to *$ is exact if and only if $g = \operatorname{coker} f$.

(*d*) Combining the two preceding properties we find that a sequence $* \to A \xrightarrow{f} B \xrightarrow{g} C \to *$ is exact if and only if $A \xrightarrow{f} B \xrightarrow{g} C$ is a short exact sequence. For this reason we usually call the first sequence short exact as well.

(*e*) From the exactness of $C_i \xrightarrow{d_{i-1}} C_i \xrightarrow{d_i} C_{i+1}$ it follows that $d_i d_{i-1} = *$.

An exact sequence is a special case of the following notion.

**2.6.5 Definition** A sequence $\ldots \to C_{i-1} \xrightarrow{d_{i-1}} C_i \xrightarrow{d_i} C_{i+1} \to \ldots$ is called a *null sequence* or a *complex* provided $d_i d_{i-1} = *$ for all relevant (=nonextremal) indices $i$.

In the following three subsections we treat a few lemmas which assert that sequences in a certain diagram are exact. They are used in the theory of derived functors (chapter 3).

**2.6.6 The nine-lemma** Consider the commutative diagram (top of page 99) in an exact category. Suppose that the three vertical sequences (columns) and also the middle horizontal sequence (row) are exact. Then the bottom row is exact if and only if the top row is exact.

*Proof.* Suppose the top row is exact. We want to prove that the bottom row is exact. For this purpose we have to show that (i) $C' \to C = \ker(C \to C'')$ and that (ii) $C \to C'' = \operatorname{coker}(C' \to C)$. If we prove (ii) and if we show that $C' \to C$ is monic we are through since then (i) follows from (ii) with the aid of lemma 2.5.1.

To show (ii) we first have to prove $C' \to C \to C'' = C' \xrightarrow{*} C''$, or, in other words, that the lower row is a null sequence. Now $B' \to C' \to C \to C'' = B' \to B \to C \to C'' = B' \to B \to B'' \to C'' =$

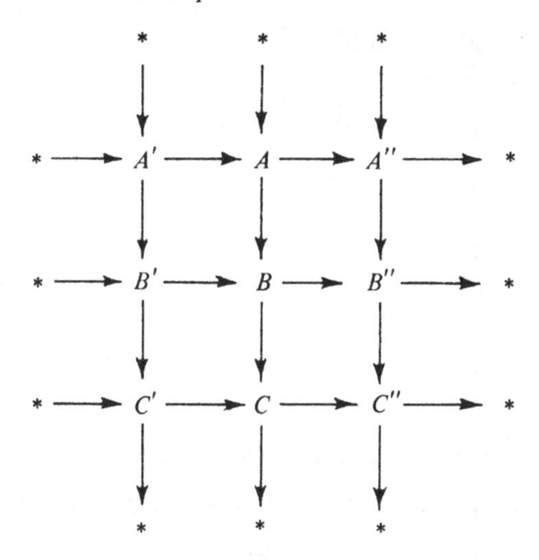


$B' \overset{*}{\to} C''$ because the middle row is a null sequence. Since $B' \to C'$ is epic we may conclude that $C' \to C \to C'' = C' \overset{*}{\to} C''$. Now suppose $C' \to C \to X = C' \overset{*}{\to} X$. Then $B' \to C' \to C \to X = B' \overset{*}{\to} X$. Commutativity yields $B' \to B \to C \to X = B' \overset{*}{\to} X$. Since the middle row is exact there is exactly one morphism $B'' \to X$ such that $B \to C \to X = B \to B'' \to X$. Now we also have $A \to B \to C = A \overset{*}{\to} C$. Furthermore $A \to A'' \to B'' \to X = A \to B \to B'' \to X = A \to B \to C \to X = A \overset{*}{\to} X$ and $A \to A''$ being epic, this implies $A'' \to B'' \to X = A'' \overset{*}{\to} X$.

Since the last column is exact there is exactly one morphism $C'' \to X$ so that $B'' \to C'' \to X = B'' \to X$. But then $B \to C \to X = B \to B'' \to X = B \to B'' \to C'' \to X = B \to C \to C'' \to X$. But $B \to C$ is epic so that $C \to X = C \to C'' \to X$. The unicity of the morphism $C'' \to X$ with this property follows from the unicity of $B'' \to X$ with the required commutativity property. Finally we have to demonstrate that $C' \to C$ is monic. For this purpose we factor $B' \to C' \to C$ as $B' \twoheadrightarrow D \rightarrowtail C$. In a similar fashion to the above one proves that $A' \to B' = \ker(B' \to D)$. This then implies that $* \to A' \to B' \to D \to *$ is exact (since $B' \to D$ is epic). It follows that $C'$ and $D$ are isomorphic. Since $D \to C$ is monic it follows that $C' \to C$ is monic.

Dually one proves that the exactness of the bottom row implies the exactness of the top row.

**2.6.7 The eight-lemma** Suppose the three columns in the following commutative diagram in an exact category, as well as the top two rows, are exact. Then the bottom row is also exact.

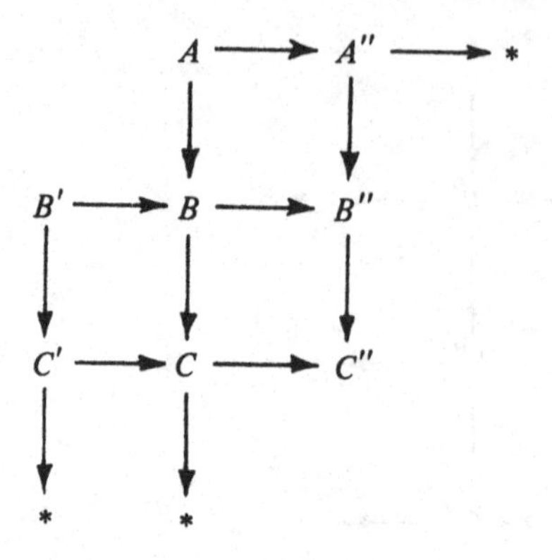


*Proof.* Consider the following epic–monic factorizations:

$$B' \to B = B' \twoheadrightarrow B_1 \rightarrowtail B;\ B \to B'' = B \twoheadrightarrow B_2 \rightarrowtail B'';$$

$$C' \to C = C' \twoheadrightarrow C_1 \rightarrowtail C;\ C \to C'' = C \twoheadrightarrow C_2 \rightarrowtail C'';$$

$$A \to B = A \twoheadrightarrow A_0 \rightarrowtail B;\ A'' \to B'' = A'' \twoheadrightarrow A_2 \rightarrowtail B''.$$

Putting this together in a diagram yields

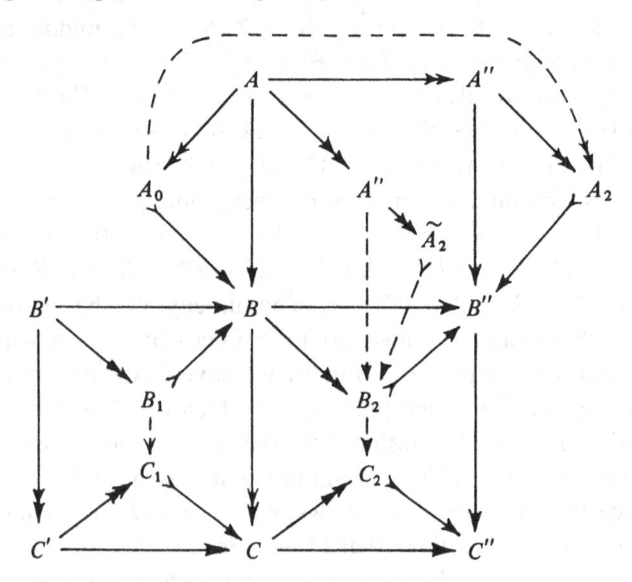


The existence of the morphisms $B_i \to C_i$ $(i = 1, 2)$, $A_0 \to A_2$ and $A'' \to B_2$ leaving the diagram commutative is guaranteed by lemma 2.5.10. Factoring $A'' \to B_2$ into $A'' \twoheadrightarrow \tilde{A}_2 \rightarrowtail B_2$ yields an epic–monic factorization for $A'' \to B''$, namely $A'' \twoheadrightarrow \tilde{A}_2 \rightarrowtail B_2 \rightarrowtail B''$. Due to lemma 2.5.10 we find that $\tilde{A}_2 \simeq A_2$. From the exactness of $A'' \to B'' \to C''$ it follows that $A_2 \to B'' = \ker(B'' \to C'')$. Then we have $A_2 \rightarrowtail B_2 \rightarrowtail B'' =$

$A_2 \rightarrowtail B'' = \ker(B'' \to C'')$. From 2.4.1($h$) it follows that $A_2 \rightarrowtail B_2 = \ker(B_2 \rightarrowtail B'' \twoheadrightarrow C'')$. Therefore we have $A_2 \rightarrowtail B_2 = \ker(B_2 \twoheadrightarrow C_2 \rightarrowtail C'') = \ker(B_2 \twoheadrightarrow C_2)$ by 2.4.1($g$). This means that $A_2 \rightarrowtail B_2 \twoheadrightarrow C_2$ is exact. Then consider the following commutative diagram.

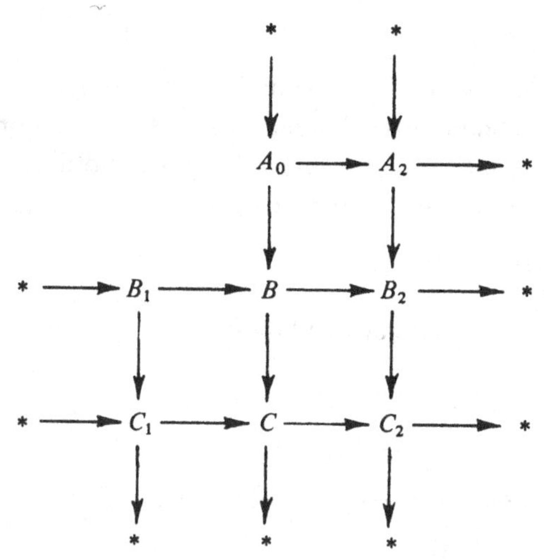

It follows, as before in lemma 2.6.6, that $C \to C_2 = \operatorname{coker}(C_1 \to C)$. This means that the bottom row in the last diagram is exact, which means that $C' \to C \to C''$ is exact.

Remark. In proving that $C \to C_2 = \operatorname{coker}(C_1 \to C)$ as we did in lemma 2.6.6, the condition that $B' \to C'$ is epic was only needed to prove that $C_1 \to C \to C_2$ is a null sequence. For the eight-lemma one may therefore replace the condition that $B' \to C'$ is epic by the condition that $C_1 \to C \to C_2$ is a null sequence.

**2.6.8 Lemma** If in the following commutative diagram the rows are exact then there is a unique $f'\colon A' \to B'$ leaving the diagram commutative. The morphism $f'$ is monic if and only if $f$ is, and epic if $f$ is.

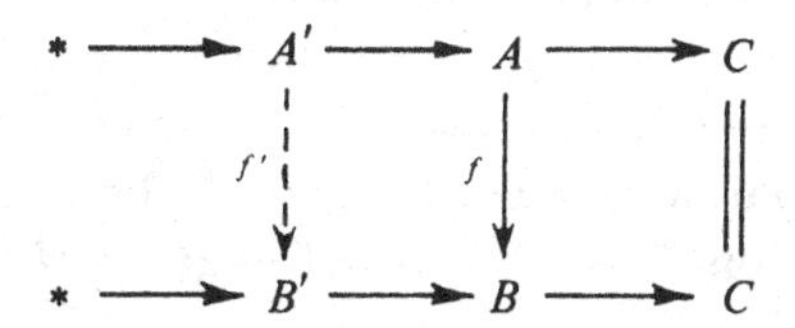

*Proof.* The existence and uniqueness of $f'\colon A' \to B'$ follows from the fact that $A' \to A \to B \to C = A' \to A \to C = A' \overset{*}{\to} C$ and that $B' \to B =$

ker $(B \to C)$. Observe that the left hand square is cartesian. If $f$ is monic the composition $A' \to B' \to B$ is monic; so $f'$ is monic. If $f'$ is monic, use the exercise in 2.4.1 to show that $f$ is monic. The last statement of the lemma follows from the dual version of the eight-lemma. The natural idea that $f$ is universally epic would need the category to be abelian (2.5.17).

**2.6.9 The snake-lemma** (Lemma of the connecting morphism.) Suppose the following commutative diagram (in an exact category) has exact columns and exact middle rows. Then the top and bottom row are exact and there is a morphism $\delta: K'' \to C'$ such that

$$K' \to K \to K'' \xrightarrow{\delta} C' \to C \to C''$$

is exact. Furthermore $\delta$ is natural and self dual.

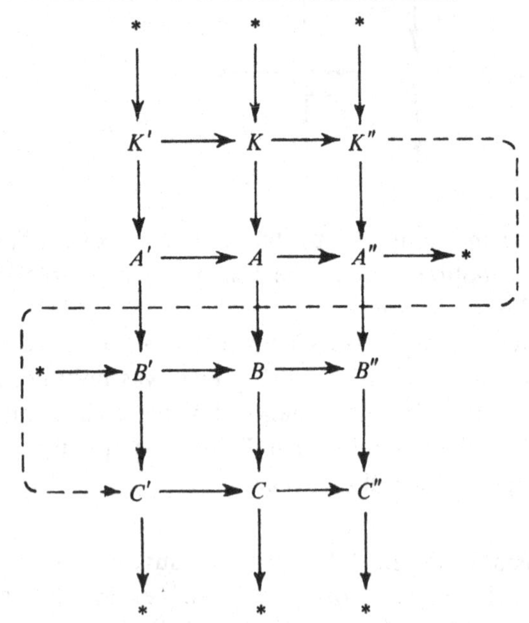


*Proof.* The bottom row is exact according to lemma 2.6.7. Dually the top row is exact. Then take $H''$ and $H'$ in such a way that $K \to K'' \to H'' \to *$ and $* \to H' \to C' \to C$ are exact. We shall establish a natural and self dual isomorphism between $H''$ and $H'$. From this it will follow that the composition $\delta: K'' \to C' = K'' \to H'' \xrightarrow{\sim} H' \to C'$ satisfies the requirements.

Take $X'$ and $X''$ in such a way that $* \to X' \to A \to B''$ and $A' \to B \to X'' \to *$ are exact. In the following commutative diagram

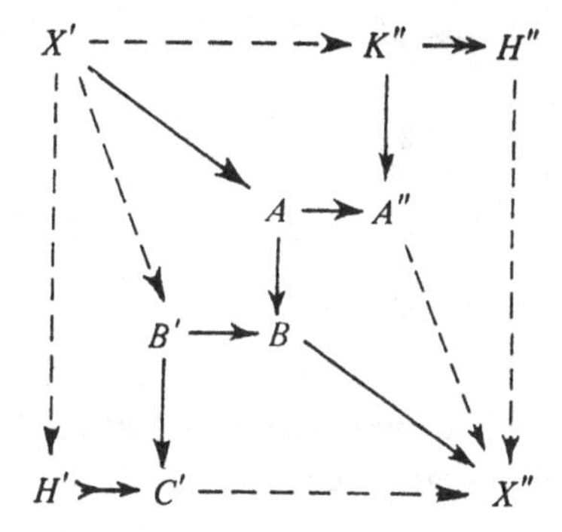


the broken arrows represent morphisms whose unique existence the reader should check using the following device (and its dual).

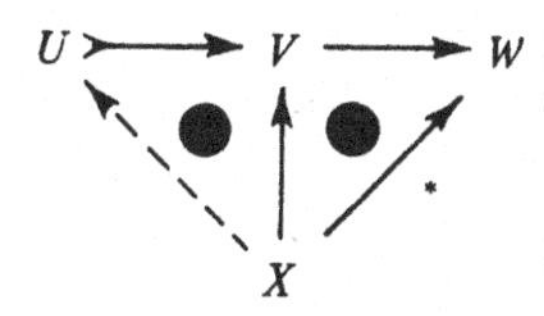


The top row is assumed to be exact. The morphism $X \to U$ is the unique morphism making the left hand triangle commutative. If the morphisms $X' \to K''$ and $X' \to H'$ are epic and $C' \to X''$ and $H'' \to X''$ are monic we would have

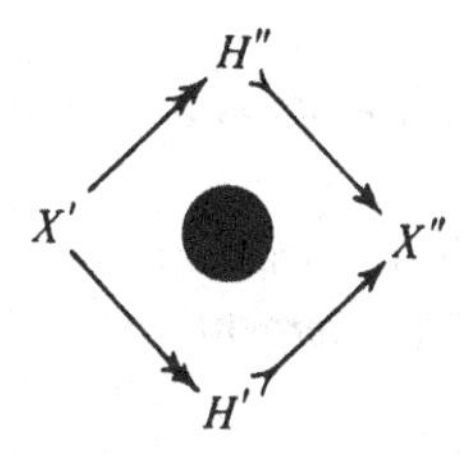


in which case we would have an isomorphism between $H'$ and $H''$ due to the uniqueness of the epic–monic factorization. So all we have to do is to show that $X' \to K''$ and $X' \to H'$ are epic (the other two assertions follow by duality). Using lemma 2.6.8 one proves that $X' \to K''$ is epic:

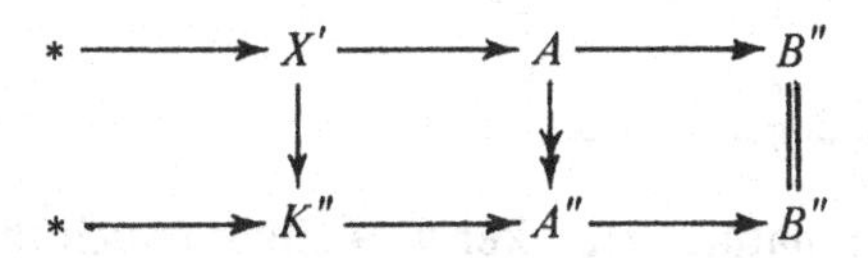


To treat $X' \to H'$ we factor $A \to B$ as $A \twoheadrightarrow D \rightarrowtail B$. Let $* \to D' \to D \to B''$ be exact. From the commutative diagram

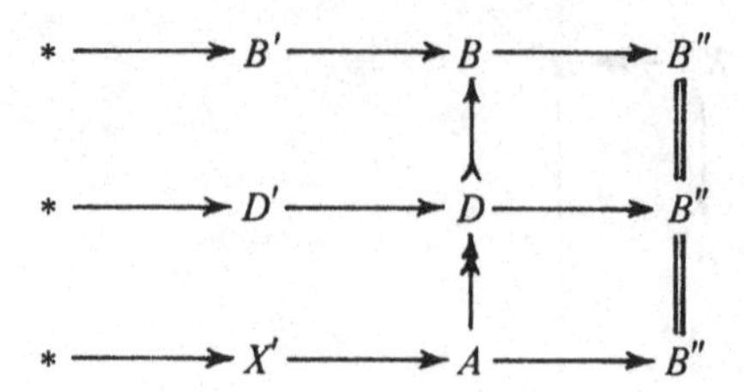


whose rows are exact we infer $X' \twoheadrightarrow D' \rightarrowtail B'$, again using 2.6.8.

From the dual version of the eight-lemma it follows that the bottom row in the following commutative diagram is exact:

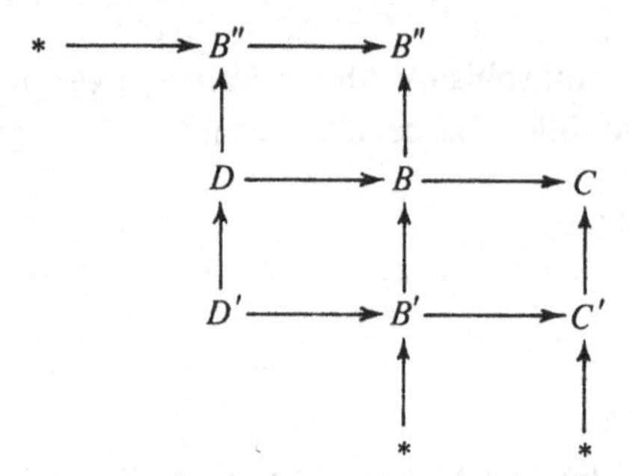


Apply lemma 2.6.8 to the commutative diagram

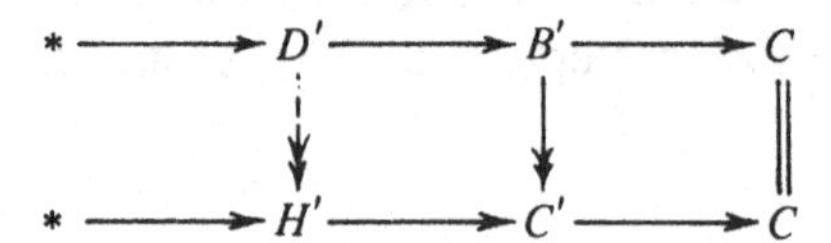


whose rows are exact. We then find a composition $X' \twoheadrightarrow D' \twoheadrightarrow H'$ which is epic. This completes the proof.

Remark. Since each step in this proof is natural the constructed morphism $K'' \to C'$, called the *connecting* or '*serpent*' *morphism*, is functorial in the obvious sense.

**Corollary** (The ker–coker sequence.) If in the following commutative diagram in an exact category

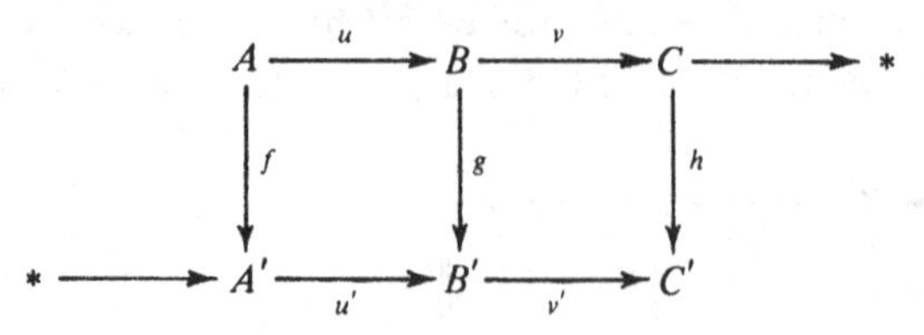


the rows are exact there is a natural map $\ker h \to \operatorname{coker} f$ such that the sequence

$$\ker f \to \ker g \to \ker h \to \operatorname{coker} f \to \operatorname{coker} g \to \operatorname{coker} h$$

is exact. It is easily seen that if $u$ is monic so is ($\ker f \to \ker g$), and if $v'$ is epic, so is ($\operatorname{coker} g \to \operatorname{coker} h$).

**Exercise** If $A \xrightarrow{f} B \xrightarrow{g} C$ is any composition of morphisms in an exact category, there exists a natural exact sequence

$$* \to \ker f \to \ker gf \to \ker g \to \operatorname{coker} f \to \operatorname{coker} gf \to \operatorname{coker} g \to *$$

**2.6.10** Let **A** be an abelian category. Suppose the sequences

$$\dot{E}: \ldots \to A_{i-1} \xrightarrow{d_{i-1}} A_i \xrightarrow{d_i} A_{i+1} \to \ldots$$

and

$$\dot{F}: \ldots \to B_{i-1} \xrightarrow{f_{i-1}} B_i \xrightarrow{f_i} B_{i+1} \to \ldots.$$

are exact in **A**. Then the sequence

$$\dot{E}\oplus\dot{F}: \ldots \to A_{i-1}\oplus B_{i-1} \xrightarrow{d_{i-1}\oplus f_{i-1}} A_i \oplus B_i \xrightarrow{d_i \oplus f_i} A_{i+1} \oplus B_{i+1} \to \ldots$$

is also exact. This is easily proved by choosing an appropriate factorization.

**2.6.11 Definition** Let **A** be abelian and let $A \xrightarrow{f} B \xrightarrow{g} C$ be a short exact sequence. We say that this sequence *splits* provided there is an isomorphism $u: B \to A \oplus C$ such that the following diagram commutes:

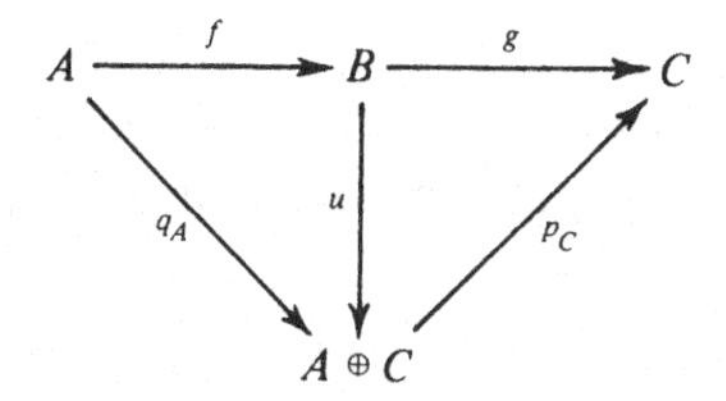


**2.6.12 Lemma** Let $A \xrightarrow{f} B \xrightarrow{g} C$ be short exact sequence in an abelian category. Then the following assertions are equivalent.

(i) The sequence splits.
(ii) There is a morphism $s: B \to A$ such that $sf = 1_A$.
(iii) There is a morphism $t: C \to B$ such that $gt = 1_C$.

*Proof.* (ii)$\Rightarrow$(i) Take $u: B \to A \oplus C$ equal to $\begin{pmatrix} s \\ g \end{pmatrix}$. Then $p_A uf = sf = 1$ and $p_C uf = gf = 0$. Hence $uf = \begin{pmatrix} 1 \\ 0 \end{pmatrix} = q_A$ and $p_C u = g$. We now prove that $u$ is an isomorphism. Since the category is abelian and hence balanced we only need to prove that $u$ is monic and epic. Let $ur = 0$. Then $0 = p_C ur = gr$. Since $f = \ker g$ there is a morphism $x$ such that $r = fx$. Hence $0 = ur = ufx = q_A x$. But $q_A$ is monic, therefore $x = 0$, and then $r = fx = 0$. Thus $u$ is monic.

To prove that $u$ is also epic let $eu = 0$. Then we have $e(q_A p_A + q_C p_C)u = e\begin{pmatrix} 1 & 0 \\ 0 & 1 \end{pmatrix} u = eu = 0$. Thus $eq_A s + eq_C g = 0$ from which we see $eq_A sf + eq_C gf = 0$ or $eq_A = 0$. So we have $eq_C g = eq_A s + eq_C g = 0$. From the fact that g is epic we find that $eq_C = 0$. From $eq_A = 0$ and $eq_C = 0$ it follows that $e = 0$ and so $u$ is epic. Dually one proves (iii)$\Rightarrow$ (i).

(i)$\Rightarrow$(ii) Now $u$ is given. Take $s = p_A u$. Then it follows that $sf = p_A uf = p_A q_A = 1$. Dually one proves (i)$\Rightarrow$(iii).

**2.6.13 Definition** Let **A** be an abelian category and let $\dot{E}$ be an exact sequence in **A**:

$$\ldots \to A_{j-1} \xrightarrow{d_{j-1}} A_j \xrightarrow{d_j} A_{j+1} \to \ldots$$

The sequence $\dot{E}$ is said to *split* provided for every $A_j$ in

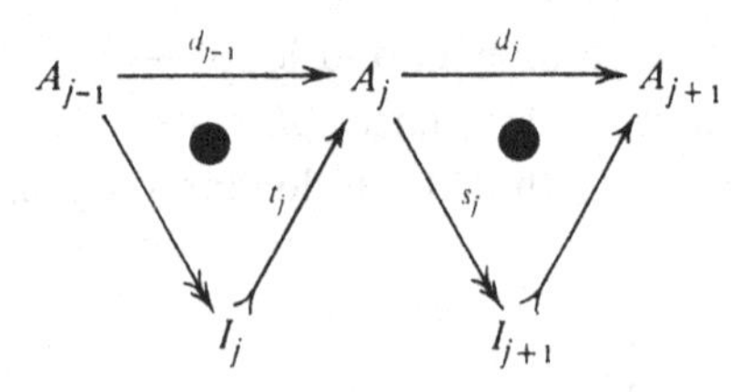

the sequence $I_j \xrightarrow{t_j} A_j \xrightarrow{s_j} I_{j+1}$ splits (except for the first and last objects in the sequence $\dot{E}$, if they exist).

**2.6.14 Lemma** Let **A** be abelian and suppose $\dot{E}$ is the exact sequence $0 \xrightarrow{d_0} A_1 \xrightarrow{d_1} \ldots \xrightarrow{d_{n-1}} A_n \xrightarrow{d_n} 0$. Then the following assertions are equivalent.

(i) The sequence splits.

(ii) There are objects $B_i$ and $C_i$ together with morphisms $p_j^B, q_j^B, p_j^C$ and $q_j^C$ such that $\ulcorner A_j, p_j^B, q_j^B, p_j^C, q_j^C \urcorner$ is a biproduct of $B_j$ and $C_j$ and such that one alternately has $d_j = 1 \oplus 0$ and $d_{j+1} = 0 \oplus 1$ for $j = 1, \ldots, n-1$. More explicitly there are exact sequences

$$\dot{F}: 0 \to B_1 \to \ldots \to B_n \to 0$$

and

$$\dot{G}: 0 \to C_1 \to \ldots \to C_n \to 0$$

such that $\dot{E} = \dot{F} \oplus \dot{G}$ while the morphisms in the sequence $\dot{F}$ are alternately 0 and 1 just as for the sequence $\dot{G}$ in the following way:

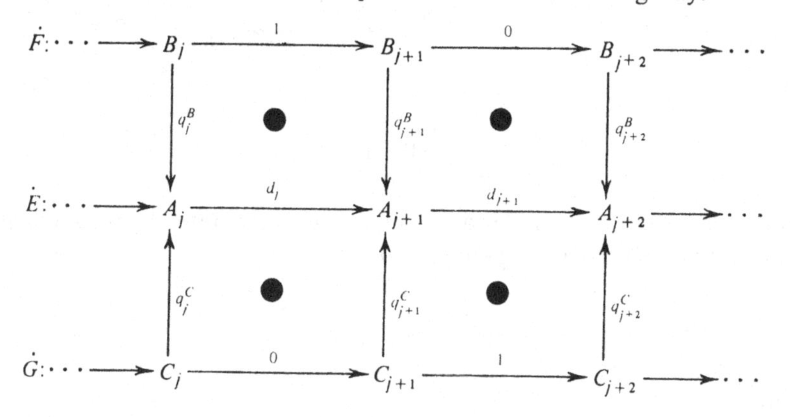

*Proof.* (i)⇒(ii) Suppose the sequence $\dot{E}$ splits. This means that for $1 \leq j \leq n$ the sequence $I_j \to A_j \to I_{j+1}$ splits (take $I_1 = I_{n+1} = 0$).

Now take $B_j = I_j$, $C_j = I_{j+1}$ for $j$ even, $B_j = I_{j+1}$, $C_j = I_j$ for $j$ odd. From

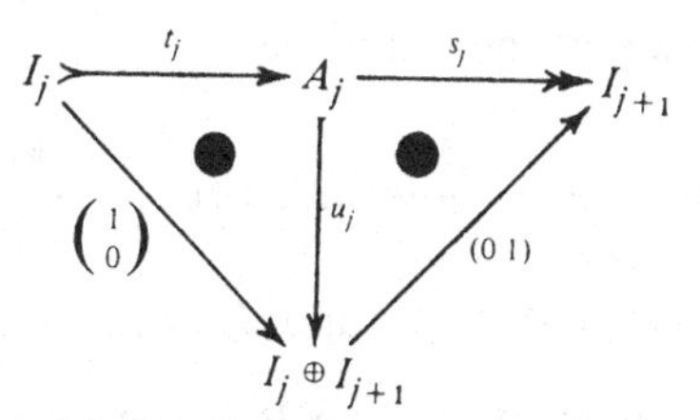

together with the fact that $u_j$ is an isomorphism, it follows that $A_j$ is a biproduct of $I_j$ and $I_{j+1}$. The projections are $(1 \;\; 0)u_j$ and $(0 \;\; 1)u_j = s_j$. The injections are $u_j^{-1}\begin{pmatrix}1\\0\end{pmatrix} = t_j$ and $u_j^{-1}\begin{pmatrix}0\\1\end{pmatrix}$. Then take $p_j^B$, $q_j^B$, $p_j^C$ and $q_j^C$ equal to these projections and injections. For even $j$ this yields the situation (see top of page 108).

Then we see that

$$d_{j-1} = q_j^B p_{j-1}^B = \begin{pmatrix}1\\0\end{pmatrix}(1 \quad 0) = \begin{pmatrix}1 & 0\\0 & 0\end{pmatrix} = 1 \oplus 0$$

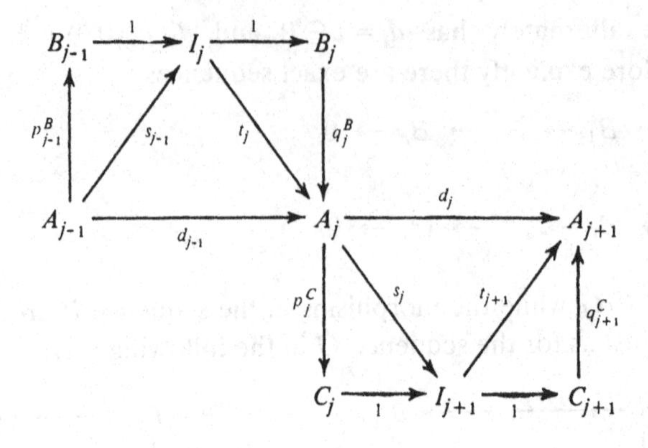


and

$$d_j = q^C_{j+1} p^C_j = \begin{pmatrix} 0 \\ 1 \end{pmatrix} (0 \quad 1) = 0 \oplus 1.$$

For $j$ odd we obtain a similar diagram with $B$ and $C$ interchanged. From this we have $\dot{E} = \dot{F} \oplus \dot{G}$ together with

$$\dot{F}: 0 \to I_2 \xrightarrow{1} I_2 \xrightarrow{0} I_4 \xrightarrow{1} \ldots$$

and

$$\dot{G}: 0 \to I_1 \xrightarrow{0} I_3 \xrightarrow{1} I_3 \xrightarrow{0} \ldots$$

(ii)$\Rightarrow$(i) We now have to prove that the sequence

$$\ldots \to B_{j-1} \oplus C_{j-1} \xrightarrow{1 \oplus 0} B_j \oplus C_j \xrightarrow{0 \oplus 1} B_{j+1} \oplus C_{j+1} \xrightarrow{0 \oplus 1} \ldots$$

splits. We first remark that $I_1 = 0 \to A_1 \rightrightarrows I_2$ and $I_n \rightrightarrows A_n \to I_{n+1} = 0$ always split. Further we see from $B_{j-1} \oplus C_{j-1} \xrightarrow{1 \oplus 0} B_j \oplus C_j$ that $B_{j-1} = B_j$ and similarly we find $C_j = C_{j+1}$. Then we may factor in the following way:

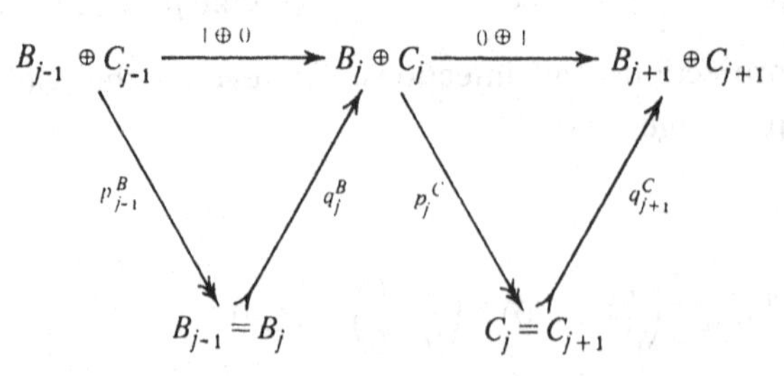

Since all sequences $B_j \xrightarrow{q_j^B} B_j \oplus C_j \xrightarrow{p_j^C} C_j$ split we are through.

## 2.7 Functors preserving extra structure

We now examine functors between categories with some extra structure. We have already seen, for instance in theorem 1.9.5, that adjoints tend to preserve certain constructions and properties. In this section a further miscellany of results in this direction is presented.

**2.7.1 Definition** Suppose $\mathbf{A}$ and $\mathbf{B}$ are (semi-) additive categories (2.3.1 and 2.3.2). A functor $T: \mathbf{A} \to \mathbf{B}$ is called *(semi-) additive* provided:

(i) $T(f+g) = Tf + Tg$ whenever $f$ and $g$ belong to the same set $(A_1, A_2)$ for $A_1$ and $A_2 \in \mathbf{A}$;

(ii) $T0 = 0$ for all null morphisms in $\mathbf{A}$.

Remark. In the additive case (ii) follows from (i).

**Examples** (*a*) Let $\mathbf{A}$ be (semi-) additive and $A \in \mathbf{A}$. Then $h_A: \mathbf{A} \to \mathbf{Ab}$ and $h^A: \mathbf{A} \to \mathbf{Ab}$ are (semi-) additive.

(*b*) For $\mathbf{A} = \mathbf{M}_R$, $\mathbf{B} = \mathbf{Ab}$ and $M \in {}_R\mathbf{M}$ the functor $-\otimes_R M: \mathbf{M}_R \to \mathbf{Ab}$ is additive (see 1.2.2(*h*)).

**2.7.2 Proposition** Let $\mathbf{A}$ and $\mathbf{B}$ be punctured categories and assume that for all objects $A \in \mathbf{A}$ there is a biproduct $A \oplus A$ in $\mathbf{A}$ and for all objects $B \in \mathbf{B}$ there is a biproduct $B \oplus B$ in $\mathbf{B}$. On account of 2.3.6 $\mathbf{A}$ and $\mathbf{B}$ are semi-additive. Let $T: \mathbf{A} \to \mathbf{B}$ be a functor. Then the following assertions are equivalent.

(i) $T$ is semi-additive.

(ii) $T$ preserves the biproducts $A \oplus A$.

If $\mathbf{A}$ and $\mathbf{B}$ are categories with biproducts then (iii) is also equivalent with (i) and (ii).

(iii) $T$ preserves biproducts.

*Proof.* The statement that $T$ preserves biproducts means that whenever $\ulcorner A_1 \oplus A_2, p_1, q_1, p_2, q_2 \urcorner$ is a biproduct in $\mathbf{A}$, then $\ulcorner T(A_1 \oplus A_2)$, $Tp_1, Tq_1, Tp_2, Tq_2 \urcorner$ is a biproduct in $\mathbf{B}$. We leave the details of the proof to the reader. For (ii)$\Rightarrow$(i) use the relation $p_1 q_2 = *$ and 2.3.5. For (i)$\Rightarrow$(ii) utilize the fact that $(Tq_1 \;\; Tq_2)$ and $\begin{pmatrix} Tp_1 \\ Tp_2 \end{pmatrix}$ are inverse to each other

$$TA_1 \oplus TA_2 \underset{\begin{pmatrix} Tp_1 \\ Tp_2 \end{pmatrix}}{\overset{(Tq_1 \;\; Tq_2)}{\rightleftarrows}} T(A_1 \oplus A_2).$$

When $\mathbf{A}$ and $\mathbf{B}$ are known to be semi-additive it is enough for $T$ to preserve products $A \prod A$ (or sums $A \coprod A$) in virtue of the exercise in 2.3.4.

**2.7.3 Proposition** Let $T: \mathbf{B} \to \mathbf{A}$ be an additive functor between the additive categories $\mathbf{A}$ and $\mathbf{B}$. Let $S: \mathbf{A} \to \mathbf{B}$ be a left adjoint of $T$. Then $S$ is additive.

*Proof.* Consider the adjunctions $\phi: (S\text{-}, \text{-})_{\mathbf{B}} \to (\text{-}, T\text{-})_{\mathbf{A}}$ and $\Phi: I_{\mathbf{A}} \to TS$. For $f_1, f_2: A \to A'$ in $\mathbf{A}$ we have

$$\begin{aligned}
\phi \ulcorner A, SA' \urcorner S(f_1+f_2) &= TS(f_1+f_2) \circ \Phi(A) \\
&= \Phi(A') \circ (f_1+f_2) \\
&= \Phi(A') \circ f_1 + \Phi(A') \circ f_2 \\
&= TSf_1 \circ \Phi(A) + TSf_2 \circ \Phi(A) \\
&= (TSf_1 + TSf_2) \circ \Phi(A) \\
&= T(Sf_1 + Sf_2) \circ \Phi(A) \\
&= \phi \ulcorner A, SA' \urcorner (Sf_1 + Sf_2).
\end{aligned}$$

Since $\phi$ is an isomorphism it follows that $S(f_1+f_2) = Sf_1 + Sf_2$. Also, if $S$ is additive, so is $T$.

**2.7.4** We want to adapt Yoneda's lemma to the special case that $\mathbf{A}$ and $\mathbf{B}$ are both additive categories with $\mathbf{A}$ a small additive category. As we know from 1.3.3, $(\mathbf{A}, \mathbf{B})$ is then a category and it is even an additive category in the following way. Let $S$ and $T \in (\mathbf{A}, \mathbf{B})$ and let $\eta, \zeta: S \to T$ be morphisms of functors. Then for $f: A \to A'$ in $\mathbf{A}$ we have

$$\begin{aligned}
Tf \circ (\eta(A) + \zeta(A)) &= Tf \circ \eta(A) + Tf \circ \zeta(A) \\
&= \eta(A') \circ Sf + \zeta(A') \circ Sf \\
&= (\eta(A') + \zeta(A')) \circ Sf.
\end{aligned}$$

Hence, if we define $(\eta + \zeta)(A) = \eta(A) + \zeta(A)$ we make $\eta + \zeta$ a morphism of functors from $S$ to $T$. With this addition $(S, T)_{(\mathbf{A}, \mathbf{B})}$ becomes an abelian group. One easily checks that the distributive laws hold and that $(\mathbf{A}, \mathbf{B})$ is an additive category. Notice that so far we have not needed additivity of the category $\mathbf{A}$. The full additive subcategory of $(\mathbf{A}, \mathbf{B})$ whose objects are the additive functors from $\mathbf{A}$ to $\mathbf{B}$ will be denoted by $(\mathbf{A}, \mathbf{B})_+$.

Now let $\mathbf{A}$ be an additive category. Let $S: \mathbf{Ab} \to \mathbf{Sets}$ be the forgetful functor and suppose $F \in (\mathbf{A}, \mathbf{Ab})_+$ and $A \in \mathbf{A}$. For $h_A \in (\mathbf{A}, \mathbf{Ab})_+$ we have $Sh_A: \mathbf{A} \to \mathbf{Sets}$. According to Yoneda's lemma $\Theta: SFA \to (Sh_A, SF)_{(\mathbf{A},\mathbf{Sets})}$ is a bijection of sets. If $A' \in \mathbf{A}$, $a \in SFA$, $f$ and $g \in (A, A')$ we have

$$\begin{aligned}\Theta(a)(A')(f+g) &= SF(f+g)(a)\\ &= F(f+g)(a) = (Ff+Fg)(a) = Ff(a)+Fg(a)\\ &= SFf(a)+SFg(a)\\ &= \Theta(a)(A')(f)+\Theta(a)(A')(g).\end{aligned}$$

Thus $\Theta(a)(A')$ is a homomorphism between abelian groups and $\Theta(a)$ becomes a morphism from $h_A$ to the functor $F$ in the category $(\mathbf{A}, \mathbf{Ab})_+$. This shows that $(h_A, F)_{(\mathbf{A},\mathbf{Ab})_+}$ may be identified with $(Sh_A, SF)_{(\mathbf{A},\mathbf{Sets})}$. Similarly one shows that $\Theta(a+b) = \Theta(a)+\Theta(b)$. Therefore $\Theta$ is indeed a homomorphism

$$\Theta: FA \cong (h_A, F)_{(\mathbf{A},\mathbf{Ab})_+}$$

between abelian groups which is functorial. In other words, the bifunctors -(-) and $(h_-, -)$ from $(\mathbf{A}, \mathbf{Ab})_+$ to $\mathbf{Ab}$ are isomorphic. This is the additive version of 1.4.2.

**2.7.5** Let $\mathbf{C}$ be an additive category. Consider the diagram

$$A \xrightarrow{f} B \underset{g_2}{\overset{g_1}{\rightrightarrows}} C.$$

Then we may easily show that $f$ is an equalizer for $g_1$ and $g_2$ if and only if $f = \ker(g_1 - g_2)$. Hence the name *difference kernel* is sometimes used for equalizer.

Dually one has for the diagram

$$A \underset{f_2}{\overset{f_1}{\rightrightarrows}} B \xrightarrow{g} C$$

that $g$ is coequalizer for $f_1$ and $f_2$ if and only if $g = \operatorname{coker}(f_1 - f_2)$. From this it follows that an additive functor between two additive categories is left (right) exact if and only if it preserves kernels (cokernels). In virtue of exercises 1.8.4 and 2.7.2 left (right) exactness means that cartesian (cocartesian) squares are preserved.

**2.7.6 Proposition** Let $\mathbf{A}$ be a punctured category. Suppose $T: \mathbf{B} \to \mathbf{A}$ is a fully faithful functor which has a left adjoint. Then:

(i) $\mathbf{B}$ is punctured and $T$ preserves null objects;

(ii) $T$ preserves null morphisms and kernels;
(iii) a morphism $f$ in **B** is monic if and only if $Tf$ is monic in **A**.

*Proof.* Let $B \in \mathbf{B}$ and $TB = A$. Then $T1_B = 1_{TB} = 1_A$ is monic in **A**. Therefore $* \to A = \ker T1_B$ in **A**. Since $T$ is full there is a morphism $g: B \to B$ in **B** such that $Tg = *_{AA}$. Now in **A** there is an equalizer (the kernel) for $1_A, *: A \rightrightarrows A$. Then it follows from 1.9.6 that there is also an equalizer $k: K \to B$ in **B** for $1_B, g: B \rightrightarrows B$ so that $Tk = \ker T1_B$. Then $TK \simeq *$. We assert that $K$ is a null object for **B**. To show this let $B' \in \mathbf{B}$. Since $T$ is fully faithful $(B', K)_{\mathbf{B}} \simeq (TB', TK)_{\mathbf{A}}$. But $TK$ is a null object, so $(B', K)$ consists of just one element. In the same way one proves that $(K, B')$ consists of a single element. This proves that $K$ is a null object in **B**.

Since $K$ and $TK$ are null objects, $T$ preserves a null object and hence every null object. This proves (i). For (ii) we refer to 1.9.5, and (iii) is clear.

**2.7.7 Theorem** Let **A** be abelian and suppose $S: \mathbf{A} \to \mathbf{B}$ is a left adjoint of $T: \mathbf{B} \to \mathbf{A}$ with $T$ fully faithful. Then we have that:
(i) **B** is additive;
(ii) $S$ and $T$ are additive;
(iii) if $T$ preserves cokernels, **B** is abelian;
(iv) if $S$ preserves kernels, **B** is abelian.

*Proof.* **A** is abelian and hence finitely right complete and finitely left complete (2.5.5 and 1.8.4). Then **B** is also finitely right and left complete (1.9.7). Moreover **B** is punctured and every null object is preserved by $T$ (2.7.6). Suppose $f, g: B_1 \to B_2$ in **B**, so that $Tf, Tg: TB_1 \to TB_2$ and hence $Tf + Tg: TB_1 \to TB_2$ are in **A**. Since $T$ is full, there is a morphism $h: B_1 \to B_2$ in **B** such that $Th = Tf + Tg$. Since $T$ is faithful, $h$ is unique. Define $f + g = h$. This makes **B** into an additive category. From 2.3.5 it follows that **B** is not only additive but its additive structure is uniquely determined. From the definition of addition in **B** it follows that $T$ is additive. From proposition 2.7.3 it follows that also $S$ (left adjoint of $T$) is additive.

We proceed to prove (iii). Since **B** is finitely right and left complete P, P°, K and K° hold in **B**. We only have to prove that N and N° hold in **B**. Let $f$ be monic in **B** and $g = \operatorname{coker} Tf$ in **A**. Since $S$ is a left adjoint it is right exact, and also additive; hence $S$ preserves cokernels, as does $T$ by assumption. Therefore $TSg = \operatorname{coker} TSTf$. According to proposition 2.7.6 $Tf$ is monic. Furthermore $\Psi: ST \to I_{\mathbf{B}}$ is an isomorphism (see 1.7.8). From this one sees that $Tf$ being monic implies that $TSTf$ is also monic. Hence $TSTf = \ker TSg$ since **A** is abelian. From the fact that $T$ is fully faithful and additive one can derive that $STf = \ker Sg$. Now the isomorphism $ST \simeq I_{\mathbf{B}}$ implies that $f$ itself is the kernel of some morphism. Consequently N holds. To show that N° also holds, let $f: B_1 \to B_2$ be epic in **B**. For the null object 0 of **B**, $T0$ is a null object of **A**. Let $c: B_2 \to 0$ be coker $f$. Then $Tc = \text{coker } Tf$ by assumption. Hence $Tf$ is epic in **A**. Therefore $Tf = \text{coker } g$ for a certain $g$. Then $STf = \text{coker } Sg$. Since $ST \simeq I_{\mathbf{B}}$, the epimorphism $f$ is a cokernel of some morphism. Thus N° holds in **B**.

Finally we prove (iv). As before we only have to prove N and N°. Let $f$ be monic in **B**. Then $Tf$ is monic in **A**. Hence $Tf = \ker g$ in **A** for a certain $g$. Then $STf = \ker Sg$ by assumption. It follows that $f$ must be the kernel of some morphism (as in (iii)). So N holds in **B**. Let $f$ be epic in **B**. In **A** we have

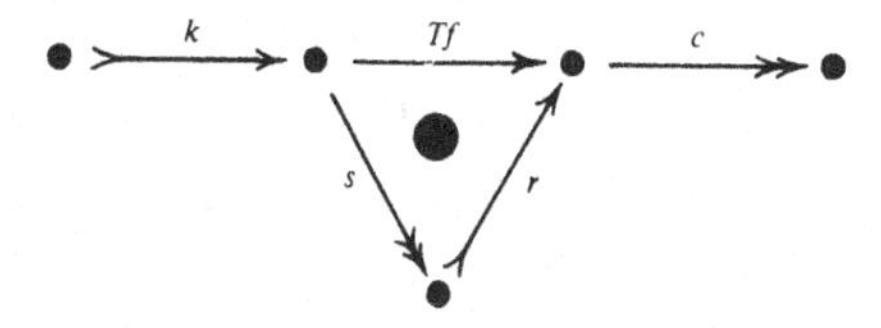


with $k = \ker Tf$, $c = \text{coker } Tf$, $s = \text{coker } k$, $r = \ker c$. Then $Sr = \ker Sc$; hence $Sr$ is monic in **B**. With $f$ also $STf$ is epic in **B**, so $Sr$ is epic in **B**. Then $Sr$ is a bimorphism in **B**. Since N holds in **B**, the category **B** is balanced and so $Sr$ is an isomorphism in **B**. As $Ss = \text{coker } Sk$, also $STf = \text{coker } Sk$. Thus $f$ must be the cokernel of some morphism (as in (iii)). Hence N° holds in **B**, and the theorem is proved.

In particular 2.7.6 and 2.7.7 apply when $T$ embeds **B** as a full reflective subcategory of **A** (1.7.10).

**2.7.8 Lemma** Let $T: \mathbf{A} \to \mathbf{B}$ be a functor between exact categories. Then:

(i) $T$ preserves kernels if and only if for every short exact sequence $A \xrightarrow{f} B \xrightarrow{g} C$ in **A** the sequence $* \to TA \xrightarrow{Tf} TB \xrightarrow{Tg} TC$ is exact in **B**;

(ii) $T$ preserves cokernels if and only if for every short exact sequence $A \xrightarrow{f} B \xrightarrow{g} C$ in **A** the sequence $TA \xrightarrow{Tf} TB \xrightarrow{Tg} TC \to *$ is exact in **B**;

(iii) $T$ preserves kernels and cokernels if and only if $T$ preserves short exact sequences.

Remark. If in addition $T$ is an additive functor between two additive categories then the condition that $T$ is left (right) exact is equivalent to the condition that for short exact sequences $A \xrightarrow{f} B \xrightarrow{g} C$ in **A** the sequence $0 \to TA \overset{Tf}{\to} TB \overset{Tg}{\to} TC$ ($TA \overset{Tf}{\to} TB \overset{Tg}{\to} TC \to 0$) is exact in **B** (see the second definition of 1.9.1, and 2.7.5).

*Proof.* (i) Let $T$ preserve kernels and let $A \overset{f}{\to} B \overset{g}{\to} C$ be short exact in **A**. This implies $f = \ker g$ so that $Tf = \ker Tg$. But then $0 \to TA \overset{Tf}{\to} TB \overset{Tg}{\to} TC$ is exact according to 2.6.4($b$). To prove the converse we first observe that $T$ preserves monomorphisms. Indeed, let $f: A \to B$ be monic and let $g: B \to C$ be the cokernel of $f$. Then $f = \ker g$ (2.5.1). Thus $A \overset{f}{\to} B \overset{g}{\to} C$ is short exact and hence $* \to TA \overset{Tf}{\to} TB \overset{Tg}{\to} TC$ is exact. Again by 2.6.4($b$), this yields $Tf = \ker Tg$ so that $Tf$ is monic. Now suppose $f: A \to B$ equals $\ker g: B \to C$ and factor $g$ as $B \overset{s}{\twoheadrightarrow} D \overset{i}{\rightarrowtail} C$. Then we have $f = \ker g = \ker s$. Since $s$ is epic we also have $\operatorname{coker} \ker s = s$ (2.5.1) or $\operatorname{coker} f = s$. This means that $A \overset{f}{\to} B \overset{s}{\to} D$ is short exact so that $* \to TA \overset{Tf}{\to} TB \overset{Ts}{\to} TD$ is exact. But then $Tf = \ker Ts$. Since $TB \overset{Tg}{\to} TC$ factors as $TB \overset{Ts}{\to} TD \overset{Ti}{\to} TC$ with $Ti$ monic, we finally find $\ker Ts = \ker Tg$ so that $Tf = \ker Tg$.

Dually one proves (ii). Then (iii) is clear.

The reader should formulate the appropriate statements for contravariant functors.

**2.7.9 Exercises** ($a$) If **C** is an additive category, for $C \in \mathbf{C}$ the functor $h_C$ takes its values in **Ab** (2.3.1). We know that the set-valued functor is left exact (1.9.3). Prove that $h_C$ is also left exact.

($b$) If **C** is an additive exact category a sequence $0 \to A \to B \to C$ is exact if and only if for every $X \in \mathbf{C}$ the sequence $0 \to h_X A \to h_X B \to h_X C$ is exact in **Ab**. By dualizing one sees that $A \to B \to C \to 0$ is exact if and only if for every $X \in \mathbf{C}$ the sequence of abelian groups $0 \to h^X C \to h^X B \to h^X A$ is exact.

($c$) The functor $- \otimes_R M: \mathbf{M}_R \to \mathbf{Ab}$ for $M \in {}_R\mathbf{M}$ has a left adjoint $h_M: \mathbf{Ab} \to \mathbf{M}_R$ (1.7.12($b$)), which is additive. By 1.9.5 and 2.7.3 the tensor product is an additive right exact functor, as we already know (see 2.7.1($b$)).

## 2.8 Special objects: projectives, injectives, generators and cogenerators

Let **C** be any category. As need arises, we shall assume that **C** is punctured, additive, exact or even abelian.

**2.8.1 Definitions** An object $Q$ in the category $\mathbf{C}$ is called *injective* provided for every monomorphism $A \to B$ and every morphism $A \to Q$ there is a morphism $B \to Q$ such that the following triangle commutes:

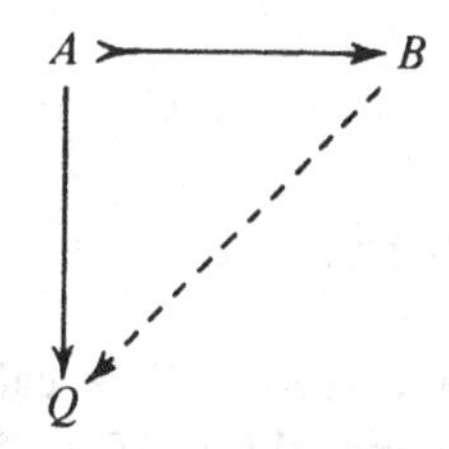


Dually an object $P$ in $\mathbf{C}$ is called *projective* provided for every epimorphism $B \to A$ and for every morphism $P \to A$ there is a morphism $P \to B$ such that the following triangle commutes:

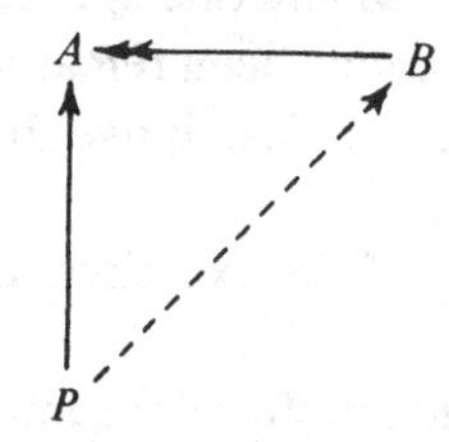


It follows directly that $Q$ is injective in $\mathbf{C}$ if and only if the functor $h^Q: \mathbf{C} \to \mathbf{Sets}$ takes a monomorphism $A \to B$ into an epimorphism $(B, Q)_{\mathbf{C}} \to (A, Q)_{\mathbf{C}}$. Dually $P$ is projective in $\mathbf{C}$ if and only if the functor $h_P: \mathbf{C} \to \mathbf{Sets}$ preserves epimorphisms.

**2.8.2 Lemma** (*a*) If $Q$ is an injective of $\mathbf{C}$ then $Q$ satisfies the condition that for every monomorphism $i: Q \rightarrowtail C$ there is a morphism $j: C \to Q$ with $ji = 1_Q$ ($C$ is a *retract* of $Q$).

(*b*) If $\mathbf{C}$ is abelian this condition implies that $Q$ is injective.

*Proof.* The assertion (*a*) follows immediately from the definition. To prove (*b*) let $u: A \to B$ be a given monomorphism and $f: A \to Q$ an arbitrary morphism. Since $\mathbf{C}$ is abelian we may complete these morphisms into a cocartesian square

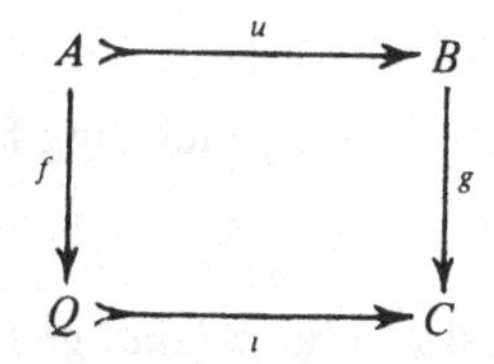


with $i$ also a monomorphism (see lemma 2.5.17). By assumption there exists a morphism $j: C \to Q$ such that $ji = 1_Q$. This implies $jgu = jif = f$.

**2.8.3 Exercises** (*a*) Let **A** be abelian. Then the object $Q$ is injective in **A** if and only if every monomorphism $Q \to A$ splits (see 2.6.11).

(*b*) Let $Q = \prod_i Q_i$ in **C**. If the objects $Q_i$ are injective, so is the object $Q$. The converse is true if the category **C** is punctured.

(*c*) If the category **C** is additive and exact, injectivity of $Q$ is equivalent to $h^Q: \mathbf{C} \to \mathbf{Ab}$ being exact.

(*d*) Dualize lemma 2.8.2 and exercises (*a*), (*b*) and (*c*).

**2.8.4 Definitions** An object $U$ of a category **C** is called a *generator* provided for every pair of distinct morphisms $f, g: A \rightrightarrows B$ there is a morphism $u: U \to A$ such that $fu \neq gu$. A collection of objects $\{U_i\}$ is called a *generating family* if for every such $f$ and $g$ there exists a morphism $u$ from some $U_i$ to $A$ with $fu \neq gu$. Dually one speaks of a *cogenerator* and a *cogenerating family*. As has already been remarked by other authors the name 'separator' would be more descriptive, but we abide with the traditional term.

The proofs of the following statements are left as exercises, as are their duals.

(*a*) If **C** is a category the representable functors form a generating family for (**C**, **Sets**). If **C** is an additive category they also do for both (**C**, **Ab**) and $(\mathbf{C}, \mathbf{Ab})_+$; cf. 2.7.4 and 1.8.2(*c*).

(*b*) The object $U \in \mathbf{C}$ is a generator if and only if $h_U$ is a faithful functor.

(*c*) Suppose the category **C** has equalizers. Then $U$ is a generator if and only if the following two conditions are satisfied:

(i) for every $C \in \mathbf{C}$ the set $(U, C)$ is nonvoid;

(ii) $k: A \rightarrowtail B$ and $h_U k: h_U A \twoheadrightarrow h_U B$ imply that $k$ is an epimorphism.

(*d*) Suppose that in an abelian category with generator $U$ we have $A \xrightarrow{f} B \xrightarrow{g} C$. Then this sequence is exact if

$$h_U A \xrightarrow{h_U f} h_U B \xrightarrow{h_U g} h_U C$$

is exact.

(*e*) If $U = \coprod U_i$ is a generator in **C** then $\{U_i\}$ is a generating family. The converse is true if **C** is punctured.

**2.8.5 Definitions** A category **C** is said to have (*enough*) *injectives* if for every object $C \in \mathbf{C}$ there is a monomorphism $C \to Q$ into an injective object $Q$ of **C**. Dually **C** is said to have (*enough*) *projectives*.

Interplay of injectivity and projectivity with the notions of generator and cogenerator is illustrated by the following easy results, of which the reader should also note the duals.

**2.8.6 Proposition** (*a*) If **C** has injectives and a cogenerator, it has an injective cogenerator.

(*b*) If **C** has products and an injective cogenerator, it has injectives.

*Proof.* (*a*) If $C \in \mathbf{C}$ is a cogenerator we have $C \rightarrowtail Q$ for some injective $Q$ since **C** has injectives. Clearly $Q$ is also a cogenerator.

For (*b*) let $C$ be an injective cogenerator of **C** and suppose $A$ is an object of **C**. Put $Q = \prod_u C_u$ where $u$ runs through the set $(A, C)$ and $C_u = C$ for every $u \in (A, C)$. Then $Q$ is injective and there is a unique morphism $i: A \rightarrow Q$ defined by $p_u i = u$. In order to prove that $i$ is monic, let $f, g: B \rightrightarrows A$. If $if = ig$, then $p_u if = p_u ig: B \rightarrow C_u$ for all $u \in (A, C)$. Since $u = p_u i$ this contradicts the fact that $C$ is a cogenerator. Hence $i$ is monic and the result is proved.

In an abelian category **A** an injective cogenerator $Q$ is succinctly characterized by the additive functor $h^Q$ being both exact and faithful. Dually one characterizes a projective generator. Thus **A** is embedded in a particularly nice way into **Ab**. We shall elaborate upon this theme in the next section.

**2.8.7 Examples** (*a*) In **Sets** every nonvoid set is a projective generator and every set containing at least two distinct elements an injective cogenerator.

(*b*) In **Top** every nonvoid space with the discrete topology is a projective generator and every space containing at least two distinct points is, equipped with the trivial topology, an injective cogenerator.

(*c*) Consider the category of subfields of $\mathbb{C}$ (the complex numbers) which are algebraic over the field of rational numbers $\mathbb{Q}$ (the setting for ordinary Galois theory). The algebraic closure $\bar{\mathbb{Q}}$ in $\mathbb{C}$ is an injective cogenerator.

(*d*) In the category $\mathbf{Ban}_1$ of real Banach spaces with linear maps of norm $\leqslant 1$ (see 1.5.3(*g*)) the field of real numbers $\mathbb{R}$ is an injective cogenerator. This is the content of the celebrated Hahn–Banach theorem [45, p. 144] which asserts that a bounded linear functional defined on a subspace can always be extended to the whole space. The same is true for $\mathbb{C}$ in the case of a complex Banach space, in view of the pendant of that result due to Bohnenblust and Sobczyk [45, p. 145].

(*e*) In the category **Gr** the free groups (1.7.12(*h*)) are easily seen to be projective. Every group is a homomorphic image of a free group and hence is a retract of such a group by 2.8.2(*a*)°. Since every subgroup of a free group is free [27, p. 95], the projectives are just the free groups.

Contrast this with the situation in **CRg**, where for instance the free ring $\mathbb{Z}[X]$ is not projective (take the epimorphism $\mathbb{Z} \to \mathbb{Q}$ of 2.1.4(*c*)).

We shall examine in more detail the case of modules, which will be important in the sequel. If $R$ is a ring and $f: M \to N$ an epimorphism in ${}_R\mathbf{M}$ and $g: R \to N$, one can choose an element $m \in M$ with $f(m) = g(1)$ since an epimorphism in this case is also a surjection (2.1.5). Then define $h: R \to M$ by $h(\rho) = \rho m$ for $\rho \in R$, which is an $R$-homomorphism such that $fh = g$. This shows that $R$, considered as a left $R$-module, is projective. If $R_i \simeq R$ in ${}_R\mathbf{M}$, the modules $\coprod_{i \in I} R_i$ are projective for every index set $I$ (2.8.3(*b*)°). These objects are just the free modules which have a free $R$-basis consisting of the images $q_j(1)$ for $q_j: R \to \coprod_{i \in I} R_i$. As in (*e*) above we can always map a free module onto an arbitrary module $M$ by lifting each $m \in M$, $m \neq 0$, to a free generator $e_m$. Thus we see that free modules are projectives in our category ${}_R\mathbf{M}$. But from 2.8.2° it follows that projectives are just direct summands of free modules. In the case of **Ab** every projective is hence free. This is also the case for the category of modules over a local ring but certainly not for the category of modules over an arbitrary ring. For instance, writing $\mathbb{Z}_k$ for $\mathbb{Z}/k\mathbb{Z}$, we have $\mathbb{Z}_{mn} \simeq \mathbb{Z}_m \times \mathbb{Z}_n$ for coprime integers $m$ and $n$. Now $\mathbb{Z}_m$ and $\mathbb{Z}_n$ are projective but not free $\mathbb{Z}_{mn}$-modules.

In general it is an important question to determine the projective objects in ${}_R\mathbf{M}$; witness the celebrated problem of Serre: if $k$ is a field, are all finitely generated projective modules over the polynomial rings $k[X_1, \ldots, X_n]$ free $(n = 1, 2, \ldots)$? See [4]; an affirmative solution has recently been given by Quillen [36]. Clearly $R$ is a projective generator for the category ${}_R\mathbf{M}$ and so is every free module. But, for instance, $\mathbb{Z}_m$ is not a generator for the category ${}_R\mathbf{M}$ (where $R = \mathbb{Z}_{mn}$) since $(\mathbb{Z}_m, \mathbb{Z}_n)_{{}_R\mathbf{M}} = 0$.

Again in general an interesting question is which projectives are generators [41]? But for homological algebra the important thing is that ${}_R\mathbf{M}$ has enough projectives: for every module $M$ there is a projective module $F$ and an epimorphism $f: F \to M$. We can take for $F$ the module with the elements of $M$ as free generators (or, more economically, a set of generators of $M$).

Turning to injective modules we first treat **Ab** where the injective objects have a neat characterization. An abelian group $D$ is called *divisible* provided for every $c \in D$ and for every integer $m \neq 0$ there exists an element $d \in D$ with $md = c$. If the element $c$ is always uniquely determined we call $D$ *uniquely divisible*. Examples of divisible abelian groups are $\mathbb{Q}$, $\mathbb{R}$, $\mathbb{Q}/\mathbb{Z}$, $\mathbb{R}/\mathbb{Z}$. The first two of these are uniquely divisible.

**2.8.8 Proposition** The injective objects in **Ab** are precisely the divisible abelian groups.

*Proof.* Let $Q$ be an injective object of **Ab**. Suppose $q \in Q$ and let $m$ be a nonnull integer. Let $f: \mathbb{Z} \rightarrowtail \mathbb{Z}$ be determined by $f(1) = m$. Let $g: \mathbb{Z} \to Q$ be given by $g(1) = q$. The injectivity of $Q$ yields a homomorphism $h: \mathbb{Z} \to Q$ with $hf = g$. Then $q = g(1) = hf(1) = h(m\,.\,1) = mh(1)$ and so $Q$ is divisible.

Now suppose conversely that $D$ is divisible. Then for a given monomorphism $f: B \to C$ and $g: B \to D$ we have to find a homomorphism $h: C \to D$ such that $hf = g$. For convenience we suppose that $B$ is a subgroup of $C$ and consider subgroups $B_i$ such that $B \subset B_i \subset C$ and homomorphisms $h_i: B_i \to D$ with $h_i|B = g$, $i \in I$. This nonvoid collection $\{B_i, h_i\}$ is partially ordered by inclusion. Every chain, i.e. totally ordered subset $\{B_j, h_j\}$, $j \in J \subset I$, has an upper bound in this ordering, viz. $\{B_j, h'\}$ with $h'(b) = h_j(b)$ for all $b \in B_j$. By Zorn's lemma there exists a maximal element, say $\ulcorner \bar{B}, \bar{h} \urcorner$, in the collection $\{B_i, h_i\}$. We shall show that $\bar{B} = C$. Then taking $h = \bar{h}$ we have found an $h: C \to D$ with $hf = g$. Now suppose $c \in C \backslash \bar{B}$. The set $\{z \in \mathbb{Z} | zc \in \bar{B}\}$ is an ideal in $\mathbb{Z}$. Let this ideal be generated by $m$. If $m \neq 0$ choose an element $d$ in the divisible group $D$ with $md = \bar{h}(mc)$. If $m = 0$ choose an arbitrary $d \in D$. Define $h''$ on the submodule of $C$ generated by $\bar{B}$ and $c$ by $h''(b + nc) = \bar{h}(b) + nd$, where $b \in \bar{B}$. Then $h''$ is clearly a homomorphism, provided it is well defined as a mapping. Suppose therefore $b + nc = b_1 + n_1 c$. It should follow that $\bar{h}(b) + nd = \bar{h}(b_1) + n_1 d$ or $\bar{h}(b - b_1) = (n_1 - n)d$. But since $(n_1 - n)c \in \bar{B}$, we may assume that $n_1 - n = km$, so that $b - b_1 = kmc$ which, in turn, implies $\bar{h}(b - b_1) = k\bar{h}(mc) = kmd = (n_1 - n)d$. So we have extended the homomorphism $\bar{h}$ on $\bar{B}$ to $h''$ on a larger subgroup, which contradicts the maximality of $\ulcorner \bar{B}, \bar{h} \urcorner$ and so $\bar{B} = C$. This establishes the proposition, which is originally due to R. Baer [3, p. 766].

The reader may prove that $\mathbb{Q}/\mathbb{Z}$ is a cogenerator observing that this group contains elements of every finite order. Since this means that **Ab** has an injective cogenerator the category **Ab** has injectives (by 2.8.6($b$)).

In order to extend this result to modules over an arbitrary ring we need a general fact.

**2.8.9 Proposition** Let $\mathbf{A} \underset{T}{\overset{S}{\rightleftarrows}} \mathbf{B}$ be an adjoint situation of additive functors between exact and additive categories, where $S$ is a left adjoint of $T$. If $S$ is exact, $T$ preserves injectives. If $S$ is in addition faithful and if $\mathbf{B}$ has an injective cogenerator, so has $\mathbf{A}$.

*Proof.* If $Q$ is injective in $\mathbf{B}$, so is $TQ$ in $\mathbf{A}$. Indeed, for $X \in \mathbf{A}$ the natural isomorphism $(X, TQ)_{\mathbf{A}} \simeq (SX, Q)_{\mathbf{B}}$ shows that $h^{TQ} \simeq h^{Q} \circ S: \mathbf{A} \to \mathbf{Ab}$. Since $S$ is exact and $h^{Q}$ is contravariant and exact, the functor $h^{TQ}$ is contravariant and exact. Hence $TQ$ is injective (see 2.8.1). Now if $Q$ is a cogenerator, the functor $h^{Q}$ is faithful (see 2.8.4($b$)°). But $S$ is also faithful. Therefore $h^{TQ}$ is faithful, making $TQ$ an injective cogenerator.

**Corollary** For every ring $R$ the category $\mathbf{M}_R$ has an injective cogenerator. Hence it has injectives.

*Proof.* There is a ring homomorphism $\mathbb{Z} \to R$, and hence a functor $S: \mathbf{M}_R \to \mathbf{Ab}$ by restriction of scalars. This functor is clearly exact and faithful. By 1.7.12($l$) it has a right adjoint $T$ determined by $TA = (R, A)_{\mathbf{Ab}}$ for $A \in \mathbf{Ab}$. Then $S$ is a left adjoint of $T$ and we may apply the previous proposition. Now $\mathbb{Q}/\mathbb{Z}$ being an injective cogenerator in $\mathbf{Ab}$ implies that $(R, \mathbb{Q}/\mathbb{Z})_{\mathbf{Ab}}$ is an injective cogenerator of $\mathbf{M}_R$.

We need to discuss a special type of injective object in the category $\mathbf{M}_R$.

**2.8.10 Definitions** A submodule $N \subset M$ is called *essential* (or *large*) if for $X \subset M$ the condition $N \cap X = (0)$ implies $X = (0)$.

Clearly this is equivalent to the fact that the monomorphism $u: N \to M$ has the property that if $vu$ is monic for some $v: M \to Y$ then $v$ itself is monic. Such a monomorphism $u$ is called an *essential extension.* In this form the concept generalizes to arbitrary categories.

It is also obvious that if $L \rightarrowtail N$ and $N \rightarrowtail M$ are both essential extensions, so is $L \rightarrowtail M$.

**2.8.11 Lemma** A module $Q$ is injective if and only if it has no proper essential extensions.

*Proof.* Any proper extension of an injective module splits (see 2.8.3($a$)), so is not essential. Suppose, conversely, that $Q$ has no proper essential extensions. We wish to show that every proper extension $Q \rightarrowtail M$ splits (which would show that $Q$ is injective by 2.8.3(*a*)). Consider, for this purpose, the collection of submodules $S \subset M$ with $S \cap Q = (0)$. This collection is ordered by inclusion. Any chain $(S_i)$ has an upper bound $\bigcup_i S_i$. By Zorn's lemma there exists a maximal element $T$ in the collection of submodules. Then the composition $Q \rightarrowtail M \to M/T$ is monic and it is essential by the maximality of $T$. By assumption $Q \rightleftarrows M/T$. But then $M \simeq Q \oplus T$, so $Q$ is injective.

**2.8.12 Proposition** Let $M$ be a submodule of an injective module $Q$. Then there exists an injective module $\bar{Q}$ with $M \subset \bar{Q} \subset Q$ and such that $M \subset \bar{Q}$ is an essential extension.

*Proof.* We order the collection of submodules $S$ of $M$ such that $M \subset S \subset Q$ with $M \subset S$ essential by inclusion. Then it is again easily seen that every chain has an upper bound in the collection. Let $\bar{Q}$ be a maximal element in the collection (Zorn's lemma). Since $M \subset \bar{Q}$, it remains to be seen (in view of our preceding lemma) that any proper extension $u: \bar{Q} \rightarrowtail N$ cannot be essential. Because $Q$ is injective there exists a morphism $v: N \to Q$ such that $vu$ equals the embedding $\bar{Q} \rightarrowtail Q$ of $\bar{Q}$ as a submodule of $Q$. If $K = \ker v$, then $K \cap \bar{Q} = (0)$. If $u$ is essential this entails $K = (0)$. Thus $v$ is an embedding of $N$ into $Q$, with $M \subset N$ essential. But then $N = \bar{Q}$ because of the maximality of the latter.

**2.8.13 Definition** If $M \rightarrowtail Q$ is essential and $Q$ is injective, the module $Q$ is called an *injective envelope* or an *injective hull* of $M$.

It is easy to show that if $i_1: M \rightarrowtail Q_1$ and $i_2: M \rightarrowtail Q_2$ are both injective envelopes of $M$ there exists an isomorphism $f$:

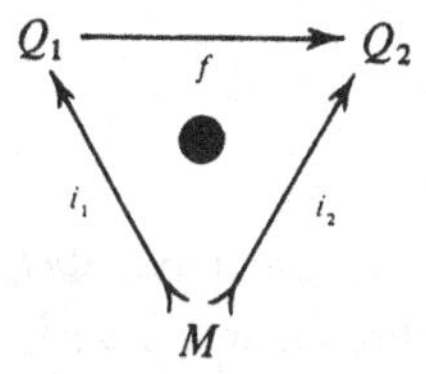


which is called an *isomorphism under M*, dual to the morphisms over in 2.1.7. We have proved that in $\mathbf{M}_R$ every module has an injective envelope. For an arbitrary ring $R$ this is not the case for the dual notion, viz. a *projective envelope* (or *projective cover*). Both notions play an important role in the standard theory of rings and modules [30], [38], [41]. These concepts may also be formulated for an arbitrary category. If every object in a category **C** has an injective (projective) envelope we say that **C** *has injective (projective) envelopes.* The following result says something about the preservation of injective envelopes in an adjoint situation.

**2.8.14 Proposition** Let $\mathbf{A} \underset{T}{\overset{S}{\rightleftarrows}} \mathbf{B}$ be an adjoint situation between exact categories with $T$ embedding $\mathbf{B}$ as a full reflective subcategory of $\mathbf{A}$. Suppose in addition that the left adjoint $S$ is exact. Then if $Q$ is an injective envelope of $TB$ in $\mathbf{A}$, the object $SQ$ is an injective envelope of $B$ in $\mathbf{B}$. Thus if $\mathbf{A}$ has injective envelopes, so has $\mathbf{B}$.

*Proof.* Let $u\colon TB \rightarrowtail Q$ be an injective envelope in $\mathbf{A}$. Applying $S$ we find $Su\colon STB \to SQ$. Since $S$ is exact it preserves monomorphisms so $Su$ is monic. From theorem 1.7.10 we know that $ST = I_{\mathbf{B}}$. To show that $SQ$ is an injective object of $\mathbf{B}$, let $v\colon X \rightarrowtail Y$ and $g\colon X \to SQ$ be morphisms in $\mathbf{B}$. We shall show in a minute that there is a morphism $j\colon TSQ \to Q$ with $Sj = 1_{SQ}$. Since $T$ preserves monomorphisms (2.7.6) we then would have the diagram

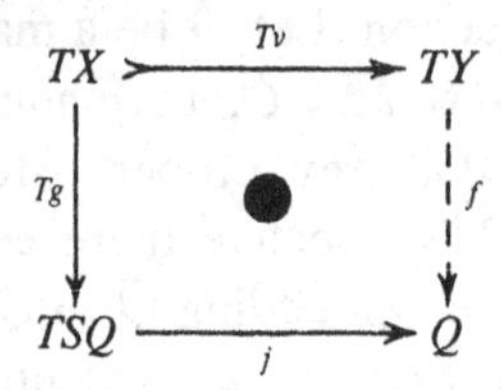


from which we infer the existence of the morphism $f$ (due to the fact that $Q$ is injective) making the diagram commute. Applying $S$ we find $Sf \circ STv = Sj \circ STg$ or $Sf \circ v = g$. Thus $SQ$ is injective. To fulfil our promise we consider the following commutative diagram:

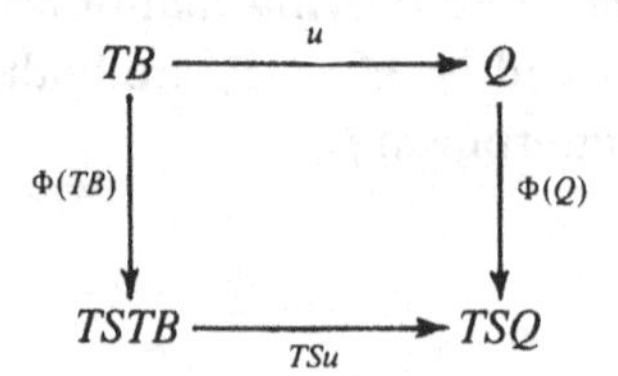


Due to the fact that $\Phi T = 1_T$ (theorem 1.7.10) we have $\Phi(Q) \circ u = TSu \circ \Phi(TB) = TSu$. Since $T$ preserves monomorphisms, $TSu$ is monic. As $u$ is an essential monomorphism, $\Phi(Q)$ is monic. Invoking lemma 2.8.2 (*a*) for the injective object $Q$ we find a morphism $j\colon STQ \to Q$ such that $j \circ \Phi(Q) = 1_Q$. Application of $S$ yields $Sj = 1_{SQ}$ since $S\Phi = 1_S$ (theorem 1.7.10).

Finally we have to show that $Su\colon B \to SQ$ is an essential monomorphism. Let $h\colon SQ \to X$ be a morphism such that $h \circ Su$ is monic. Then $Th \circ TSu = Th \circ \Phi(Q) \circ u$ is monic in $\mathbf{A}$. Since $u$ is an essential extension $Th \circ \Phi(Q)$ must be monic. But then $STh \circ S\Phi(Q)$ is monic ($S$ is exact) and this morphism equals $h$ because $S\Phi = 1_S$. So $Su$ is indeed an essential monomorphism.

## 2.9 Grothendieck categories

For certain purposes of homological algebra abelian categories are not quite good enough. So we add a few axioms satisfied by $\mathbf{M}_R$. Before doing so we first continue the discussion started in 2.1.7.

Let $\mathbf{A}$ be an exact category and let $A$ be a fixed object of $\mathbf{A}$. Suppose we have a class of monomorphisms $\{m_i: A_i \to A\}_{i\in I}$. Assuming that the sum $\coprod_{i\in I} A_i$ exists we may consider the following diagram:

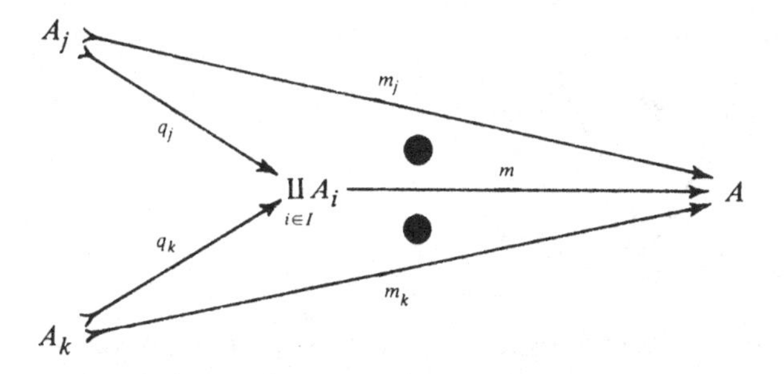


We then may factor $m$ through an epimorphism and monomorphism (in a unique way, up to isomorphism):

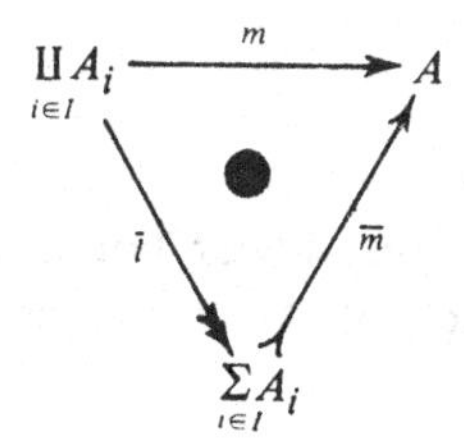


The notation $\sum A_i$ is chosen to bring out an analogy with an abelian group $A$ and its subgroups $A_i$. Put

$$\bar{l}q_j = \bar{q}_j : A_j \to \textstyle\sum_{i\in I} A_i$$

for every $j \in I$. It follows easily that $\bar{q}_j$ is monic. If $m_{jk}: A_j \rightarrowtail A_k$ is a morphism over $A$ then it follows readily that

$$A_j \overset{m_{jk}}{\rightarrowtail} A_k \overset{q_k}{\rightarrowtail} \textstyle\sum A_i = A_j \overset{q_j}{\rightarrowtail} \sum A_i.$$

Endowing the class of all subobjects of $A$ with the partial ordering introduced in 2.1.7 one may prove that $\sum_{i\in I} A_i$ is the least upper bound of the subobjects $\{A_i\}_{i\in I}$ in the class of all subobjects of $A$. When the index class $I$ is a set consisting of only two elements we denote $\sum_{i=1,2} A_i$ by $A_1 + A_2$.

Now assume that the diagram $D$ given by all the monomorphisms in the collection $\{A_i \xrightarrow{m_i} A\}_{i\in I}$ has an infimum. Factoring the morphism $\inf D \xrightarrow{p} A$ through an epimorphism and a monomorphism yields

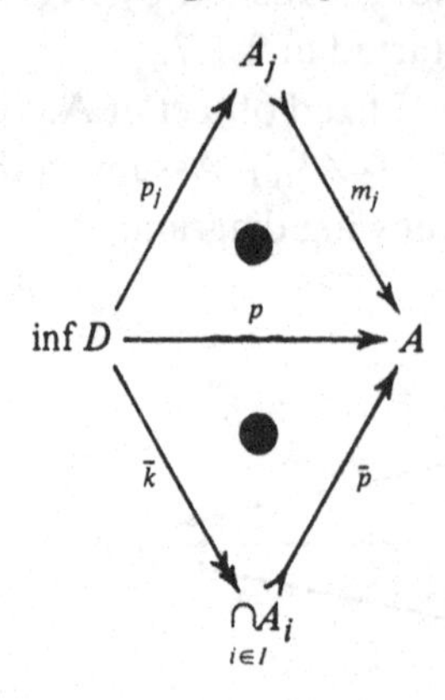


Using the epic–monic factorization of $\inf D \xrightarrow{p_j} A_j$ one proves that there is a monomorphism $\bigcap_{i\in I} A_i \xrightarrow{\bar{p}_j} A_j$ such that $\bar{p}_j\bar{k} = p_j$ for every $j \in I$ and one may also show easily that if $m_{jk}\colon A_j \rightarrowtail A_k$ is a morphism over $A$ then

$$\bigcap_{i\in I} A_i \xrightarrow{\bar{p}_j} A_j \xrightarrow{m_{jk}} A_k = \bigcap_{i\in I} A_i \xrightarrow{\bar{p}_k} A_k.$$

One readily proves that $\bigcap_{i\in I} A_i$ is the greatest lower bound of the subobjects $\{A_i\}_{i\in I}$ in the class of all subobjects of $A$. This explains the notation $\bigcap_{i\in I} A_i$.

From $\bigcap_{i\in I} A_i \xrightarrow{\bar{p}_j} A_j \xrightarrow{\bar{q}_j} \sum A_i$ we obtain a single morphism $\bigcap_{i\in I} A_i \to \sum A_i$. Denote coker $(A_i \xrightarrow{m_i} A)$ by $c_i\colon A \to A/A_i$. Then consider $c\colon A \to \prod_{i\in I}(A/A_i)$ defined by $\bar{p}_i c = c_i$ for all $i \in I$ and denote $\ker c$ by $\bar{c}\colon K \to A$. Putting things together into one diagram we obtain

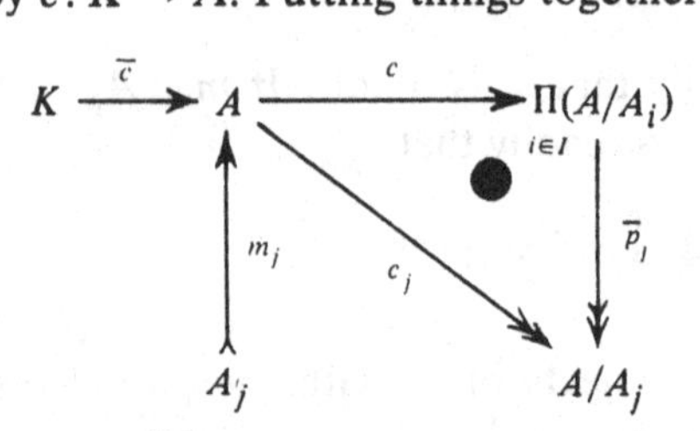


It follows easily from this diagram that there is a monomorphism $\bar{p}_j'\colon K \to A_j$ with $m_j\bar{p}_j' = \bar{c}$ for each $j \in I$ and if $m_{jk}\colon A_j \rightarrowtail A_k$ is a morphism over $A$ then $K \xrightarrow{\bar{p}_j'} A_j \xrightarrow{m_{jk}} A_k = K \xrightarrow{\bar{p}_k'} A_k$. One then proves also without trouble that $K$ is the greatest lower bound of the subobjects $\{A_i\}_{i\in I}$ in the class of all subobjects of $A$. Now it follows that the subobjects $K$ and $\bigcap_{i\in I} A_i$ coincide.

**2.9.1 Lemma** Let **A** be an abelian category. Consider the diagram $D$ in **A** given by $\{A_i \overset{m_i}{\rightarrowtail} A\}_{(i=1,2)}$. With the notations as introduced above the following square is bicartesian:

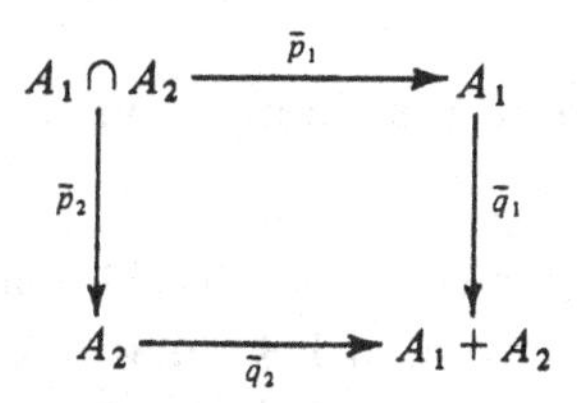


*Proof.* We shall first prove that the square is cartesian. Suppose $u_i: X \to A_i$ $(i=1,2)$ such that $\bar{q}_1 u_1 = \bar{q}_2 u_2$. We have to show that there is a unique morphism $u: X \to A_1 \cap A_2$ such that $\bar{p}_i v = u_i (i=1,2)$. It follows from the given situation that there is a unique morphism $u: X \to \inf D$ making the following diagram commutative:

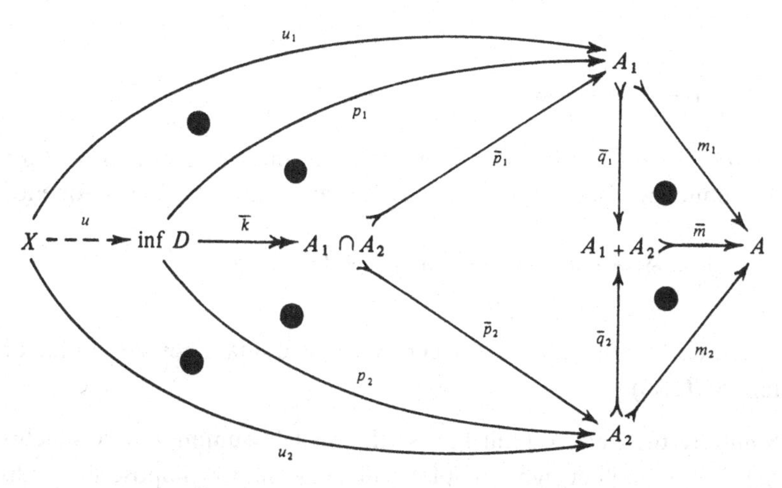


Then $u_i = p_i u = \bar{p}_i \bar{k} u$ $(i=1,2)$. So we take $v = \bar{k}u$. The morphism $v$ is unique since $\bar{p}_i \tilde{v} = u$ $(i=1,2)$ implies $v = \tilde{v}$, due to the fact that $\bar{p}_i$ is monic $(i=1,2)$.

Now that we have proved that the above square is cartesian we may infer (lemma 2.5.16) that in the sequence

$$A_1 \cap A_2 \xrightarrow{\binom{\bar{p}_2}{\bar{p}_2}} A_1 \oplus A_2 \xrightarrow{(-\bar{q}_1\ \bar{q}_2)} A_1 + A_2$$

we have

$$\begin{pmatrix}\bar{p}_1\\ \bar{p}_2\end{pmatrix} = \ker\,(-\bar{q}_1\;\;\bar{q}_2).$$

Since the morphism $(\bar{q}_1\;\;\bar{q}_2)$ is epic (this morphism is just the morphism denoted by $\bar{l}$ earlier) the morphism $(-\bar{q}_1\;\;\bar{q}_2)$ is also epic. Therefore

$$\operatorname{coker}\begin{pmatrix}\bar{p}_1\\ \bar{p}_2\end{pmatrix} = (-\bar{q}_1\;\;\bar{q}_2).$$

Thus the sequence is short exact. From the exercise in 2.6.1 it follows that the square is bicartesian.

Since **A** is abelian it has coequalizers. If we assume, in addition, that **A** has infinite sums it follows from the theorem in 1.8.4 that **A** is right complete. Suppose further that $I$ is a directed set such that for all $i \in I$ there is a short exact sequence $\dot{E}_i\colon A_i \to B_i \to C_i$ in **A** and assume that, whenever $i \geqslant j$, there is a morphism $\dot{E}_i \to \dot{E}_j$ in the category of short exact sequences such that

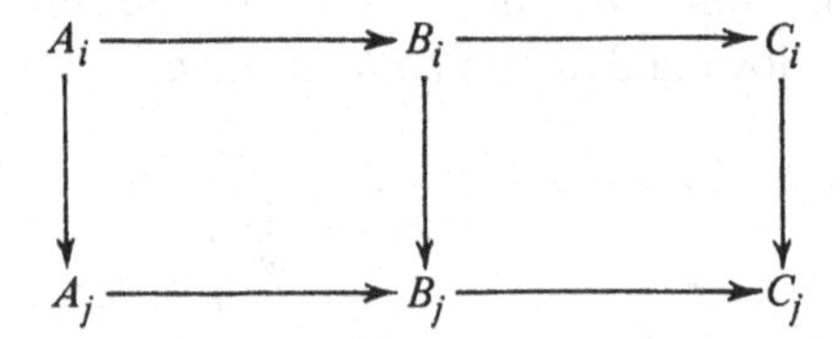


is commutative. In other words, we have an inductive system of short exact sequences (see 1.6.1). It is easy to verify that the limit sequence

$$\varinjlim A_i \to \varinjlim B_i \to \varinjlim C_i \to 0$$

is exact, i.e. the functor $\varinjlim$ is right exact (its domain being the functor category $(\mathbf{I}, \mathbf{A})$).

Similarly the inverse limit $\varprojlim$ is left exact (assuming left completeness). On an abelian right complete category we now impose the additional axiom that the functor $\varinjlim$ is exact. For historical reasons this axiom is always referred to as AB5. In view of the above this means that we are stipulating that $\varinjlim$ preserves monomorphisms.

**2.9.2 Lemma** In a right complete abelian category satisfying AB5, suppose $m_i\colon A_i \rightarrowtail A$ is an inductive system of subobjects. If $X \rightarrowtail A$ is a subobject then $\sum (X \cap A_i) = X \cap \sum A_i$.

*Proof.* For each $i$ we consider the following cartesian square (all morphisms are monic).

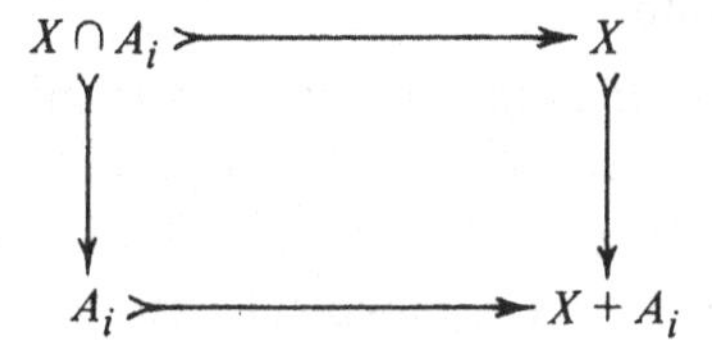

Taking direct limits and using the fact that $\varinjlim$ is left exact we may conclude from 2.7.5 that we get a cartesian square. It follows further from AB5 that $\varinjlim A_i$ is again a subobject and one easily checks that $\varinjlim A_i = \sum A_i$ and similarly for the subobjects $X \cap A_i$ and $X + A_i$. In the commutative square

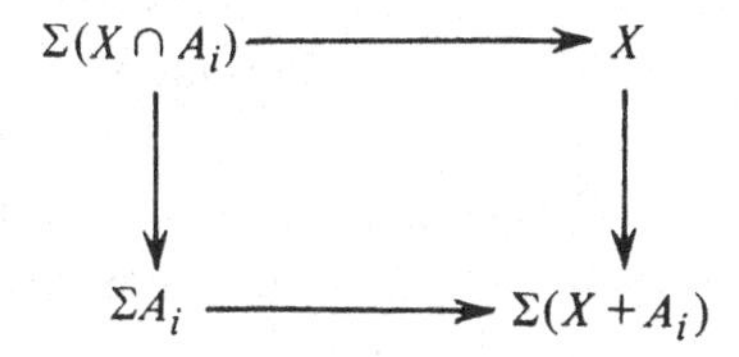

because of AB5 all morphisms are still monic and in fact all the objects are still subobjects of $A$. From the fact that the square

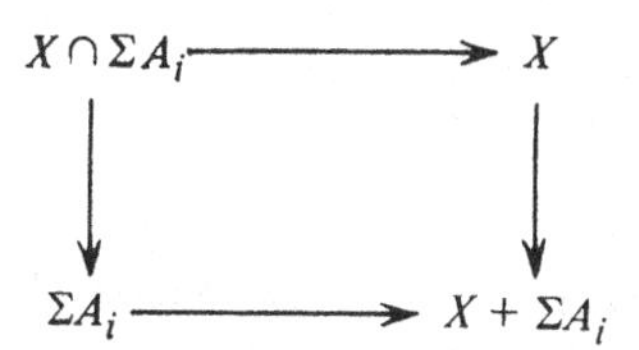

is also cartesian it follows that the subobjects $X \cap \sum A_i$ and $\sum (X \cap A_i)$ are the same, since $\sum (X + A_i)$ and $X + \sum A_i$ are easily seen to determine isomorphic subobjects of $A$. The condition of the lemma is in fact equivalent to AB5, as the reader may prove for himself. A thorough discussion of different forms of AB5 and related axioms is given in [33, p. 50 seq.].

Sometimes a category is called *locally small* if for each object the class of subobjects is a set.

**Exercise** Suppose **A** is an abelian category with a generator $U$. Prove that **A** is locally small by showing that the map which associates to a subobject $m: B \rightarrowtail A$ the subset of $(U, A)$ consisting of those morphisms $f: U \to A$ which factor through $m$ is injective.

**2.9.3 Definition** An abelian category is called a *Grothendieck category* if it is right complete, satisfies the axiom AB5 (i.e. the functor $\varinjlim$ is exact) and possesses a generator.

The principal example, as the reader may convince himself, is the category $\mathbf{M}_R$, for any ring $R$. Not every author makes precisely these stipulations for a Grothendieck category, but ours is the most convenient definition. We shall prove that a Grothendieck category can be embedded in a nice way in a category of modules. First, however, we embark upon some preliminary considerations.

Let $J: \mathbf{U} \to \mathbf{C}$ be a functor from a small category $\mathbf{U}$ to a right complete category $\mathbf{C}$. Define $\bar{J}: \mathbf{C} \to (\mathbf{U}^\circ, \mathbf{Sets})$ by $\bar{J}C(U) = (JU, C)_\mathbf{C}$, $C \in \mathbf{C}$, $U \in \mathbf{U}$. Notice that for each $C \in \mathbf{C}$ the functor $\bar{J}C = h^C \circ J: \mathbf{U}^\circ \to \mathbf{Sets}$ is a familiar one; it was denoted by $G^C$ in theorem 1.7.7° and elsewhere. We shall construct a left adjoint of $\bar{J}$. Recall the Yoneda embedding $h^{\cdot}: \mathbf{U} \to (\mathbf{U}^\circ, \mathbf{Sets})$ which is full and faithful (1.4.7). This induces a functor $- \circ h^{\cdot} = T_1: ((\mathbf{U}^\circ, \mathbf{Sets}), \mathbf{C}) \to (\mathbf{U}, \mathbf{C})$ which has a left adjoint $S_1: (\mathbf{U}, \mathbf{C}) \to ((\mathbf{U}^\circ, \mathbf{Sets}), \mathbf{C})$ such that $T_1 S_1 = I_{(\mathbf{U},\mathbf{C})}$ as is proved in the Kan extension theorem (1.8.6). We claim that the Kan extension of $J$ is a functor $S_1 J: (\mathbf{U}^\circ, \mathbf{Sets}) \to \mathbf{C}$ which is left adjoint to $\bar{J}$.

First observe that for the representable functor $h^U$, $U \in \mathbf{U}$ and $C \in \mathbf{C}$, we have

$$\begin{aligned}(S_1 J(h^U), C)_\mathbf{C} &= ((S_1 J \circ h^{\cdot})U, C)_\mathbf{C} \\ &= ((T_1 S_1 J)U, C)_\mathbf{C} = (JU, C)_\mathbf{C} \\ &= \bar{J}C(U) \\ &\simeq (h^U, \bar{J}C)_{(\mathbf{U}^\circ, \mathbf{Sets})} \qquad \text{(Yoneda's lemma).}\end{aligned}$$

Thus the behaviour of $S_1 J$ on the full subcategory of the representable functors from $\mathbf{U}^\circ$ to $\mathbf{Sets}$ is that of a left adjoint to $\bar{J}$. To construct the Kan extension of $J$ we have to consider for each $F: \mathbf{U}^\circ \to \mathbf{Sets}$ the category $\mathbf{H}_F$ whose objects are morphisms $\alpha: h^U \to F$ in $(\mathbf{U}^\circ, \mathbf{Sets})$ and whose morphisms are $v: U \to U'$ such that $\alpha' \circ h^v = \alpha$. The functor $J_F: \mathbf{H}_F \to \mathbf{C}$ is defined by $J_F \ulcorner U, \alpha \urcorner = JU$ and $J_{Fv} = Jv$. Then $S_1 J(F)$ is the supremum of the diagram $J_F$ in $\mathbf{C}$. Now the chain of identifications

$$\begin{aligned}(S_1 J(F), C)_\mathbf{C} &= (\sup J_F, C)_\mathbf{C} \\ &= (\sup JU, C)_\mathbf{C} \\ &= \inf\,(JU, C)_\mathbf{C} \\ &\simeq \inf\,(h^U, \bar{J}C)_{(\mathbf{U}^\circ,\mathbf{Sets})} \\ &= (\sup h^U, \bar{J}C)_{(\mathbf{U}^\circ,\mathbf{Sets})} \\ &= (F, \bar{J}C)_{(\mathbf{U}^\circ,\mathbf{Sets})}\end{aligned}$$

proves that $S_1J$ is a left adjoint of $\bar{J}$. Here we have used the fact that $F = \sup h^U$ as pointed out in 1.8.2($c$).

Now suppose in the above that **U** and **C** are additive categories and $J$ is an additive functor between them. Replacing **Sets** by **Ab** we may carry through the same programme where now all the functors occurring will be additive (compare 2.8.4($a$)). In particular, take for **C** an abelian right complete category **A** and let $U \in \mathbf{A}$. Take for **U** the full subcategory of **A** consisting of the single object $U$ and whose morphisms are the elements of the endomorphism ring $R = (U, U)_{\mathbf{A}}$. The functor $J$ is just the embedding of **U** in **A**. Notice that the category of contravariant additive functors $(\mathbf{U}^\circ, \mathbf{Ab})_+$ may be identified with $\mathbf{M}_R$ by associating to such a functor $F$ the abelian group $FU$. Then for a morphism $\rho: U \to U$ in **U** the morphism $F\rho$ is an endomorphism of $FU$ and this yields a ring-homomorphism of $R$ into $(FU, FU)_{\mathbf{Ab}}$ making $FU$ into a right $R$-module. Further details of this identification are readily supplied. Then, since $\bar{J}A$ is determined by $(\bar{J}A)U = (JU, A)_{\mathbf{A}} = (U, A)_{\mathbf{A}}$, our $\bar{J}$ is just $(U, \text{-})_{\mathbf{A}} = h_U$, considered now as an additive functor from **A** to $\mathbf{M}_R$. Call this functor $T$.

**2.9.4 Theorem** Let **A** be a Grothendieck category with a generator $U$ and denote the ring $(U, U)_{\mathbf{A}}$ by $R$. Then the following assertions hold.

(i) $T$ is fully faithful.
(ii) $T$ has an additive left adjoint $S$.
(iii) $S$ is exact.

*Proof.* That $T$ has a left adjoint was seen in the previous discussion; this is additive by proposition 2.7.3. We proceed to prove that $T$ is full and faithful.

Now $T = (U, \text{-})_{\mathbf{A}}$, where the right $R$-module structure on $TA$, $A \in \mathbf{A}$, is given by the action of $R$ on $U$ in the following way. Let $f \in TA = (U, A)_{\mathbf{A}}$ and let $\rho \in R = (U, U)_{\mathbf{A}}$. Then $f\rho$ equals the morphism $U \xrightarrow{\rho} U \xrightarrow{f} A$. The fact that $T$ is faithful, i.e. the injectivity of $(A, A')_{\mathbf{A}} \to (TA, TA')_{\mathbf{M}_R}$, is tantamount to $U$ being a generator. To prove surjectivity suppose $\phi: TA \to TA'$ in $\mathbf{M}_R$. Write $L = (U, A)_{\mathbf{A}} = TA$ and put $U_l = U$ for $l \in L$. The morphism $f: \coprod_{l \in L} U_l \to A$ determined by $fq_l = l$ is clearly epic because $U$ is a generator. For each $l \in L$ the map $\phi(l)$ is a morphism from $U$ to $A'$ and there exists a unique $g: \coprod_{l \in L} U_l \to A'$ such that $gq_l = \phi(l)$ for every $l \in L$. We wish to prove that $g$ is the null morphism on $K = \ker f$. In that case $g$ factors through $f = \operatorname{coker} \ker f$ (2.5.1; $f$ is epic); i.e. there is a morphism $s: A \to A'$ such that $sf = g$.

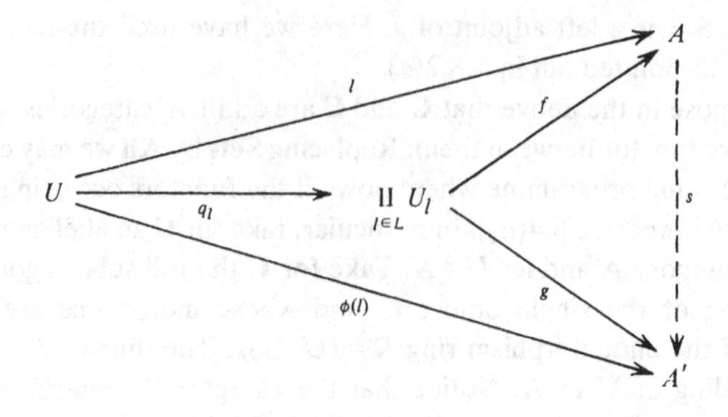


From the diagram it transpires that $sl = \phi(l)$ for every $l \in (U, A)_{\mathbf{A}}$ which means that $\phi = Ts$, thus establishing the fact that $T$ is full.

To prove the assertion that $g$ is the null morphism on $K = \ker f$, consider the collection of finite subsets $i$ of the index set $L$. These subsets form a directed set $I$ by inclusion. For each $i \in I$ there is a morphism $q_i: \coprod_{l\in i} U_l \to \coprod_{l\in L} U_l = B$, given by $q_i q_l' = q_l$ for $l \in i$. Here $q_l (q_l'$respectively) is the injection of $U_l$ into $\coprod_{l \in L} U_l$ (into $\coprod_{l\in i} U_l$ respectively), while we shall soon denote the respective projections by $p_l$ and $p_l'$. We write $b_i: B_i \rightarrowtail B$, or just $B_i$, for im $q_i$. Then the subobjects $B_i$ of $B$ form an inductive system for which we have $\sum_{i\in I} B_i = B$ (as is easily proved). If we write $K_i$ for $B_i \cap K$, lemma 2.9.2 yields

$$\sum_{i\in I} K_i = \sum_{i\in I} (B_i \cap K) = \left(\sum_{i\in I} B_i\right) \cap K = B \cap K = K.$$

Let $\coprod_{l\in i} U_l \overset{c_i}{\twoheadrightarrow} B_i \overset{b_i}{\rightarrowtail} B$ be the epic-monic factorization of $q_i$. Putting $k_i: K_i \rightarrowtail B_i = \ker fb_i$ and $k_i': K_i' \rightarrowtail \coprod_{l\in i} U_l = \ker fq_i$, one finds a unique morphism $k_i'': K_i' \to K_i$ with $k_i k_i'' = ck_i'$, completing a cartesian square. From 2.5.17 one concludes that $k_i''$ is epic.

We wish to show that $K \overset{k}{\to} B \overset{g}{\to} A'$ is the null morphism, but since $K = \sum_{i\in I} K_i$, it is enough to do so for all compositions $K_i \overset{k_i}{\to} B_i \overset{b_i}{\to} B \overset{g}{\to} A'$ or also for all $gb_i k_i k_i'' = gq_i k_i'$. Or even, since $U$ is a generator, for every $gq_i k_i' w$ with $w: U \to K_i'$. Put $k_i' w = t: U \to \coprod_{l\in i} U_l$. Now recall, and this is the heart of the proof, that the set $i$ is finite and hence $\coprod_{l\in i} U_l = \bigoplus_{l\in i} U_l$ so that identity morphism on this object is given by $\sum_{l\in i} q_l' p_l'$. Hence

$$0 = fq_i t = fq_i(\textstyle\sum_{l\in i} q_l' p_l')t = \sum_{l\in i} fq_l p_l' t = \sum_{l\in i} l p_l' t.$$

Similarly $gq_it = \sum_{l\in i} \phi(l)p_i't$.

But $p_i't \in (U, U)_{\mathbf{A}} = R$ and $\phi(l)$ is a homomorphism of right $R$-modules, so that $gq_it = \phi(fq_it) = 0$. This completes the proof of assertion (i).

Since $TA = TA_1$ certainly entails $A = A_1$ we have, thus far, proved that $T$ is a fully faithful embedding of $\mathbf{A}$ in $\mathbf{M}_R$ with a left adjoint $S$. As follows from the discussion just preceding theorem 1.7.10 this means that $T$ embeds $\mathbf{A}$ in $\mathbf{M}_R$ as a full reflective subcategory of $\mathbf{M}_R$. Furthermore the left adjoint $S$ may be chosen in such a way that $ST = I_{\mathbf{A}}$ and $\Psi$: $ST \to I_{\mathbf{A}}$ is the identity.

To prove assertion (iii) we first remark that $S$ is right exact according to the theorems 1.9.5 and 1.9.2. To show that $S$ is left exact we have to prove that $S$ preserves kernels (2.7.5). For this purpose it is sufficient to prove that $S$ preserves monomorphisms (see 2.7.8).

First assume $M$ is a finitely generated $R$-module and $s: M \to L_0$ a monomorphism into a free module $L_0$, which we may as well assume to be finitely generated. Map a finitely generated free module $L_1$ onto $M$. Repeat the process for the kernel of this map thus finding an exact sequence of free modules $L_2 \xrightarrow{g} L_1 \xrightarrow{f} M \to 0$. Here $L_j$ equals $\coprod_{i\in I_j} R_i = \coprod_{i\in I_j} TU_i$ for $j = 0, 1, 2$ and for certain index sets $I_j$, of which the first two are finite. The modules $R_i$ are copies of the free module $R$ on a single generator and $U_i \simeq U$. The functor $S$ is right continuous (theorem 1.9.5) so $SL_j = \coprod_{i\in I_j} STU_i = \coprod_{i\in I_j} U_i$. Because of finiteness for $j = 0, 1$ we have $\coprod_{i\in I_j} U_i = \bigoplus_{i\in I_j} U_i$ so that for these two indices we obtain $TSL_j = L_j$ ($T$ is left continuous by 1.9.5). Consider the following commutative diagram:

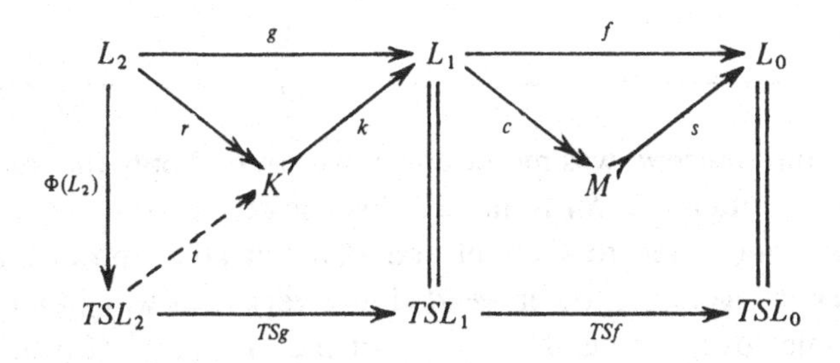


The top row is exact, so that $K \overset{k}{\rightarrowtail} L_1 \overset{c}{\twoheadrightarrow} M$ is short exact. The bottom row is a null sequence. So there exists a morphism $t$: $TSL_2 \to K$ with $kt = TSg$. From $TSg \circ \Phi(L_2) = g$ we then conclude $kt \circ \Phi(L_2) = kr$ and since $k$ is monic it follows that $t \circ \Phi(L_2) = r$. Hence $t$ is epic which implies that the bottom row is exact. But $T = h_U$, so exercise ($d$) in 2.8.4 asserts that $SL_2 \xrightarrow{Sg} SL_1 \xrightarrow{Sf} SL_0$ is exact. On the other hand, $c = \text{coker } g$ and $S$, being right exact, preserves cokernels so that $Sc = \text{coker } Sg$. This means that $Ss$ is monic.

Next suppose $s: M \to L$ is a monomorphism into a free module where now $M$ is arbitrary. For each finitely generated submodule $M_i$ of $M$ the image $s(M_i)$ is contained in a finitely generated free submodule $L_1 \subset L$. Then the restrictions $s_i = s|M_i: M_i \to L_i$ form an inductive system of monomorphisms with $\varinjlim s_i = s$. Since $S$ is right continuous we have $Ss = \varinjlim Ss_i$, where each $Ss_i$ is monic as we just proved. Since, however, **A** is a Grothendieck category we may conclude that $Ss$ is monic by AB5.

To finish the proof, finally suppose $u: M \to N$ is an arbitrary monomorphism in $\mathbf{M}_R$. Let $t$ be a homomorphism of a free module $L$ onto $N$. Completing to a cartesian square and taking $k' = \ker t'$ we obtain a commutative diagram (see exercise in 2.4.1)

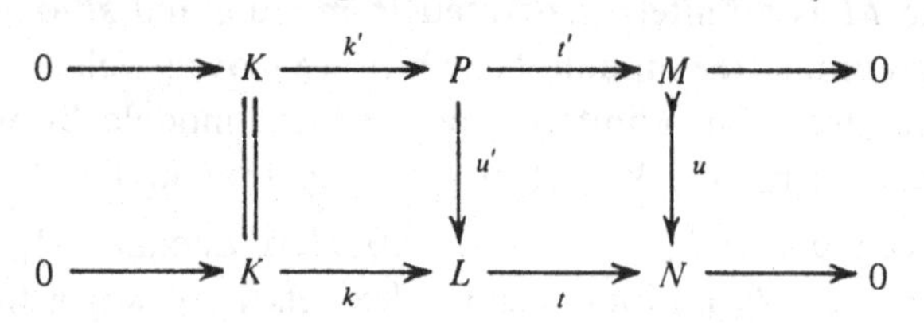

with exact rows.

From 2.1.2° it follows that $u'$ is monic. Applying the right exact functor $S$ we obtain the commutative diagram with exact rows

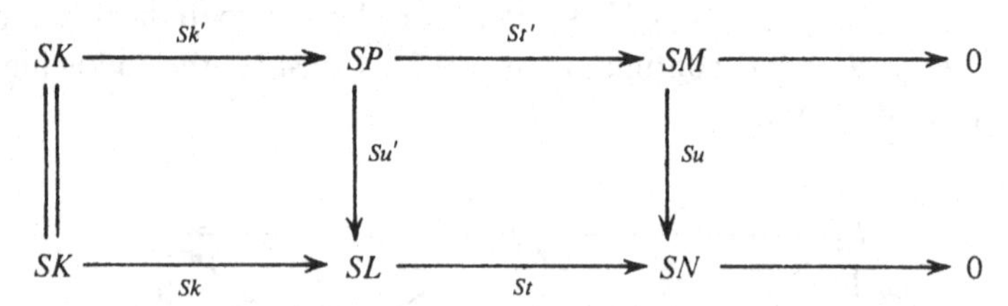

Now $Su'$ is monic since $u'$ was monic and $L$ was free. Apply the dual of lemma 2.6.8 to obtain that $Su$ is monic. This ends the proof.

The above result, due to Gabriel and Popescu [16], shows that a Grothendieck category can be embedded in a very neat way in a category of modules over a ring. This is one of the highlights of a theory inaugurated in Gabriel's thesis [14], where the current catchwords are embedding theorems, localization in categories and torsion theories. We shall not pursue these matters here. Good recent references are [24] and [40].

Our purpose is to derive a few useful properties of Grothendieck categories using the Gabriel–Popescu theorem.

**2.9.5 Corollary** Every Grothendieck category **A** is left complete, has injective envelopes and has an injective cogenerator.

*Proof.* The category $\mathbf{M}_R$ is left complete since it has equalizers and products (theorem in 1.8.4). Then using theorem 1.9.7 we infer that **A** is left complete.

The second assertion is a consequence of proposition 2.8.14 in which **A** is now $\mathbf{M}_R$ and **B** is our **A**. The category $\mathbf{M}_R$ has injective envelopes by 2.8.13.

To prove the last assertion, let $U$ be a generator of **A**. Then the subobjects $U_i$ of $U$ form a set (see exercise in 2.9.2) and hence the same holds for the quotient objects $C_i$. Let $v_i: C_i \rightarrowtail Q_i$ be an injective envelope of $C_i$. Then we will show that $\prod Q_i$ is an injective cogenerator of $A$. Injectivity being clear it is enough to prove that if $f: X \to Y$ is a nonnull morphism then there exists a morphism $t_i: Y \to Q_i$, for some $i$, with $t_i f \neq 0$. In the first place there exists a morphism $g: U \to X$ such that $fg \neq 0$ since $U$ is a generator. We look at the epic–monic factorization of $fg$.

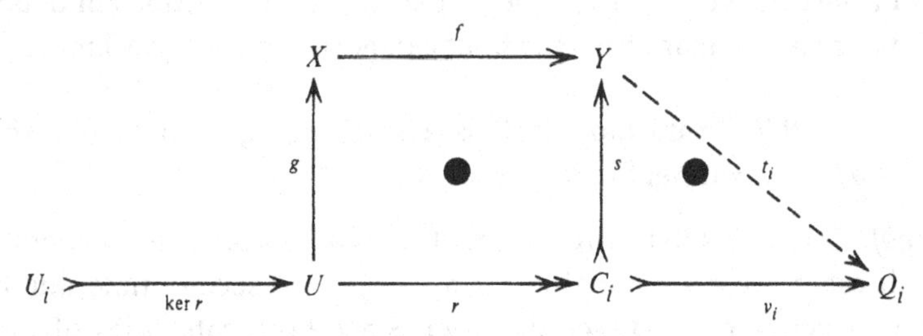


The existence of $t_i$ with $t_i s = v_i$ is implied by injectivity of $Q_i$. If $t_i f$ were null we would have $t_i fg = 0$ or also $v_i r = 0$. Since $v_i$ is monic this would imply $r = 0$, so then $sr = fg = 0$ which is not the case. This finishes the proof of the corollary.

Remark. If the generator $U$ happens to be projective then the functor $T = h_U$ is exact.

Consider $\Delta: \mathbf{A} \to \mathbf{A}^\circ$ where **A** is a Grothendieck category. Let $U \in \mathbf{A}$ be an injective cogenerator, then $\Delta U$ is a projective generator in $\mathbf{A}^\circ$. Defining $T_1: \mathbf{A}^\circ \to \mathbf{M}_R$ by $T_1 = h_{\Delta U}$, we have achieved that $T_1$ is exact. But then $T_1 \circ \Delta: \mathbf{A} \to \mathbf{M}_R$ is contravariant and exact and clearly full and faithful. So we have obtained

**2.9.6 Corollary** For any Grothendieck category **A** there is a contravariant functor $T': \mathbf{A} \to \mathbf{M}_R$ which is full, faithful and exact, defined by $T' = h^U$ where $U$ is an injective cogenerator of **A**.

Continuing from here, one may prove the historically important theorem of Mitchell which asserts that every small abelian category admits a full and faithful exact covariant functor into a full module category over a ring [13, p. 101], [29, p. 151].

We end the section with a few remarks on categories of functors. If **C** is a small category then the category of functors (**C**, **A**) tends to inherit properties of the target category **A**. For additivity this has already been remarked in 2.7.4. Other constructions are also carried out objectwise. Assume for instance that **A** has kernels. Then if $\eta: S \to T$ is a morphism in (**C**, **A**), $K = \ker \eta$ as defined on an object $C$ is $KC = \ker \eta(C)$ and in an obvious manner we may verify that $K$ indeed satisfies the requirements for a kernel. Similarly the existence of cokernels and biproducts in **A** is transferred to (**C**, **A**). If **A** is abelian, then (**C**, **A**) is abelian, since the axioms are verified in **A**. Also if **A** is complete, respectively satisfies AB5, so does (**C**, **A**). Now if **C** is also additive we can carry out the same programme in the full subcategory of additive functor $(\mathbf{C}, \mathbf{A})_+$. In order to appreciate what is involved the reader may find it instructive to verify at least part of these assertions in detail and also to make similar remarks about categories of contravariant functors.

**2.9.7 Proposition** If **C** is a small category both (**C**, **Ab**) and $(\mathbf{C}, \mathbf{Ab})_+$ are Grothendieck categories.

*Proof.* It remains to be proved that these categories have a generator. In both cases the representable functors $h_C$ (the second time additively) form a generating set; see 2.8.4($a$). Since both categories of functors have sums, 2.8.4($e$) provides us with a generator.

# 3

# Homological algebra

## 3.1 Extensions

Let **A** be an abelian category.

**3.1.1 Definitions** Let $\dot{E}: 0 \to A \xrightarrow{f} B \xrightarrow{g} C \to 0$ be an exact sequence in **A**. This sequence (or sometimes $B$) is then called an *extension of $C$ by $A$*.

Example. We may take the case $A \oplus C \simeq B$. Then $B$ is called a *split extension* of $C$ by $A$ (see 2.6.11).

Let $\dot{E}$ and $\dot{F}$ be exact sequences. As usual,

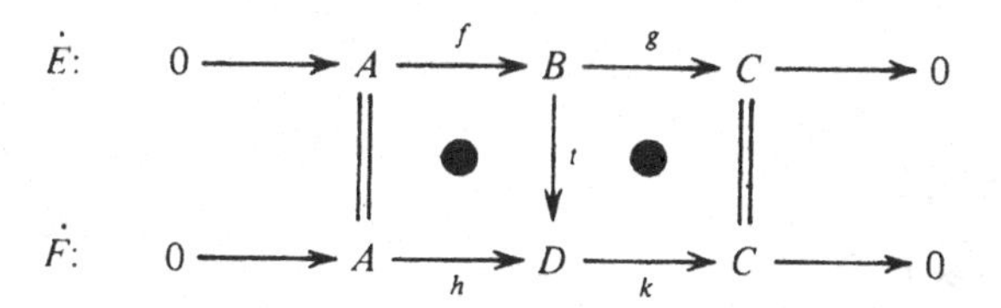

the symbol ● indicates that the corresponding square is commutative. The lines between the objects $A$ (and $C$) indicate the corresponding identity morphism. We call $\dot{E}$ and $\dot{F}$ *equivalent* in this situation and we denote this by $\dot{E} \sim \dot{F}$.

**3.1.2 Lemma** (i) In the above diagram $t$ is an isomorphism.

(ii) The relation $\sim$ is an equivalence relation.

*Proof.* From the ker–coker sequence – corollary in 2.6.9 – we see that $\ker t = \operatorname{coker} t = 0$, hence $t$ is a bimorphism. Since the abelian category **A** is balanced (see 2.5.3) $t$ has an inverse $t^{-1}$ and $t^{-1}h = f$, $gt^{-1} = k$. The symmetry is then evident, as are reflexivity and transivity.

Let $A$ and $C$ be objects of **A**. The equivalence classes of extensions of $C$ by $A$ form a class denoted by $\operatorname{Ext}(C, A)$. If **A** is small this class is actually a set. This is often true under less stringent conditions; see 3.1.24. An element of $\operatorname{Ext}(C, A)$, i.e. an equivalence class of extensions, is also called an extension of $C$ by $A$. Note that $\operatorname{Ext}(C, A)$ is nonempty because there is always the split extension of $C$ by $A$. Even when $\operatorname{Ext}(C, A)$ is only a class, we can distinguish subclasses of elements and, for instance, speak of exactness of a complex of such objects as we shall do in 3.1.21 seq.

**3.1.3 Lemma** Given the extension $\dot{E}: 0 \to A \xrightarrow{a} B \xrightarrow{b} C \to 0$ and $f: A \to X$ in **A** there exists a short exact sequence $\dot{F}$ and a morphism $h$ so that

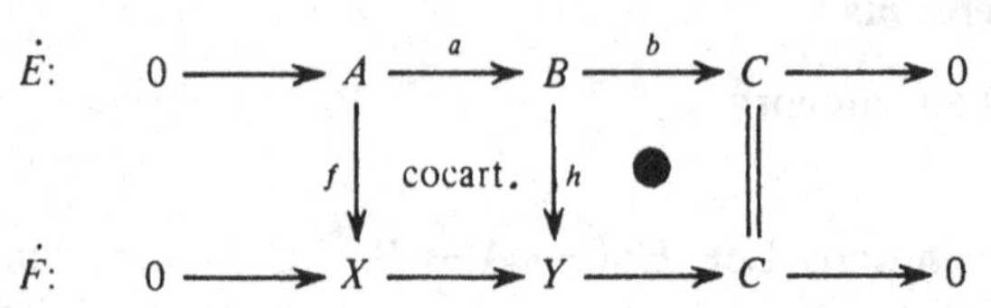


*Proof.* Using lemma 2.5.5° and exercise (b) in 1.8.4 we may complete the diagram

$$\begin{array}{ccc} A & \xrightarrow{a} & B \\ {\scriptstyle f}\downarrow & & \\ X & & \end{array}$$

to a cocartesian square

$$\begin{array}{ccc} A & \xrightarrow{a} & B \\ {\scriptstyle f}\downarrow & & \downarrow{\scriptstyle h} \\ X & \xrightarrow[x]{} & Y \end{array}$$

Since $a$ is a monomorphism in an abelian category, it is universally monic and so $x$ is monic. From the exercise in 2.4.1° there is a unique $y: Y \to C$ with $y = \operatorname{coker} x$ and $yh = b$. From 2.5.1 it follows that $x = \ker y$. Hence $X \xrightarrow{x} Y \xrightarrow{y} C$ is a short exact sequence.

**3.1.4 Lemma** Suppose both $\dot{E}$ and $\dot{E}'$ are representatives from the same class in $\operatorname{Ext}(C, A)$ and let $\dot{F}$ and $\dot{F}'$ be related to $\dot{E}$ and $\dot{E}'$ respectively as described in the diagram below. Then $\dot{F} \sim \dot{F}'$. (See diagram on top of page 137.)

*Proof.* The fact that the lower left square is cocartesian together with the fact that the two morphisms $h'ta, x'f: A \rightrightarrows Y'$ are equal yields a unique morphism $s: Y \to Y'$ with $sh = h't$ and $sx = x'$. From this we

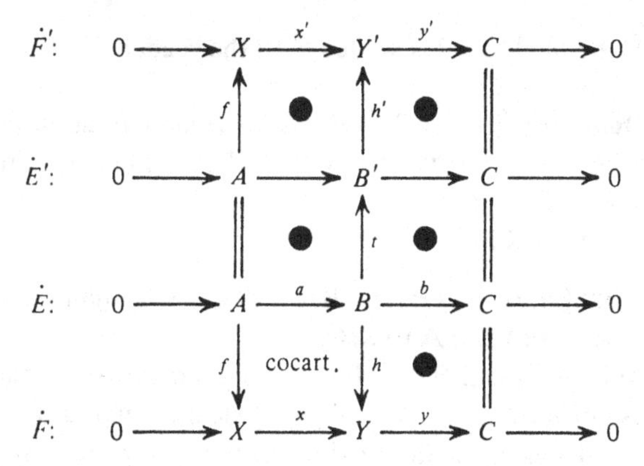


have $y'sx = y'x' = 0 = yx$ and $y'sh = y'h't = b = yh$ so that $y's = y$ (universal property of a cocartesian square). Hence we have

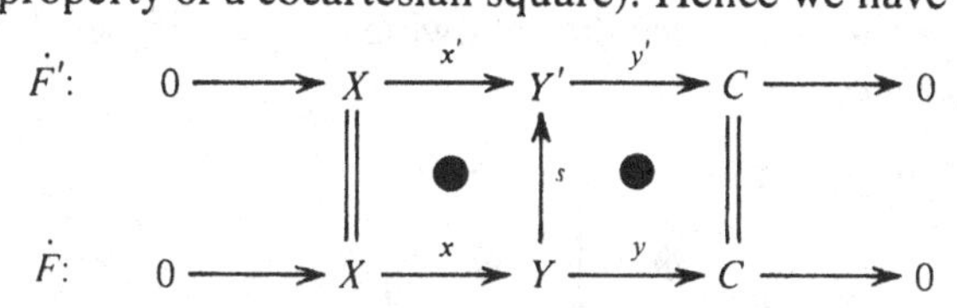


**3.1.5 Definition** Let $f: A \to X$. Take $\dot{E}$ and $\dot{F}$ as in 3.1.3. Then, denoting the equivalence classes of $\dot{E}$ and $\dot{F}$ by $\mathscr{E}$ and $\mathscr{F}$ respectively, we write $f_*\mathscr{E} = \mathscr{F}$. This definition is justified by lemma 3.1.4. Clearly, if $g: X \to Z$ then $(gf)_*\mathscr{E} = g_*(f_*\mathscr{E})$.

Changing notation, we define for $f: B \to B'$

$$\operatorname{Ext}(A, \text{-}): \mathbf{A} \to \mathbf{Cl}$$

by $\operatorname{Ext}(A, \text{-})B = \operatorname{Ext}(A, B)$ and $\operatorname{Ext}(A, \text{-})f = f_*$.

**Cl** denotes the category of classes. Actually this is not a bona fide category as the collection of morphisms between two classes may not be a set. Nevertheless we will use the familiar terminology and call Ext $(A, \text{-})$ a functor. If **A** is small there is no trouble as we then have a functor $\operatorname{Ext}(A, \text{-}): \mathbf{A} \to \mathbf{Sets}$.

Dualizing the previous lemma we obtain

**3.1.3° Lemma** For the extension $\dot{E}: 0 \to A \xrightarrow{a} B \xrightarrow{b} C \to 0$ and $g: Z \to C$ in **C** there exists a short exact sequence $\dot{F}$ and a morphism $h$ so that

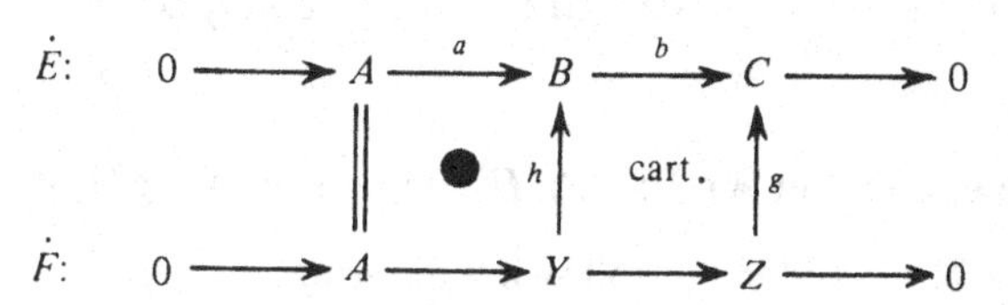

The statement of the dual to 3.1.4 we leave to the reader.

**3.1.5° Definition** Let $g: Z \to C$. Take $\dot{E}$ and $\dot{F}$ as in 3.1.3°. Then we define $g^*\mathscr{E} = \mathscr{F}$. Thus $g^*\mathscr{E} \in \text{Ext}(Z, A)$. Furthermore we define, in case now $g: B' \to B$,

$$\text{Ext}(\text{-}, B): \mathbf{A} \to \mathbf{Cl}$$

by $\text{Ext}(\text{-}, B)A = \text{Ext}(A, B)$ and $\text{Ext}(\text{-}, B)g = g^*$. If $\mathbf{A}$ is small $\text{Ext}(\text{-}, B)$ is a contravariant functor from $\mathbf{A}$ to **Sets**.

The functors $\text{Ext}(A, \text{-})$ and $\text{Ext}(\text{-}, B)$ are very similar to $h_A$ and $h^B$. Just as in the case of $h_A$ and $h^B$ we are really dealing with a bifunctor $\text{Ext}(\text{-}, \text{-}): \mathbf{A}^\circ \times \mathbf{A} \to \mathbf{Cl}$ as we shall see shortly. First we prove two lemmas.

**3.1.6 Lemma** Consider the following diagram in which $\dot{E}$ and $\dot{F}$ are exact sequences.

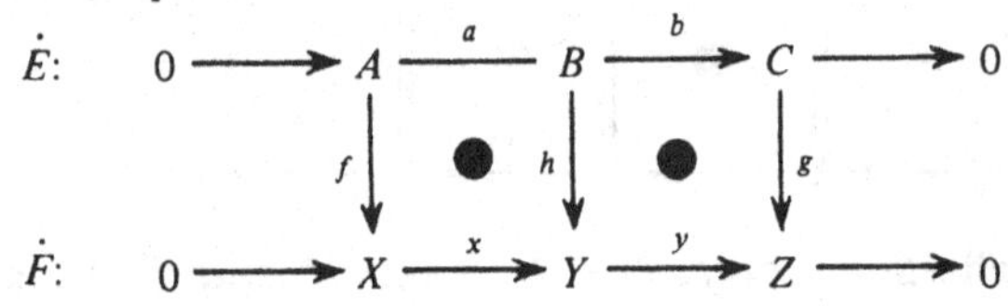


Then $f_*\mathscr{E} = g^*\mathscr{F}$.

*Proof.* Consider the diagram

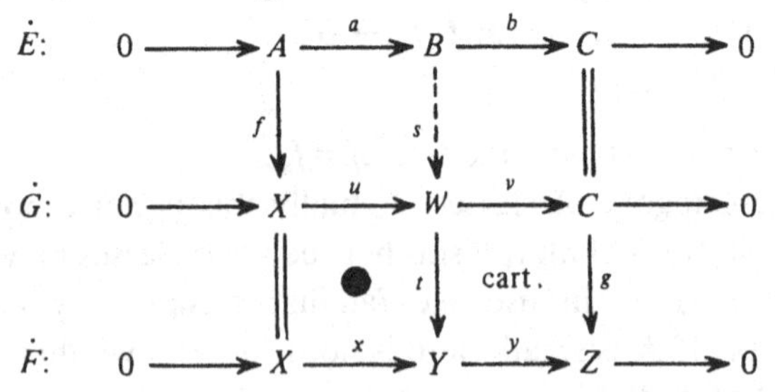


in which the middle row is a constructed representative of $g^*\mathscr{F}$. Due to the fact that the lower right square is cartesian and that $gb = yh$ there is a unique $s: B \to W$ such that $vs = b$ and $ts = h$. Then $vsa = ba = 0 = vuf$ and $tsa = ha = xf = tuf$. From the universal property of the cartesian square we find $sa = uf$. Thus all squares commute. It now follows from 3.1.4 ($\dot{E}' = \dot{E}$) that $\dot{G} \in f_*\mathscr{E}$. Since also $\dot{G} \in g^*\mathscr{F}$ we may conclude $f_*\mathscr{E} = g^*\mathscr{F}$.

**3.1.7 Lemma** Let $\mathscr{E} \in \text{Ext}(A, B)$. Then $f_*g^*\mathscr{E} = g^*f_*\mathscr{E}$.

*Proof.* Take $\dot{E} \in \mathscr{E}$. Consider the following diagram:

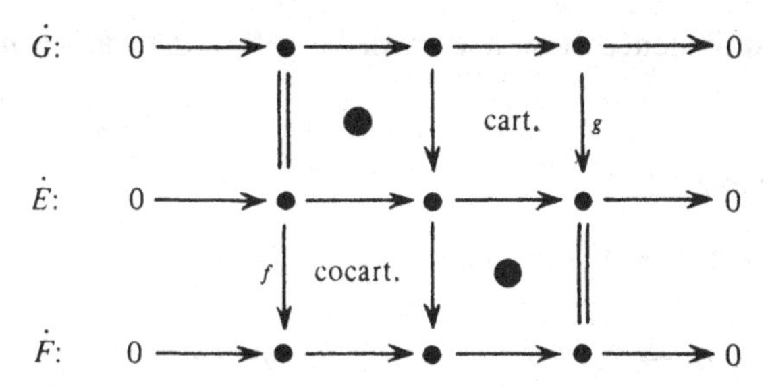


Thus $\dot{G}=g^*\mathscr{E}$ and $\dot{F}=f_*\mathscr{E}$. But according to our previous lemma $f_*g^*\mathscr{E}=g^*f_*\mathscr{E}$.

From now on we will denote $f_*\mathscr{E}$ by $f\mathscr{E}$ and $g^*\mathscr{E}$ by $\mathscr{E}g$. Our last lemma then says $f(\mathscr{E}g)=(f\mathscr{E})g$. This associative law allows us to delete brackets. Since also $f(g\mathscr{E})=(fg)\mathscr{E}$ and $(\mathscr{E}f)g=\mathscr{E}(fg)$, here too brackets are superfluous.

Now define the bifunctor $\operatorname{Ext}(\text{-},\text{-})\colon \mathbf{A}^\circ\times\mathbf{A}\to\mathbf{Cl}$ by $\operatorname{Ext}(\text{-},\text{-})\ulcorner A^\circ, B\urcorner=\operatorname{Ext}(A,B)$ and $\operatorname{Ext}(\text{-},\text{-})\ulcorner g^\circ, f\urcorner=\operatorname{Ext}(f,g)$ by $\operatorname{Ext}(f,g)\mathscr{E}=f\mathscr{E}g$.

We next introduce long extensions.

**3.1.8 Definitions** Consider the exact sequence

$$\dot{E}^n: 0\to B\to Y_1\to\ldots\to Y_n\to A\to 0.$$

Then $n$ is called the *length* of $\dot{E}^n$ and $\dot{E}^n$ is called an *$n$-fold extension* of $A$ by $B$. If the following diagram of $n$-fold extensions commutes

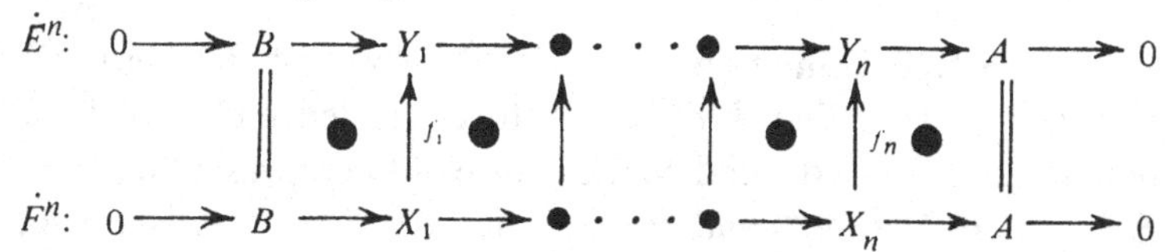


$\dot{E}^n$ and $\dot{F}^n$ are called *directly equivalent*. This is meant as a symmetric relation. Two $n$-fold extensions $\dot{E}^n$ and $\dot{F}^n$ of $A$ by $B$ are called *equivalent* if there is a sequence $\dot{E}^n_0, \dot{E}^n_1, \ldots, \dot{E}^n_k$ of $n$-fold extensions such that:

(i) $\dot{E}^n_0=\dot{E}^n$ and $\dot{E}^n_k=\dot{F}^n$;

(ii) $\dot{E}^n_i$ is directly equivalent with $\dot{E}^n_{i+1}$ for $i=0,1,\ldots,k-1$.

This equivalence relation will be denoted by $\sim$.

The collection consisting of equivalence classes of $n$-fold extensions of $A$ by $B$ we call $\operatorname{Ext}^n(A,B)$ so that $\operatorname{Ext}^1(A,B)=\operatorname{Ext}(A,B)$. Again $\operatorname{Ext}^n(A,B)\neq\varnothing$ because there is always the split extension

$$0\to B\xrightarrow{1} B\to 0\to\ldots\to 0\to A\xrightarrow{1} A\to 0.$$

Sometimes an equivalence class itself will be referred to as an $n$-fold extension.

**3.1.9 Lemma** Let

$$\dot{E}^n: 0 \to C \xrightarrow{y_0} Y_1 \to \ldots \to Y_n \xrightarrow{y_n} B \to 0$$

and

$$\dot{F}^m: 0 \to B \xrightarrow{x_0} X_1 \to \ldots \to X_m \xrightarrow{x_m} A \to 0$$

be exact ($m \geqslant 1$ and $n \geqslant 1$). Then the sequence

$$0 \to C \xrightarrow{y_0} Y_1 \to \ldots \to Y_n \xrightarrow{y_n} X_1 \xrightarrow{x_1} \ldots \to X_m \xrightarrow{x_m} A \to 0$$

is also exact. Denote this sequence by $\dot{E}^n\dot{F}^m$.

*Proof.* Clearly the following diagram is an epic–monic factorization of $x_0y_n$.

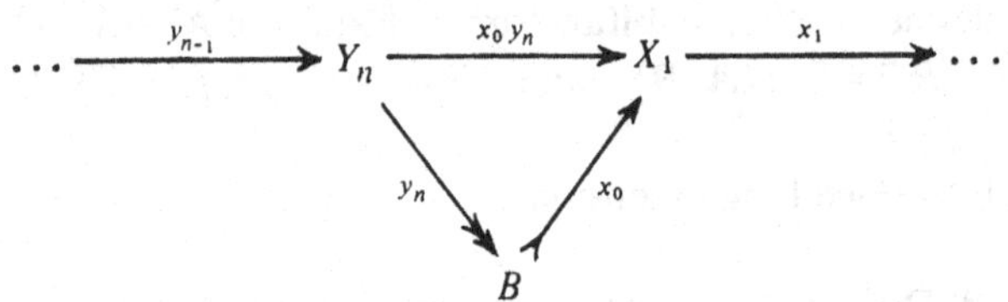


Exactness of $\dot{F}^m$ at $X_1$ means that $\ker x_1 = x_0$. But $x_0 = \operatorname{im} x_0y_n$ so that $\dot{E}^n\dot{F}^m$ is exact at $X_1$. Dually $\dot{E}^n\dot{F}^m$ is exact at $Y_n$. Exactness at other objects is inherited from $\dot{E}^n$ and $\dot{F}^m$.

**3.1.10 Definition** For $\dot{E}^m \in \mathscr{E} \in \operatorname{Ext}^m(A, B)$ and $\dot{F}^n \in \mathscr{F}^n \in \operatorname{Ext}^n(B, C)$ we define $\mathscr{F}^n\mathscr{E}^m$ to be the equivalence class of $\dot{F}^n\dot{E}^m$. This definition does not depend on the choice of the representatives, since for $\dot{E}^m$ and $\dot{E}'^m$ in $\mathscr{E}^m$ and for $\dot{F}^n$ and $\dot{F}'^n$ in $\mathscr{F}^n$ it is obvious that $\dot{F}^n\dot{E}^m \sim \dot{F}^n\dot{E}'^m$ and $\dot{F}'^n\dot{E}'^m \sim \dot{F}'^n\dot{E}'^m$. The class $\mathscr{F}\mathscr{E}^{m+n} = \mathscr{F}^n\mathscr{E}^m$ is called the *Yoneda composition of $\mathscr{E}^m$ and $\mathscr{F}^n$* which defines a pairing

$$\operatorname{Ext}^n(B, C) \times \operatorname{Ext}^m(A, B) \to \operatorname{Ext}^{m+n}(A, C)\ [44].$$

Two immediate properties are the following.

(i) For $\mathscr{E}^l \in \operatorname{Ext}^l(A, B)$, $\mathscr{F}^m \in \operatorname{Ext}^m(B, C)$ and $\mathscr{G}^n \in \operatorname{Ext}^n(C, D)$ (with all exponents at least 1) we have $(\mathscr{G}^n\mathscr{F}^m)\mathscr{E}^l = \mathscr{G}^n(\mathscr{F}^m\mathscr{E}^l) \in \operatorname{Ext}^{l+m+n}(A, D)$.

(ii) $\mathscr{E}^l \in \operatorname{Ext}^l(A, B)$ may be considered as the Yoneda composition of $l$ extensions of length 1.

*Proof.* We only prove (ii). Consider a representative $\dot{E}^l$ of $\mathscr{E}^l$ and factorize:

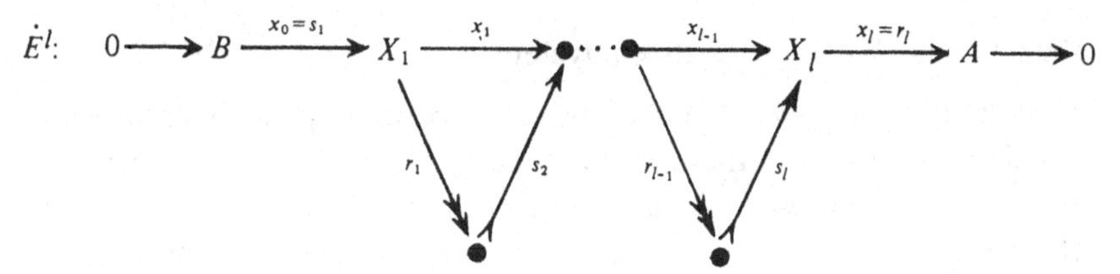


Then $\dot{E}^l = \dot{F}_1\dot{F}_2 \ldots \dot{F}_l$ with $\dot{F}_i\colon . \underset{s_i}{\rightarrow} . \underset{r_i}{\rightarrow} .$

One often calls this *splicing into short exact sequences.*

We now want to repeat the constructions and definitions given in 3.1.3 and following for long extensions.

**3.1.11 Lemma** Let

$$\dot{E}^n\colon 0 \rightarrow B \overset{x_0}{\rightarrow} X_1 \overset{x_1}{\rightarrow} \ldots \overset{x_{n-1}}{\longrightarrow} X_n \overset{x_n}{\rightarrow} A \rightarrow 0$$

be exact and suppose $f\colon B \rightarrow U$ $(n \geqslant 1)$. Then there is an exact sequence $\dot{F}^n$ and a morphism $h$ such that

$$\begin{array}{lccccccccccccccc}
\dot{E}^n\colon & 0 & \longrightarrow & B & \xrightarrow{x_0} & X_1 & \xrightarrow{x_1} & X_2 & \longrightarrow & \ldots & \longrightarrow & X_n & \xrightarrow{x_n} & A & \longrightarrow & 0 \\
 & & & {\scriptstyle f}\downarrow & \text{cocart.} & \downarrow{\scriptstyle h} & & \| & & \bullet & & \| & \bullet & \| & & \\
\dot{F}^n\colon & 0 & \longrightarrow & U & \xrightarrow{u} & V & \xrightarrow{v} & X_2 & \longrightarrow & \ldots & \longrightarrow & X_n & \longrightarrow & A & \longrightarrow & 0
\end{array}$$

*Proof.* Decompose $\dot{E}^n = \dot{E}_1\dot{E}_2^{n-1}$ with $E_1$ an extension of coker $x_0 = X$ by $B$ of length 1. Then use lemma 3.1.3 to construct $\dot{F}_1$ with

$$\begin{array}{lccccccccc}
\dot{E}_1\colon & 0 & \longrightarrow & B & \longrightarrow & X_1 & \longrightarrow & X & \longrightarrow & 0 \\
 & & & {\scriptstyle f}\downarrow & \text{cocart.} & \downarrow{\scriptstyle h} & & \| & & \\
\dot{F}_1\colon & 0 & \longrightarrow & U & \longrightarrow & V & \longrightarrow & X & \longrightarrow & 0
\end{array}$$

Take $\dot{F}_2^{n-1} = \dot{E}_2^{n-1}$. Then $\dot{F}^n = \dot{F}_1\dot{F}_2^{n-1}$ satisfies the requirements.

**3.1.12 Lemma** Consider the following commutative diagram:

$$\begin{array}{lccccccccccccccc}
\dot{F}'^n\colon & 0 & \longrightarrow & U & \xrightarrow{y_0} & Y_1 & \xrightarrow{y_1} & Y_2 & \longrightarrow & \ldots & \longrightarrow & Y_n & \longrightarrow & A & \longrightarrow & 0 \\
 & & & {\scriptstyle f}\uparrow & \bullet & \uparrow{\scriptstyle v_1} & \bullet & \uparrow{\scriptstyle v_2} & & & & \uparrow{\scriptstyle v_n} & \bullet & \| & & \\
\dot{E}'^n\colon & 0 & \longrightarrow & B & \longrightarrow & X_1' & \longrightarrow & X_2' & \longrightarrow & \ldots & \longrightarrow & X_n' & \longrightarrow & A & \longrightarrow & 0 \\
 & & & \| & \bullet & \uparrow{\scriptstyle t_1} & \bullet & \uparrow{\scriptstyle t_2} & & & & \uparrow{\scriptstyle t_n} & \bullet & \| & & \\
\dot{E}^n\colon & 0 & \longrightarrow & B & \xrightarrow{x_0} & X_1 & \longrightarrow & X_2 & \longrightarrow & \ldots & \longrightarrow & X_n & \longrightarrow & A & \longrightarrow & 0 \\
 & & & {\scriptstyle f}\downarrow & \text{cocart.} & \downarrow{\scriptstyle h} & \bullet & \| & & & & \| & \bullet & \| & & \\
\dot{F}^n\colon & 0 & \longrightarrow & U & \xrightarrow{u} & V & \xrightarrow{v} & X_2 & \longrightarrow & \ldots & & X_n & \longrightarrow & A & \longrightarrow & 0
\end{array}$$

Then $\dot{F}^n$ and $\dot{F}'^n$ are directly equivalent.

*Proof.* Due to the fact that the lower left corner square is cocartesian and since $v_1t_1x_0 = y_0f$ there is a unique $s: V \to Y_1$ with $sh = v_1t_1$ and $su = y_0$. But then $y_1sh = y_1v_1t_1 = v_2t_2vh$ and $y_1su = y_1y_0 = 0 = v_2t_2vu$ which implies $y_1s = v_2t_2v$ (universal property of a cocartesian square). Thus we have

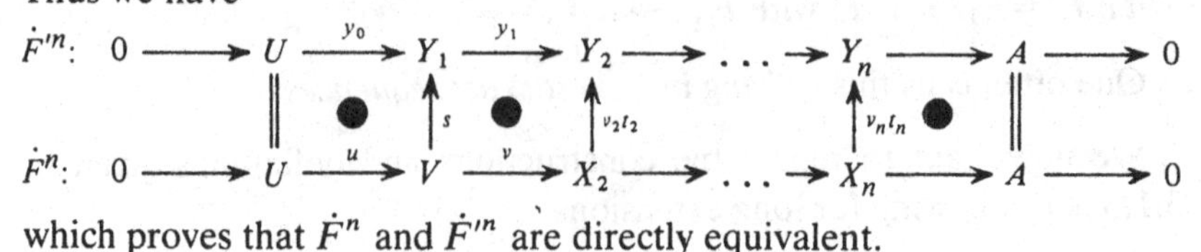

which proves that $\dot{F}^n$ and $\dot{F}'^n$ are directly equivalent.

**Corollary.** Assume the following two diagrams are given:

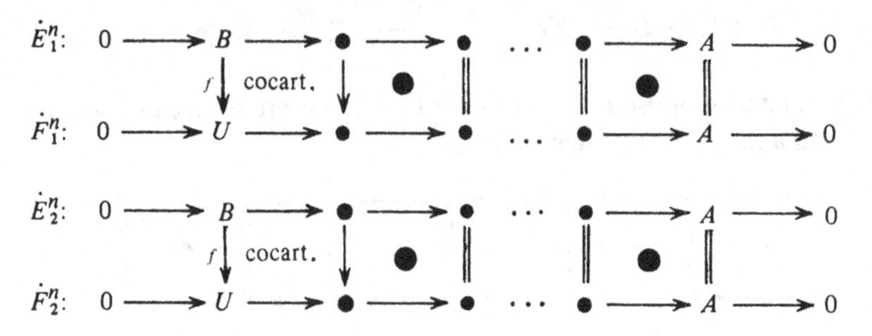

Then $\dot{F}_1^n$ and $\dot{F}_2^n$ are (directly) equivalent if $\dot{E}_1^n$ and $\dot{E}_2^n$ are (directly) equivalent.

*Proof.* If $\dot{E}_1^n$ and $\dot{E}_2^n$ are directly equivalent we are in the case of lemma 3.1.12. If $\dot{E}_1^n$ and $\dot{E}_2^n$ are just equivalent, lemma 3.1.11 allows us to construct extensions $\dot{F}^n$ to each intermediate sequence $\dot{E}^n$ and proceed step by step.

**3.1.13 Definition** For $\dot{E}^n \in \mathscr{E}^n \in \mathrm{Ext}^n\,(A, B)$ $(n \geqslant 1)$ and for $f: B \to X$ we define $f\mathscr{E}^n$ to be the equivalence class to which $\dot{F}^n$ as constructed in 3.1.11 belongs. It follows from the corollary of lemma 3.1.12 that this is well defined. Thus $f\mathscr{E}^n \in \mathrm{Ext}^n\,(A, X)$. The class $f\mathscr{E}^n$ is called the *Yoneda composition of f and* $\mathscr{E}^n$.

Remark. If $\dot{E}'^n \in \mathscr{E}^n$ and if for a certain exact sequence $\dot{F}'^n$ and morphisms $v_i$

$$\begin{array}{lccccccccccccc}
\dot{E}'^n: & 0 & \longrightarrow & B & \longrightarrow & X'_1 & \longrightarrow & X'_2 & \longrightarrow \dots \longrightarrow & X'_n & \longrightarrow & A & \longrightarrow & 0 \\
 & & & \downarrow f & \bullet & \downarrow v_1 & \bullet & \downarrow v_2 & & \downarrow v_n & \bullet & \| & & \\
\dot{F}'^n: & 0 & \longrightarrow & U & \longrightarrow & Y_1 & \longrightarrow & Y_2 & \longrightarrow \dots \longrightarrow & Y_n & \longrightarrow & A & \longrightarrow & 0
\end{array}$$

then $\dot{F}'^n \in f\mathscr{E}^n$. This follows from lemma 3.1.12.

Furthermore we define $\mathrm{Ext}^0(A, C) = (A, C)$. Then this Yoneda composition is a pairing $\mathrm{Ext}^0(B, X) \times \mathrm{Ext}^n(A, B) \to \mathrm{Ext}^n(A, X)$.

We now dualize as we did in 3.1.3°. Let

$$\dot{E}^n : 0 \to B \xrightarrow{x_0} X_1 \xrightarrow{x_1} X_2 \to \ldots \to X_n \to A \to 0$$

be exact and let $g: Z \to A$. Then there is an exact sequence $\dot{F}^n$ and a morphism $h$ such that

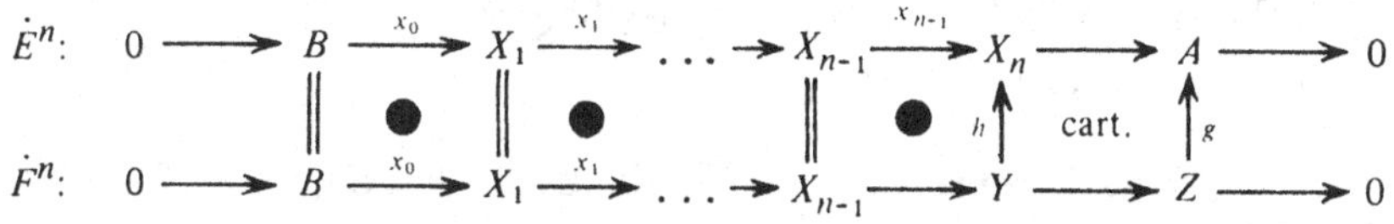

$$\begin{array}{lccccccccccccc}
\dot{E}^n: & 0 & \longrightarrow & B & \xrightarrow{x_0} & X_1 & \xrightarrow{x_1} & \ldots \to & X_{n-1} & \xrightarrow{x_{n-1}} & X_n & \longrightarrow & A & \longrightarrow 0 \\
 & & & \| & \bullet & \| & \bullet & & \| & \bullet & \uparrow h & \text{cart.} & \uparrow g & \\
\dot{F}^n: & 0 & \longrightarrow & B & \xrightarrow{x_0} & X_1 & \xrightarrow{x_1} & \ldots \to & X_{n-1} & \longrightarrow & Y & \longrightarrow & Z & \longrightarrow 0
\end{array}$$

For $\dot{E}^n \in \mathscr{E}^n \in \mathrm{Ext}^n(A, B)$ and for $g: Z \to A$ as above we define $\mathscr{E}g$ as the equivalence class to which $\dot{F}^n$ belongs. Hence $\mathscr{E}^n g \in \mathrm{Ext}^n(Z, B)$. Again, $\mathscr{E}^n g$ is called the *Yoneda composition of* $\mathscr{E}^n$ *and* $g$. This Yoneda composition is a pairing $\mathrm{Ext}^n(A, B) \times \mathrm{Ext}^0(Z, A) \to \mathrm{Ext}^n(Z, B)$. As before, the definition of $\mathscr{E}g$ is unambiguous, and the dual of the above remark allows us to recognize all the representatives of $\mathscr{E}g$.

We have thus defined a Yoneda pairing

$$\mathrm{Ext}^n(B, C) \times \mathrm{Ext}^m(A, B) \to \mathrm{Ext}^{m+n}(A, C) \qquad (m, n \geqslant 0)$$

(for $m = n = 0$: take the ordinary composition in **A**). We claim that $(\mathscr{G}^k \mathscr{F}^m)\mathscr{E}^n = \mathscr{G}^k(\mathscr{F}^m \mathscr{E}^n)$.

**3.1.14 Lemma** The Yoneda pairing is associative.

*Proof.* The cases $k = m = n = 0$; $k > 0, m > 0, n > 0$; $k = 0, m = 1, n = 0$; $k = m = 0$, $n = 1$ and $k = 1$, $m = n = 0$ have already been proved. The cases $k = m = 0$, $n > 1$; $k = 0$, $m > 0$, $n > 0$; $k > 1$, $m = n = 0$; $k > 0$, $m > 0$, $n = 0$ and $k > 0$, $m = 0$, $n > 0$ follow directly from the pertaining diagrams (some cases are dual to each other). The only remaining case is $k = 0$, $m > 1$, $n = 0$. We reduce it to the other cases as follows. According to 3.1.10 property (ii) we may splice $\mathscr{F}^m$ into $m$ sequences with length 1: $\mathscr{F}^m = \mathscr{F}_1 \mathscr{F}_2 \ldots \mathscr{F}_m$. Now we have

$$\begin{aligned}
(f\mathscr{F}^m)g &= (f(\mathscr{F}_1 \ldots \mathscr{F}_m))g \\
&= ((f\mathscr{F}_1 \ldots \mathscr{F}_{m-1})\mathscr{F}_m)g \\
&= (f\mathscr{F}_1 \ldots \mathscr{F}_{m-1})\mathscr{F}_m g \\
&= f(\mathscr{F}_1 \ldots \mathscr{F}_m g) \\
&= f(\mathscr{F}^m g).
\end{aligned}$$

**3.1.15 Definitions** We define $\text{Ext}^n(\text{-},\text{-})\colon \mathbf{A}^\circ \times \mathbf{A} \to \mathbf{Cl}$ by $\text{Ext}^n(\text{-},\text{-})\ulcorner A^\circ, B\urcorner = \text{Ext}^n(A,B)$ and $(\text{Ext}^n(\text{-},\text{-})\ulcorner g^\circ, f\urcorner)\mathscr{E} = f\mathscr{E}g$. By restriction we obtain the (covariant) functor $\text{Ext}^n(A,\text{-})\colon \mathbf{A} \to \mathbf{Cl}$ and the contravariant functor $\text{Ext}^n(\text{-},B)\colon \mathbf{A} \to \mathbf{Cl}$, generalizing for $n > 1$ the functor $\text{Ext} = \text{Ext}^1$ of 3.1.5 and 3.1.5°.

Notation.

$\text{Ext}^n(A,\text{-})f = \text{Ext}^n(\text{-},\text{-})\ulcorner 1_A^\circ, f\urcorner = \text{Ext}^n(A,f)$

$\text{Ext}^n(\text{-},B)f = \text{Ext}^n(\text{-},\text{-})\ulcorner g^\circ, 1_B\urcorner = \text{Ext}^n(g,B)$

$\text{Ext}^n(\text{-},\text{-})\ulcorner g^\circ, f\urcorner = \text{Ext}^n(g,f)$

**3.1.16 Lemma** Let $\dot{E}^n \in \mathscr{E}^n \in \text{Ext}^n(A,B)$, $\dot{F}^n \in \mathscr{F}^n \in \text{Ext}^n(C,D)$, $n \geq 1$ and suppose

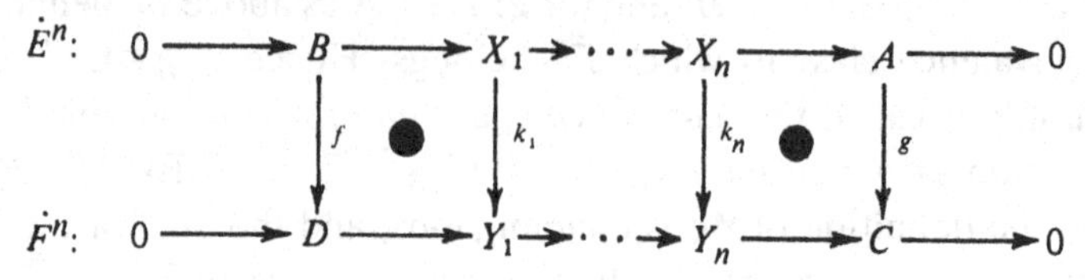

$\dot{E}^n$: 0 → B → $X_1$ → ⋯ → $X_n$ → A → 0
f
$k_1$
$k_n$
g
$\dot{F}^n$: 0 → D → $Y_1$ → ⋯ → $Y_n$ → C → 0


Then $f\mathscr{E}^n = \mathscr{F}^n g$.

*Proof.* We first observe that for $g = 1_A$ the assertion is explained in the remark after definition 3.1.13.

Now to prove the general case we may splice and use lemma 2.5.10, inserting morphisms $h_i$ in the following way:

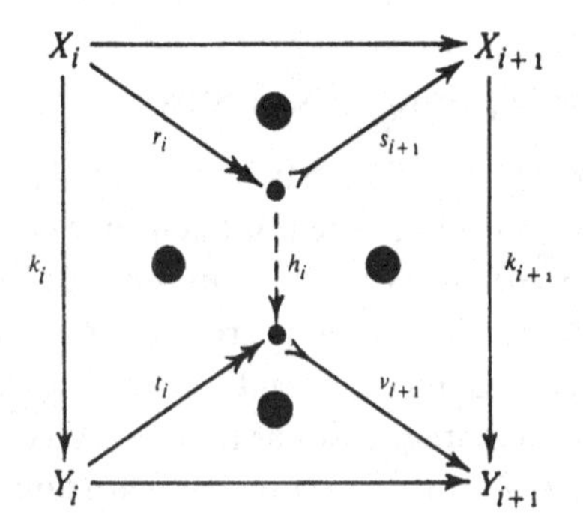


Take $\dot{E}_i\colon 0 \to . \underset{s_i}{\to} . \underset{r_i}{\to} . \to 0$ and $\dot{F}_i\colon 0 \to . \underset{v_i}{\to} . \underset{t_i}{\to} . \to 0$ in such a way that $\dot{E}^n = \dot{E}_1\dot{E}_2 \ldots \dot{E}_n$ and $\dot{F}^n = \dot{F}_1\dot{F}_2 \ldots \dot{F}_n$, as was done in 3.1.10 property (ii). Now according to 3.1.6 we have $\mathscr{F}_i h_i = h_{i-1}\mathscr{E}_i$ for $i = 1, \ldots, n$ ($h_0 = f$ and $h_n = g$). Then $\mathscr{F}^n g = \mathscr{F}_1 \ldots \mathscr{F}_n h_n = h_0 \mathscr{E}_1 \ldots \mathscr{E}_n = f\mathscr{E}^n$.

We now want to construct a punctured category $\mathbf{A}_e$ in which the morphisms are the extension classes $\text{Ext}^n(A,B)$ and the composition is the Yoneda composition. We first prove a lemma.

**3.1.17 Lemma** (i) For every $\mathrm{Ext}^n(A, B)$, with $n \geq 1$, there is exactly one $\mathscr{E}^n \in \mathrm{Ext}^n(A, B)$ such that $0_{BB}\mathscr{E}^n = \mathscr{E}^n$. This class of equivalent $n$-fold extensions of $B$ by $A$ is called the *null* of $\mathrm{Ext}^n(A, B)$. For $n = 0$, $\mathscr{E} = 0_{AB}$, the null morphism of $(A, B)$.

(ii) Let $\mathscr{E}^n$ be the null of $\mathrm{Ext}^n(A, B)$. Let $\mathscr{F}^m \in \mathrm{Ext}^m(B, C)$, $m \geq 0$. Then $\mathscr{F}^m\mathscr{E}^n$ is the null of $\mathrm{Ext}^{m+n}(A, C)$. Furthermore let $\mathscr{G}^k \in \mathrm{Ext}^k(D, A)$. Then $\mathscr{E}^n\mathscr{G}^k$ is the null of $\mathrm{Ext}^{n+k}(D, B)$.

(iii) If $\mathscr{E}^n \in \mathrm{Ext}^n(A, B)$ has a representing sequence that splits then $\mathscr{E}^n$ is the null of $\mathrm{Ext}^n(A, B)$. For $n = 1$ in fact every representative of the null class splits.

(iv) For $\dot{E}^n \in \mathscr{E}^n \in \mathrm{Ext}^n(A, B)$ with

$$\dot{E}^n : 0 \to B \xrightarrow{f} X_1 \to \ldots \to X_n \xrightarrow{g} A \to 0$$

$f\mathscr{E}^n$ and $\mathscr{E}^n g$ are null.

Remark. For the construction of $\mathbf{A_e}$ (iii) and (iv) are not needed.

*Proof.* Let $\dot{F}^n \in \mathscr{F}^n \in \mathrm{Ext}^n(A, B)$ and let $\dot{G}^n \in \mathscr{G}^n \in \mathrm{Ext}^n(C, D)$ with $n \geq 1$. From the diagram

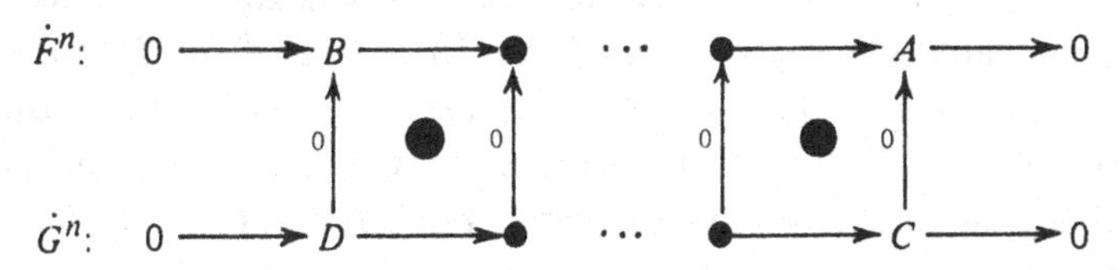


it follows (see 3.1.16) that $0_{DB}\mathscr{G}^n = \mathscr{F}^n 0_{CA}$. A similar relation holds for the case $n = 0$.

Now we prove (i). Let $\mathscr{F} \in \mathrm{Ext}^n(A, B)$ and take $\mathscr{E} = 0_{BB}\mathscr{F}$. Then $0_{BB}\mathscr{E} = 0_{BB}0_{BB}\mathscr{F} = 0_{BB}\mathscr{F} = \mathscr{E}$. Suppose we also have $0_{BB}\mathscr{G} = \mathscr{G}$ for a certain $\mathscr{G} \in \mathrm{Ext}^n(A, B)$. Then, according to the remark above, we have $\mathscr{G} = 0_{BB}\mathscr{G} = \mathscr{G}0_{AA} = 0_{BB}\mathscr{E} = \mathscr{E}$.

Next we prove (ii). Let $\mathscr{E}$, $\mathscr{F}$ and $\mathscr{G}$ be as indicated in (ii). Then we have $0_{CC}\mathscr{F}\mathscr{E} = \mathscr{F}0_{BB}\mathscr{E} = \mathscr{F}\mathscr{E}$. Hence $\mathscr{F}\mathscr{E}$ is null, and so of course is $\mathscr{E}\mathscr{G}$.

To prove (iii) (see also 2.6.14) we remark that a splitting sequence of length $m$ may be considered as the Yoneda composition of $m$ splitting sequences (3.1.10 property (ii)). So according to (ii) we only have to consider the case $n = 1$. From the commuting diagram

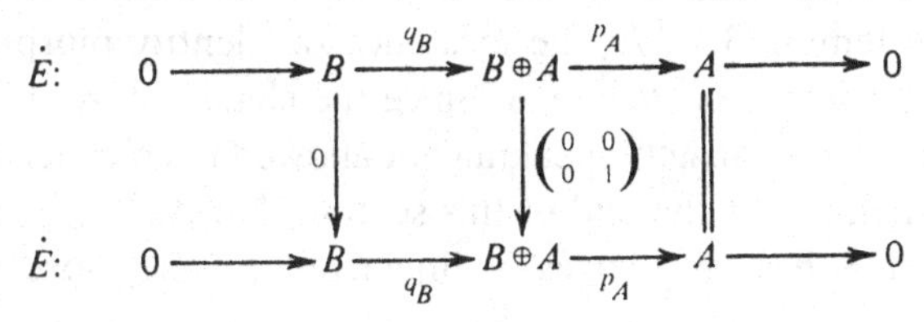

it follows that $\dot{E}$ represents the null extension class. Suppose conversely (for $n = 1$) that a short exact sequence $\dot{E}$ acts as a null. According to (i) $\dot{E}$ must be equivalent (and here this means directly equivalent) to a split sequence. But then $\dot{E}$ splits.

Finally we prove (iv). Again first consider the case $n = 1$:

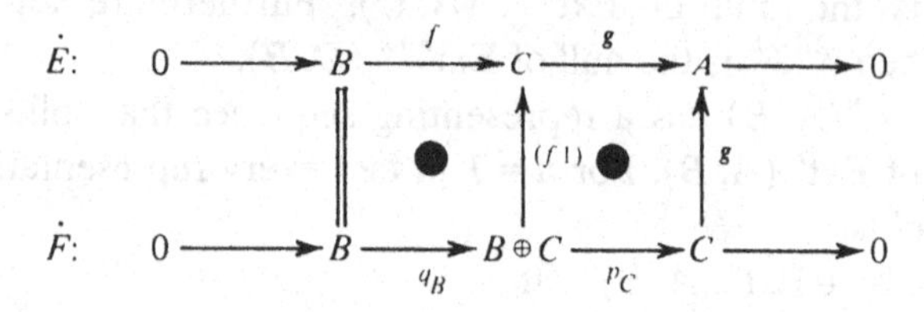

We see that $\mathscr{E}g = \mathscr{F}$ which is the null class. Dually we prove $f\mathscr{E}$ to be the null class.

For $n > 1$ we consider $\dot{E}^n$ decomposed into $n$ extensions of length one: $\dot{E}^n = \dot{E}_1\dot{E}_2 \ldots \dot{E}_n$. We then see that $f\mathscr{E}_1$ and $\mathscr{E}_n g$ are null. Use (ii).

**3.1.18 Definition** A category $\mathbf{A_e}$ is defined as follows. Objects are pairs $\ulcorner A, m\urcorner$ with $A \in \mathbf{A}$ and $m$ a nonnegative integer. The morphisms are given by $(\ulcorner A, m\urcorner, \ulcorner B, n\urcorner)_{\mathbf{A_e}} = \mathrm{Ext}^{m-n}(A, B)$ for $m \geqslant n$ and equal $\{0_{AB}^{m-n}\}$ for $m < n$. This morphism $0_{AB}^{m-n}$ is formally adjoined for every pair $\ulcorner\ulcorner A, m\urcorner, \ulcorner B, n\urcorner\urcorner$ with $m < n$. Composition of morphisms is taken in the following way. For $f: \ulcorner A, l\urcorner \to \ulcorner B, m\urcorner$ and $g: \ulcorner B, m\urcorner \to \ulcorner C, n\urcorner$ in $\mathbf{A_e}$ and for $l \geqslant m \geqslant n$ the morphism $gf: \ulcorner A, l\urcorner \to \ulcorner C, n\urcorner$ in $\mathbf{A_e}$ is given by the Yoneda composition of $g$ and $f$. Then $gf \in \mathrm{Ext}^{l-n}(A, C)$. When the condition $l \geqslant m \geqslant n$ is not fulfilled $gf$ is defined to be the null of $(\ulcorner A, l\urcorner, \ulcorner C, n\urcorner)$. More specifically, for $l > n$ it is the null defined in lemma 3.1.17, for $l = n$ the null element of $(A, C)$, and for $l < n$ the null which is formally adjoined. Furthermore we remark that we have been a little careless in defining $(\ulcorner A, m\urcorner, \ulcorner B, n\urcorner)_{\mathbf{A_e}} = \mathrm{Ext}^{m-n}(A, B)$. This would imply that $(\ulcorner A, m+1\urcorner, \ulcorner B, n+1\urcorner)_{\mathbf{A_e}} = (\ulcorner A, m\urcorner, \ulcorner B, n\urcorner)_{\mathbf{A_e}}$; this is not so, as their domains and codomains are not equal. To be precise the domains and the codomains of each morphism should be specified. However, we shall not always be fussy about this. The associativity for the composition follows from the associativity of the Yoneda composition (lemma 3.1.14) and from lemma 3.1.17. The existence of identity morphisms follows from the existence of such in $\mathbf{A}$. Since the classes of morphisms are not always sets we are strictly speaking not allowed to consider $\mathbf{A_e}$ as a category. We shall see at the end of this section that $\mathbf{A_e}$ is a genuine category provided $\mathbf{A}$ has projectives or injectives. Furthermore $\mathbf{A_e}$ is punctured with null object $\ulcorner 0, 0 \urcorner$ where the first 0 is the null object of the abelian category $\mathbf{A}$. For $(\ulcorner 0,0\urcorner, \ulcorner A, n\urcorner) = \{0\}$ is evident; and $(\ulcorner A, n\urcorner, \ulcorner 0, 0\urcorner) = \{0\}$ is evident for $n = 0$, while for $n > 0$ this follows from the fact that for $\mathscr{E} \in \operatorname{Ext}^n(A, 0) = (\ulcorner A, n\urcorner, \ulcorner 0, 0\urcorner)_{\mathbf{A_e}}$ we have $\mathscr{E} = 1_{00}\mathscr{E} = 0_{00}\mathscr{E} =$ the null extension according to lemma 3.1.17(i). It is clear that the null morphisms form a null family (see 2.2.1).

**3.1.19 Lemma** Let $A \oplus B$ be a biproduct of $A$ and $B$ with projections and injections $p_A$, $p_B$, $q_A$ and $q_B$. Then $\ulcorner A \oplus B, n\urcorner$ with $p_A, p_B, q_A$ and $q_B$ is a biproduct in $\mathbf{A_e}$ of $\ulcorner A, n\urcorner$ and $\ulcorner B, n\urcorner$ for $n \geqslant 0$.

*Proof.* We first show that $\ulcorner A \oplus B, n\urcorner$ is a sum with injections $q_A$ and $q_B$. Therefore let $f: \ulcorner A, n\urcorner \to \ulcorner C, m\urcorner$ and $g: \ulcorner B, n\urcorner \to \ulcorner C, m\urcorner$ be two morphisms in $\mathbf{A_e}$. We have to show that there is exactly one morphism $h: \ulcorner A \oplus B, n\urcorner \to \ulcorner C, m\urcorner$ in $\mathbf{A_e}$ such that $hq_A = f$ and $hq_B = g$. For $n \leqslant m$ this should be evident. Suppose therefore $n > m$ and take $k = n - m$. Let $f = \mathscr{F}^k$ and $g = \mathscr{G}^k$:

$$\dot{F}^k: 0 \to C \to X_1 \to \ldots \to X_k \to A \to 0$$

$$\dot{G}^k: 0 \to C \to Y_1 \to \ldots \to Y_k \to B \to 0.$$

We take $\dot{H}^k = \dot{F}^k \oplus \dot{G}^k$ (see 2.6.10) and consider $\mathscr{H} \in \operatorname{Ext}^k(A \oplus B, C \oplus C)$:

$$\begin{array}{llcccccccc}
\dot{F}^k: & 0 & \longrightarrow & C & \longrightarrow & \ldots & \longrightarrow & A & \longrightarrow & 0 \\
 & & & \downarrow \binom{1}{0} & \bullet & & \bullet & \downarrow \binom{1}{0} = q_A & & \\
\dot{H}^k: & 0 & \longrightarrow & C \oplus C & \longrightarrow & \ldots & \longrightarrow & A \oplus B & \longrightarrow & 0 \\
 & & & \uparrow \binom{0}{1} & \bullet & & \bullet & \uparrow \binom{0}{1} = q_B & & \\
\dot{G}^k: & 0 & \longrightarrow & C & \longrightarrow & \ldots & \longrightarrow & B & \longrightarrow & 0
\end{array}$$

According to 3.1.16 we have $\binom{0}{1}\mathscr{G} = \mathscr{H}\binom{0}{1}$ and $\binom{1}{0}\mathscr{F} = \mathscr{H}\binom{1}{0}$. Hence $(1\ 1)\mathscr{H}q_A = (1\ 1)\mathscr{H}\binom{1}{0} = (1\ 1)\binom{1}{0}\mathscr{F} = 1\mathscr{F} = \mathscr{F}$ and $(1\ 1)\mathscr{H}q_B = (1\ 1)\mathscr{H}\binom{0}{1} = (1\ 1)\binom{0}{1}\mathscr{G} = \mathscr{G}$. Now take $h = (1\ 1)\mathscr{H} \in \operatorname{Ext}^k(A \oplus B, C)$.

Then $h$ satisfies $hq_A = f$ and $hq_B = g$. We now have to show that $h$ is unique. Suppose therefore $\mathscr{K} \in \operatorname{Ext}^k(A \oplus B, C)$ also satisfies $\mathscr{K}q_A = f = \mathscr{F}$ and $\mathscr{K}q_B = g = \mathscr{G}$. We then have (for some $\dot{F}' \in \mathscr{F}$):

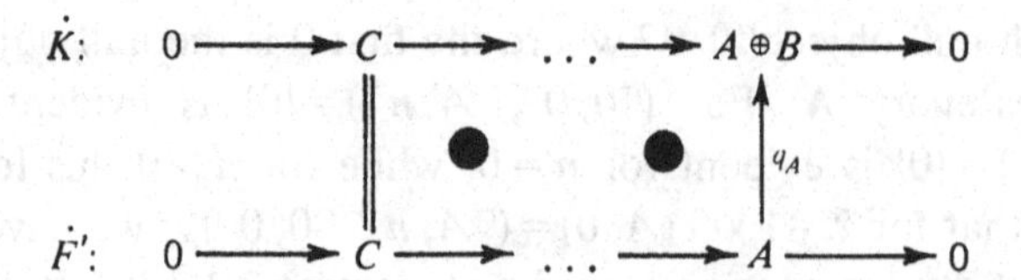


and (for some $\dot{G}' \in \mathscr{G}$):

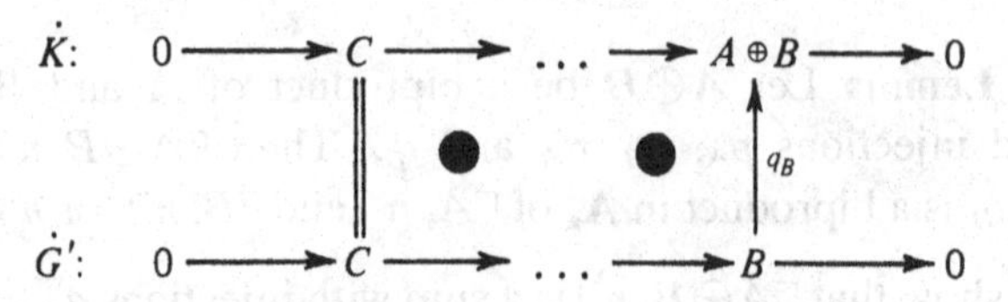


But we also have the extension $\dot{F}' \oplus \dot{G}'$ yielding

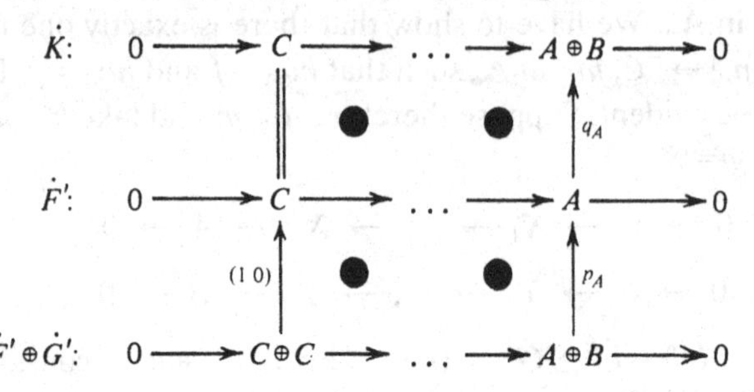


So that

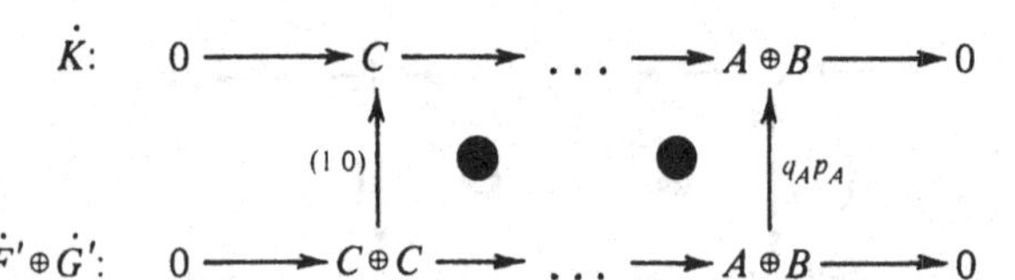


and similarly

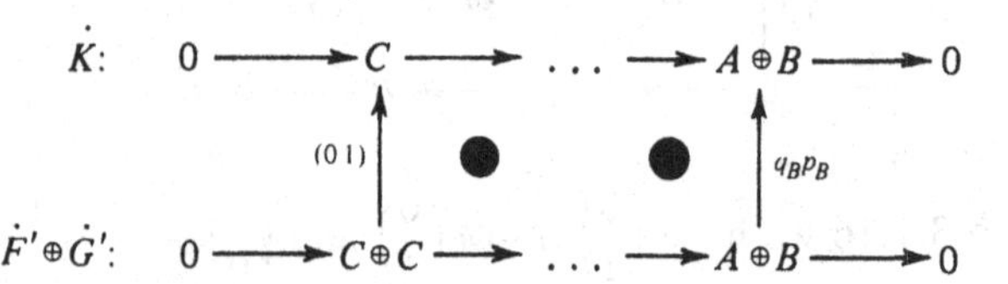


Now identifying the objects and the horizontal morphisms of the last two diagrams, we add the vertical morphisms obtaining

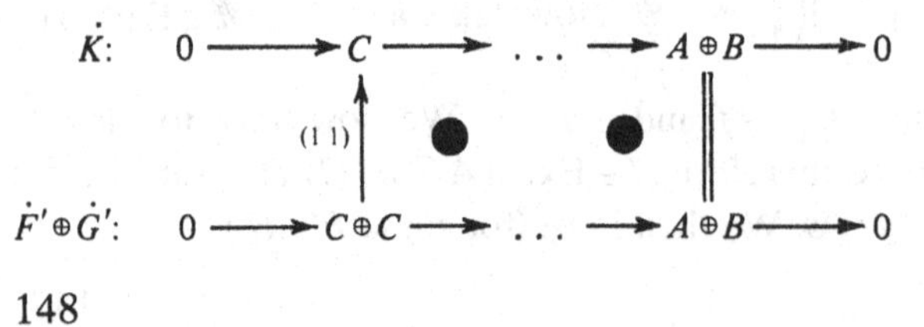

If we can show that $\dot{F}'\oplus\dot{G}'\in\mathscr{H}$ then it follows from the last diagram that $\mathscr{H}=(1\ \ 1)\mathscr{H}=h$ which proves the uniqueness of $h$. It is not hard to show that $\dot{F}\oplus\dot{G}\sim\dot{F}'\oplus\dot{G}$ and $\dot{F}'\oplus\dot{G}\sim\dot{F}'\oplus\dot{G}'$. Thus we have proved that $\ulcorner A\oplus B, n\urcorner$ is a sum with injections $q_A$ and $q_B$. To prove that $\ulcorner A\oplus B, n\urcorner$ is a product with projections $p_A$ and $p_B$ one cannot dualize since the category $\mathbf{A_e}$ is not self dual. However parts of the proof dualize correctly and it is then easy to finish. We may thus conclude that $\ulcorner A\oplus B, n\urcorner$ is a biproduct.

**3.1.20 Theorem** The category $\mathbf{A_e}$ is additive.

*Proof.* It follows from the last lemma and from theorem 2.3.6 that $\mathbf{A_e}$ is semi-additive with respect to a suitable addition, which is unique. This unique addition takes the following form (see 2.3.6): for $\mathscr{F}^n$ and $\mathscr{G}^n$ in $\text{Ext}^n(A, B)$ the extension class $\mathscr{F}^n+\mathscr{G}^n$ equals $(1\ \ 1)(\mathscr{F}^n\oplus\mathscr{G}^n)\binom{1}{1}$. This may be seen as follows, using the addition of 2.3.6. Take the upper row of the diagram exhibited there. With the objects replaced by the type of objects in our category $\mathbf{A_e}$ this row looks like

$$\ulcorner A, r\urcorner \xrightarrow{\binom{1}{1}} \ulcorner A\oplus A, r\urcorner \xrightarrow{(f\ g)} \ulcorner B, s\urcorner$$

where $f$ and $g$ are morphisms $f, g: \ulcorner A, r\urcorner \rightrightarrows \ulcorner B, s\urcorner$; $f$ and $g$ are elements of $\text{Ext}^n(A, B)$ with $n=r-s\geq 0$. Let $f=\mathscr{F}^n$ and $g=\mathscr{G}^n$. For $n>0$ the morphism $(f\ \ g)$ is just what was called $h$ in the previous lemma 3.1.19. So we have $\mathscr{F}^n+\mathscr{G}^n=f+g=(f\ \ g)\binom{1}{1}=(1\ \ 1)(\mathscr{F}^n\oplus\mathscr{G}^n)\binom{1}{1}$. This yields the following diagram:

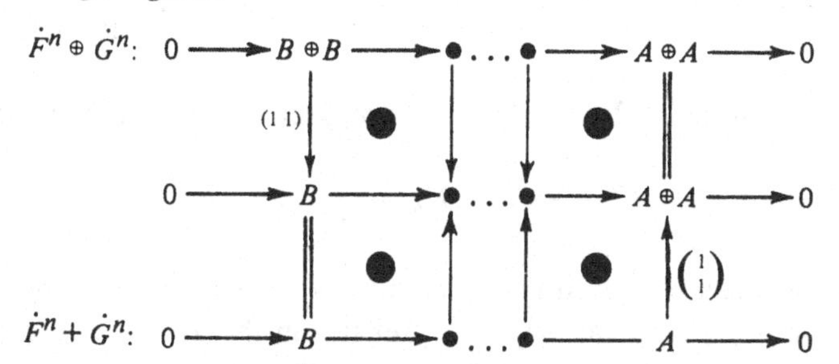


For the case $n=0$ (which means $r=s$) $f$ and $g$ are just elements of $(A, B)_{\mathbf{A}}$ and we may replace the diagram

$$\ulcorner A, r\urcorner \xrightarrow{\binom{1}{1}} \ulcorner A\oplus A, r\urcorner \xrightarrow{(f\ g)} \ulcorner B, r\urcorner$$

by

$$A \xrightarrow{\binom{1}{1}} A \oplus A \xrightarrow{(f\ g)} B$$

which shows that the addition in this case coincides with the addition in $(A, B)_{\mathbf{A}}$. Since the identity morphisms in the category $\mathbf{A}_e$ have additive inverses we may conclude that $\mathbf{A}_e$ is not only semi-additive but even additive (see the remark after theorem 2.3.6). The addition just discussed is often called the *Baer addition* after its discoverer in $\mathrm{Ext}^1$ [2]. The bifunctors $\mathrm{Ext}^n(\text{-},\text{-})\ulcorner A^\circ, B\urcorner = \mathrm{Ext}^n(A, B)$ with $\mathrm{Ext}^n(\text{-},\text{-})\ulcorner g^\circ, f\urcorner \mathscr{E} = f\mathscr{E}g$ defined in 3.1.15 are hence additive and can be taken as $\mathrm{Ext}^n(\text{-},\text{-})\colon \mathbf{A}^\circ \times \mathbf{A} \to \mathbf{Ab}$. Then also the covariant restriction

$$\mathrm{Ext}^n(A, \text{-})\colon \mathbf{A} \to \mathbf{Ab}$$

and the contravariant one

$$\mathrm{Ext}^n(\text{-}, B)\colon \mathbf{A} \to \mathbf{Ab}$$

for $A$ and $B \in \mathbf{A}$ are additive.

**3.1.21 Lemma** Let $\dot{E}\colon 0 \to A \xrightarrow{f} B \xrightarrow{g} C \to 0$ be a short exact sequence or, in other words, let $\dot{E} \in \mathscr{E} \in \mathrm{Ext}(C, A)$. Let $X \in \mathbf{A}$. Let $\mathscr{E}$- be defined by $(\mathscr{E}\text{-})h = \mathscr{E}h$ for $h \in (X, C)$. Then the following sequence in $\mathbf{Ab}$ is exact

$$0 \to \mathrm{Ext}^0(X, A) \xrightarrow{\mathrm{Ext}^0(X, f)} \mathrm{Ext}^0(X, B) \xrightarrow{\mathrm{Ext}^0(X, g)} \mathrm{Ext}^0(X, C)$$

$$\xrightarrow{\mathscr{E}\text{-}} \mathrm{Ext}^1(X, A) \xrightarrow{\mathrm{Ext}^1(X, f)} \mathrm{Ext}^1(X, B) \xrightarrow{\mathrm{Ext}^1(X, g)} \mathrm{Ext}^1(X, C).$$

*Proof.* According to 2.7.9($a$) the following sequence is exact in $\mathbf{Ab}$:

$$0 \to (X, A)_{\mathbf{A}} \xrightarrow[h \times f]{} (X, B)_{\mathbf{A}} \xrightarrow[h \times g]{} (X, C)_{\mathbf{A}}.$$

To prove exactness at the remaining points we have to show that 'im $d_i = \ker d_{i+1}$' and we may interpret this in the group-theoretic way since we work in $\mathbf{Ab}$. We first observe that the sequence is a null sequence. To begin with, $gf = 0$. Then according to 3.1.17(iv) we have $f\mathscr{E} = 0$ and $\mathscr{E}g = 0$. From this it follows that $(\mathrm{Ext}^1(X, f) \circ \mathscr{E}\text{-})h = f\mathscr{E}h = 0$ for $h \in \mathrm{Ext}^0(X, C)$ so that $\mathrm{Ext}^1(X, f) \circ \mathscr{E}\text{-} = 0$. A similar argument yields $\mathrm{im}\, d_i \subset \ker d_{i+1}$ at the remaining point. To prove the converse we first look at $\text{Ext}^0(X, C)$. Let $h \in \ker(\mathscr{E}\text{-})$, i.e. $h \in (X, C)$ and $\mathscr{E}h = 0$. We have to show that $h \in \text{im}\,\text{Ext}^0(X, g)$. Construct $\dot{F} \in \mathscr{E}h$:

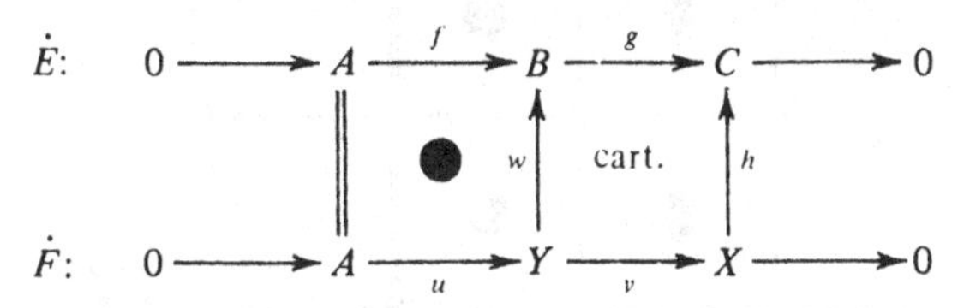


From 3.1.17(iii) it follows that $\dot{F}$ splits. Hence there is a morphism $t: X \to Y$ with $vt = 1$ (2.6.12). Now we have

$$h = hvt = gwt = \text{Ext}^0(X, g)wt \in \text{im}\,\text{Ext}^0(X, g).$$

Next we consider exactness at $\text{Ext}^1(X, A)$. Let $\mathscr{F} \in \ker \text{Ext}^1(X, f)$, i.e. $\mathscr{F} \in \text{Ext}^1(X, A)$ with $f\mathscr{F} = 0$. Take a representative for $\mathscr{F}$, say $\dot{F}: 0 \to A \xrightarrow{a} Y \xrightarrow{b} X \to 0$. We proved above that if $\mathscr{E}h = 0$, there is a morphism $s: X \to B$ (namely $wt$) such that $h = gs$. Dually one has that if $f\mathscr{F} = 0$ there is a morphism $u: Y \to B$ with $ua = f$. Then $gua = gf = 0$. Since $b = \text{coker}\, a$ there is a morphism $r: X \to C$ such that $rb = g$. This yields the diagram

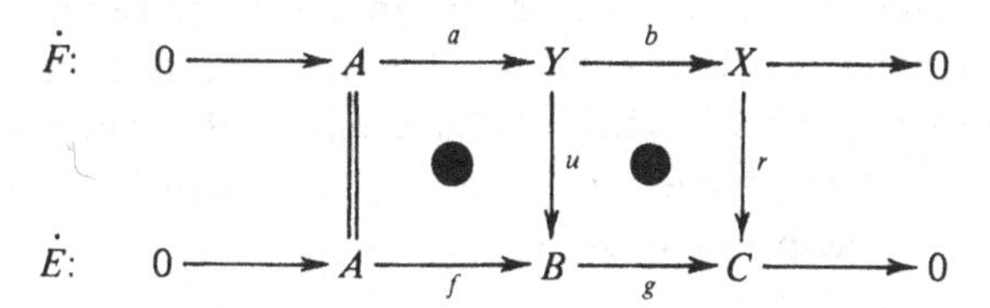


Hence $\mathscr{F} = \mathscr{E}r = (\mathscr{E}\text{-})r \in \text{im}(\mathscr{E}\text{-})$.

Finally we prove exactness at $\text{Ext}^1(X, B)$. Let $\mathscr{F} \in \ker \text{Ext}^1(X, g)$, i.e. $\mathscr{F} \in \text{Ext}^1(X, B)$ with $g\mathscr{F} = 0$. Take $\dot{F} \in \mathscr{F}$ and construct a sequence $\dot{G}' \in g\mathscr{F}$. Then use the fact that $\dot{G}'$ splits:

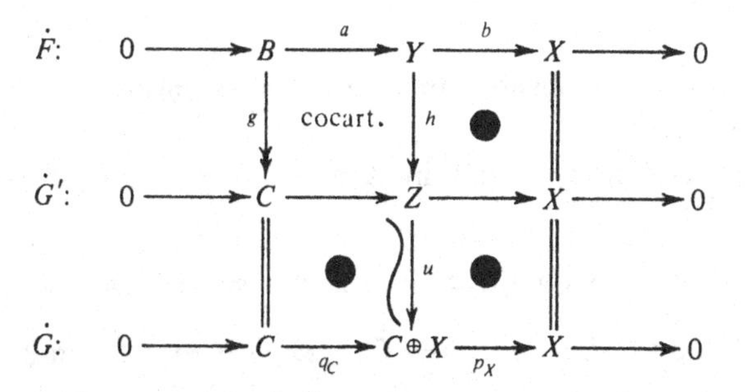


Due to 2.1.2 $h$ is epic. Take $y = uh$. Then $y$ is epic. Let $k: K \to Y$ be the kernel of $p_C y$. Since $p_C yaf = p_C q_C gf = 0$, there is a unique $h: A \to K$ with $kh = af$. In a diagram:

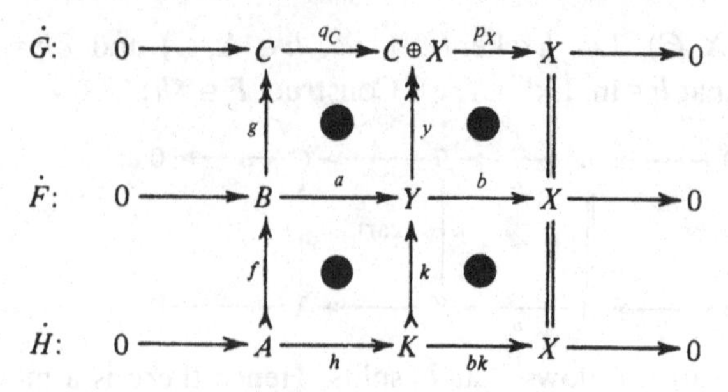


Since $a$ and $f$ are monic, so is $h$. We want to show that $\dot{H}$ is exact. Once this is proved we are through since $\dot{H} \in \mathscr{H} \in \operatorname{Ext}^1(X, A)$ and $\mathscr{F} = f\mathscr{H} = \operatorname{Ext}^1(X, f)\mathscr{H} \in \operatorname{im} \operatorname{Ext}^1(X, f)$. To prove exactness of $\dot{H}$ we first show $h = \ker bk$. In the first place $bkh = baf = 0$. Suppose next that $v: V \to K$ with $bkv = 0$. Since $a = \ker b$ we have $w: V \to B$ with $kv = aw$. Then $gw = p_C q_C gw = p_C yaw = p_C ykv = 0$. Since $f = \ker g$ there is a unique $x: V \to A$ with $fx = w$. But then $khx = afx = aw = kv$. Since $k$ is monic it follows that $hx = v$. Since $h$ is monic such a morphism $x$ is unique.

Next we prove that $bk$ is epic. By 2.5.1° we may then conclude that $\dot{H}$ is exact. Suppose therefore $z: X \to Z$ is such that $zbk = 0$. From 2.5.1° it follows that $p_C y = \operatorname{coker} k$ $(k = \ker p_C y)$. Hence there is a morphism $s: C \to Z$ such that $sp_C y = zb$. Then we have $sp_C y = zb = zp_X y$. Since $y$ is epic it follows that $sp_C = zp_X$ so that $z = zp_X q_X = sp_C q_X = 0$. Therefore $bk$ is epic and exactness has been established.

We want to extend the sequence in 3.1.21. For this purpose we need

**3.1.22 Lemma** Let $A, B$ and $C$ be objects of $\mathbf{A}$, $k \geq 1$, $l \geq 1$ and let $\mathscr{E}^l \in \operatorname{Ext}^l(B, C)$ and $\mathscr{F}^k \in \operatorname{Ext}^k(A, B)$. Then the following assertions are equivalent.

(i) $\mathscr{E}^l \mathscr{F}^k = 0$.

(ii) There exists a morphism $t$ in $\mathbf{A}$ and an extension $\mathscr{H}^k$ so that $\mathscr{F}^k = t\mathscr{H}^k$ and $\mathscr{E}^l t = 0$.

(iii) There exists a morphism $s$ in $\mathbf{A}$ and an extension $\mathscr{G}^l$ so that $\mathscr{E}^l = \mathscr{G}^l s$ and $s\mathscr{F}^k = 0$.

*Proof.* We start by proving (ii)⇒(iii) for the special case $k = l = 1$. Assume $\mathscr{F} = t\mathscr{H}$, $\mathscr{E}t = 0$, $\dot{E} \in \mathscr{E}$, $\dot{E}: 0 \to C \xrightarrow{f} D \xrightarrow{g} B \to 0$. From $\mathscr{E}t = 0$ it follows by the previous lemma 3.1.21 that $t = gh$ for a certain $h \in \operatorname{Ext}^0(X, D)$ (if we assume $t: X \to B$). Take $\mathscr{K} = h\mathscr{H}$ so that we have $\mathscr{F} = t\mathscr{H} = gh\mathscr{H} = g\mathscr{K}$. Consider the following diagram which is obtained by first choosing $\dot{K} \in \mathscr{K}$ and then taking $\dot{F} = g\dot{K}$.

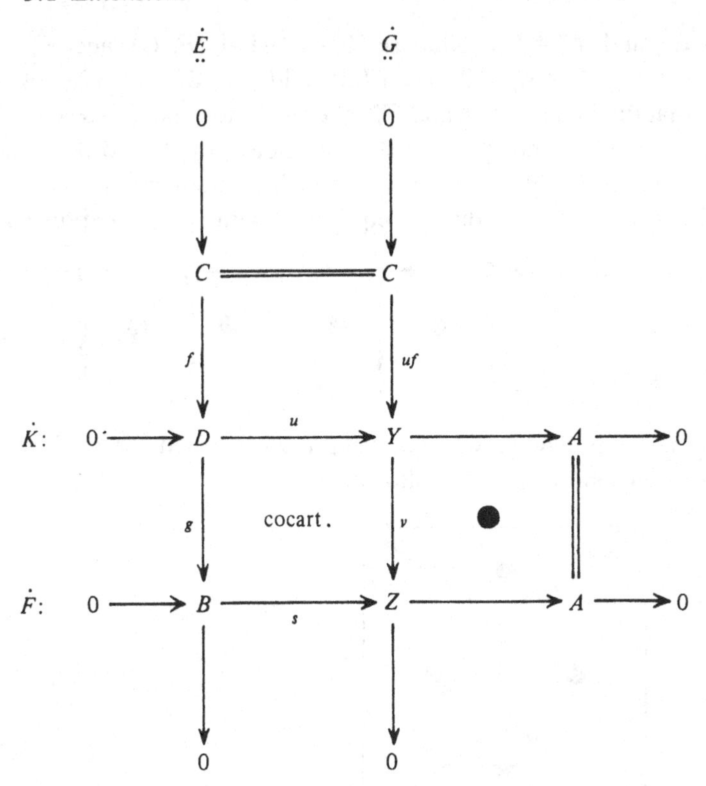


We will show that $\dot{G}$ is an exact sequence. Then we would have $\dot{G} \in \mathscr{G} \in \operatorname{Ext}(Z, C)$ with $\mathscr{E} = \mathscr{G}s$ while $s\mathscr{F} = 0$ (according to 3.1.17(iv)) which would prove (iii). Since the cocartesian square is bicartesian by exercise 2.5.18(*b*)° we know by the exercise after definition 2.4.1 that $uf = \ker v$. The morphism $v$ being epic we have $v = \operatorname{coker} uf$ and $\dot{G}$ is exact. Dually one proves (iii)$\Rightarrow$(ii) for $k = l = 1$.

To prove the equivalence of (ii) and (iii) for all values of $k$ and $l$ we use induction in the following way. We will prove the assertion $P(n)$: for all $A$, $B$, $C$, $\mathscr{E}^l$ and $\mathscr{F}^k$ as above the equivalence (ii)$\Leftrightarrow$(iii) holds for $k + l \leqslant n$ and moreover (i)$\Rightarrow$(iii) for $k + l < n$. Notice that the dual of $P(n)$ is $P(n)$ itself. Since $k \geqslant 1$ and $l \geqslant 1$, the assertions $P(0)$ and $P(1)$ are true while we just proved $P(2)$.

Now we treat the induction step from $n$ to $n+1$ ($n \geqslant 2$). We start by proving (i)$\Rightarrow$(iii) for $k + l < n + 1$ or $k + l \leqslant n$. Suppose $\mathscr{E}^l\mathscr{F}^k = 0$. Take $\dot{E}^l \in \mathscr{E}^l$, $\dot{F}^k \in \mathscr{F}^k$ and $\dot{K}^l \in \mathscr{E}^l 0_{BB}$. Then $\dot{K}^l\dot{F}^k \sim 0 \sim \dot{E}^l\dot{F}^k$. Hence there are exact sequences $\dot{H}_0^{l+k}, \dot{H}_1^{l+k}, \ldots, \dot{H}_p^{l+k}$ with $\dot{H}_0^{l+k} = \dot{K}^l\dot{F}^k$ and $\dot{H}_p^{l+k} = \dot{E}^l\dot{F}^k$ in such a way that $\dot{H}_i$ is directly equivalent with $\dot{H}_{i+1}$ for $i = 0, \ldots, p-1$. We may write $\dot{H}_i^{l+k} = \dot{E}_i^l\dot{F}_i^k$. We take $\dot{E}_0^l = \dot{K}^l$, $\dot{F}_0^k = \dot{F}^k$,

$\dot{E}_p^l = \dot{E}^l$ and $\dot{F}_p^k = \dot{F}^k$. Suppose $\dot{E}_i^l \in \mathscr{E}_i^l \in \operatorname{Ext}^l(B_i, C)$ and $\dot{F}_i^k \in \mathscr{F}_i^k \in \operatorname{Ext}^k(A, B_i)$, $B_0 = B_p = B$. Then (iii) holds for $\mathscr{E}_0^l$ and $\mathscr{F}_0^k$. Let $m < p$ such that (iii) holds for $\mathscr{E}_m^l$ and $\mathscr{F}_m^k$. We shall show that (iii) also holds for $\mathscr{E}_{m+1}^l$ and $\mathscr{F}_{m+1}^k$, thus proving that (iii) holds for $\mathscr{E}_p^l$ and $\mathscr{F}_p^k$. It would follow that (iii) holds for $\mathscr{E}^l$ and $\mathscr{F}^k$ which we wanted to prove.

The fact that $\dot{H}_m^{l+k}$ is directly equivalent with $\dot{H}_{m+1}^{l+k}$ is expressed by

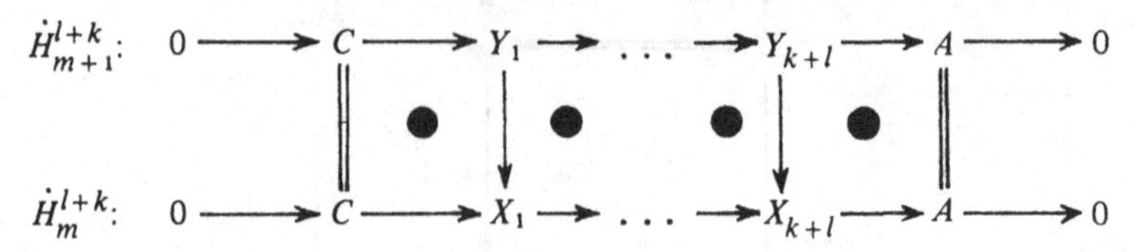


or by a similar diagram with vertical arrows reversed. By 2.5.10 we may insert a morphism $w$ in the following way:

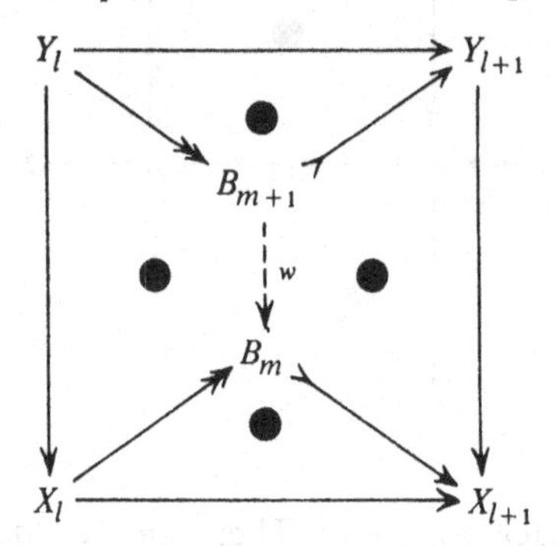


Thus we have $\mathscr{E}_m^l w = \mathscr{E}_{m+1}^l$ and $\mathscr{F}_m^k = w\mathscr{F}_{m+1}^k$.

Since (iii) holds for $\mathscr{E}_m^l$ and $\mathscr{F}_m^k$ there is a morphism $s$ and an extension class $\mathscr{G}^l$ such that $\mathscr{E}_m^l = \mathscr{G}^l s$ and $s\mathscr{F}_m^k = 0$. But then $\mathscr{E}_{m+1}^l = \mathscr{G}^l sw$ and $sw\mathscr{F}_{m+1}^k = 0$ so that (iii) holds for $\mathscr{E}_{m+1}^l$ and $\mathscr{F}_{m+1}^k$.

In the case when the direct equivalence between $\dot{H}_{m+1}$ and $\dot{H}_m$ is given by a diagram as shown above but with vertical arrows reversed we want to use condition (ii). But (iii)$\Leftrightarrow$(ii) for $\mathscr{E}_m^l$, $\mathscr{F}_m^k$ because $P(n)$ is assumed. Hence we can proceed similarly to prove (ii) for $\mathscr{E}_{m+1}^l$ and $\mathscr{F}_{m+1}^k$ and thence (iii).

Next we prove (ii)$\Rightarrow$(iii) for $k + l \leq n + 1$. Let $\mathscr{F}^k = t\mathscr{H}^k$ and $\mathscr{E}^l t = 0$. The case $k = l = 1$ has been handled. First assume $k > 1$ (and afterwards $k = 1$ and $l > 1$). We split $\mathscr{H}$ as $\mathscr{H}_1\mathscr{H}_2 \ldots \mathscr{H}_k$ with $\mathscr{H}_i$ having length one. Then (ii) holds for $\mathscr{E}^l$ and $\mathscr{F}' = t\mathscr{H}_1$. According to $P(n)$ then (iii) holds for $\mathscr{E}^l$ and $\mathscr{F}'$ (since $l + 1 < k + 1 \leq n + 1$), i.e. $\mathscr{E}^l = \mathscr{G}^l s$ and $s\mathscr{F}' = 0$. But then $s\mathscr{F}^k = st\mathscr{H}^k = s\mathscr{F}'\mathscr{H}_2 \ldots \mathscr{H}_k = 0$. Hence (iii) holds for $\mathscr{E}^l$ and $\mathscr{F}^k$. For the case $k = 1$ and $l > 1$ we split $\mathscr{E}^l$ as $\mathscr{E}_1\mathscr{E}_2 \ldots \mathscr{E}_l$ with $\mathscr{E}_i$ having length one. Then (i) holds for $\mathscr{E}_1 \ldots \mathscr{E}_{l-1}$ and $\mathscr{E}_l t$. Since we have already proved (i)$\Rightarrow$(iii) for $k+l \leqslant n$ we know that (iii) holds for $\mathscr{E}_1 \ldots \mathscr{E}_{l-1}$ and $\mathscr{E}_l t$. This means there is an extension class $\mathscr{G}'^{l-1}$ and a morphism $s'$ such that $\mathscr{E}_1 \ldots \mathscr{E}_{l-1} = \mathscr{G}'^{l-1}s'$ and $s'\mathscr{E}_l t = 0$. But this says that (ii) holds for $s'\mathscr{E}_l$ and $\mathscr{F}^k$. But then (iii) holds for $s'\mathscr{E}_l$ and $\mathscr{F}^k$, so there is an extension class $\mathscr{G}''$ and a morphism $s$ so that $s'\mathscr{E}_l = \mathscr{G}''s$ and $s\mathscr{F}^k = 0$. Then it follows that $\mathscr{E}^l = \mathscr{E}_1 \ldots \mathscr{E}_l = \mathscr{G}'^{l-1}s'\mathscr{E}_l = \mathscr{G}'^{l-1}\mathscr{G}''s$. So (iii) holds for $\mathscr{E}^l$ and $\mathscr{F}^k$. Thus we proved (ii)$\Rightarrow$(iii) for $\mathscr{E}^l$ and $\mathscr{F}^k$ in the case $k+l \leqslant n+1$. Dually one proves the converse. Hence $P(n)$ is proved for all $n$. Since (iii)$\Rightarrow$(i) is evident the lemma now has been established.

**3.1.23 Theorem** Suppose $\dot{E}: 0 \to A \xrightarrow{f} B \xrightarrow{g} C \to 0$ is a short exact sequence, $\dot{E} \in \mathscr{E} \in \operatorname{Ext}(C, A)$. Let $X$ be an object of $\mathbf{A}$. Then the infinite sequence

$$0 \to \operatorname{Ext}^0(X, A) \xrightarrow{\operatorname{Ext}^0(X,f)} \operatorname{Ext}^0(X, B) \xrightarrow{\operatorname{Ext}^0(X,g)} \operatorname{Ext}^0(X, C)$$

$$\xrightarrow{\mathscr{E}\text{-}} \operatorname{Ext}^1(X, A) \to \ldots$$

$$\xrightarrow{\mathscr{E}\text{-}} \operatorname{Ext}^n(X, A) \xrightarrow{\operatorname{Ext}^n(X,f)} \operatorname{Ext}^n(X, B) \xrightarrow{\operatorname{Ext}^n(X,g)} \operatorname{Ext}^n(X, C)$$

$$\xrightarrow{\mathscr{E}\text{-}} \operatorname{Ext}^{n+1}(X, A) \to \ldots$$

is exact in $\mathbf{Ab}$, where $\mathscr{E}$- is defined as $(\mathscr{E}\text{-})\mathscr{F}^n = \mathscr{E}\mathscr{F}^n$ for $\mathscr{F}^n \in \operatorname{Ext}^n(X, C)$.

Remark. For each $n$ the mapping $\mathscr{E}$- is actually a different one. So a more correct notation would have been $(\mathscr{E}\text{-})_n$.

*Proof.* Just as in 3.1.21 we prove that the sequence is a null sequence. We then only have to show that $\ker d_{i+1} \subset \operatorname{im} d_i$. This we do by reducing the theorem to 3.1.21, using the last lemma. For instance look at the exactness at $\operatorname{Ext}^{n+1}(X, A)$ for $n \geqslant 1$. Suppose $\mathscr{F}^{n+1} \in \ker \operatorname{Ext}^{n+1}(X, f)$. Thus $\mathscr{F}^{n+1} \in \operatorname{Ext}^{n+1}(X, A)$ with $f\mathscr{F} = 0$. Decompose $\mathscr{F}^{n+1}$ into $n+1$ extensions each with length one: $\mathscr{F}^{n+1} = \mathscr{F}_1\mathscr{F}_2 \ldots \mathscr{F}_{n+1}$. From $(f\mathscr{F}_1)\mathscr{F}_2 \ldots \mathscr{F}_{n+1} = 0$ it follows (by 3.1.22) that there exists a morphism $t$ and an extension class $\mathscr{H}$ so that $f\mathscr{F}_1 t = 0$ and $\mathscr{F}_2 \ldots \mathscr{F}_{n+1} = t\mathscr{H}^n$. Then the exactness at $\operatorname{Ext}^1(Y, A)$ of the sequence for the appropriate $Y$ means a morphism $h$ exists with $\mathscr{F}_1 t = \mathscr{E}h$. This entails

$$\mathscr{F}^{n+1} = \mathscr{F}_1 \ldots \mathscr{F}_{n+1} = \mathscr{F}_1 t\mathscr{H}^n = \mathscr{E}h\mathscr{H}^n = (\mathscr{E}\text{-})h\mathscr{H}^n \in \operatorname{im}(\mathscr{E}\text{-}).$$

In the same way the exactness at $\operatorname{Ext}^n(X, B)$ is reduced to exactness at $\operatorname{Ext}^0(Y, B)$ and exactness at $\operatorname{Ext}^n(X, C)$ to exactness at $\operatorname{Ext}^1(Y, C)$. Our proof follows Mitchell [29, ch. VIII]. Very recently R. Fritsch has given a different argument.

**3.1.23° Theorem** Suppose $\dot{E} \in \mathscr{E} \in \operatorname{Ext}(C, A)$ is a short exact sequence $0 \to A \xrightarrow{f} B \xrightarrow{g} C \to 0$. Let $Y \in \mathbf{A}$. Then the infinite sequence

$$0 \to \operatorname{Ext}^0(C, Y) \xrightarrow{\operatorname{Ext}^0(g, Y)} \operatorname{Ext}^0(B, Y) \xrightarrow{\operatorname{Ext}^0(f, Y)} \operatorname{Ext}^0(A, Y)$$

$$\xrightarrow{-\mathscr{E}} \operatorname{Ext}^1(C, Y) \to \dots$$

$$\xrightarrow{-\mathscr{E}} \operatorname{Ext}^n(C, Y) \xrightarrow{\operatorname{Ext}^n(g, Y)} \operatorname{Ext}^n(B, Y) \xrightarrow{\operatorname{Ext}^n(f, Y)} \operatorname{Ext}^n(A, Y)$$

$$\xrightarrow{-\mathscr{E}} \operatorname{Ext}^{n+1}(C, Y) \to \dots$$

is exact in **Ab**. Here $-\mathscr{E}$ is defined by $(-\mathscr{E})\mathscr{F}^n = \mathscr{F}^n\mathscr{E}$ for $\mathscr{F}^n \in \operatorname{Ext}^n(A, Y)$.

**3.1.24** We draw some conclusions from the theory. First observe that $\operatorname{Ext}^1(P, A) = 0$ for a projective object $P$ since every epimorphism $B \twoheadrightarrow P$ splits and hence represents the null class. Furthermore an extension $\dot{F}^n$ of length $n$ of $P$ by $A$ can be written as a composition $\dot{F}^n = \dot{F}_1\dot{F}_2 \dots \dot{F}_n$ of extensions of length 1 with $\dot{F}_n$ representing an element of $\operatorname{Ext}^1(P, I_{n-1}) = 0$. Hence $\operatorname{Ext}^n(P, A) = 0$ for $n \geqslant 1$ and dually $\operatorname{Ext}^n(C, Q) = 0$ for $n \geqslant 1$ when $Q$ is injective.

We shall now prove that in a category with projectives the objects $\operatorname{Ext}^n(C, A)$ are actually sets and hence legitimate abelian groups. Indeed let $P \twoheadrightarrow C$ be an epimorphism with $P$ projective. Call the kernel of this epimorphism $K$. The long exact sequence of 3.1.23° applied to the short exact sequence $0 \to K \to P \to C \to 0$ yields exactness of

$$\dots \to \operatorname{Ext}^n(P, A) \to \operatorname{Ext}^n(K, A) \to \operatorname{Ext}^{n+1}(C, A)$$
$$\to \operatorname{Ext}^{n+1}(P, A) \to \operatorname{Ext}^{n+1}(K, A) \to \dots$$

for $n \geqslant 0$. Since $\operatorname{Ext}^{n+1}(P, A) = 0$ for $n \geqslant 0$ we find $\operatorname{Ext}^0(K, A) \twoheadrightarrow \operatorname{Ext}^1(C, A)$. The former being just the abelian group $(K, A)_{\mathbf{A}}$ we see that $\operatorname{Ext}^1(C, A)$ is a set. Since $C$ is arbitrary $\operatorname{Ext}^1(K, A)$ is an abelian group, while $\operatorname{Ext}^2(C, A) \simeq \operatorname{Ext}^1(K, A)$. By induction we achieve our aim. Dually, using 3.1.23, we prove that the objects $\mathrm{Ext}^n(C, A)$ are abelian groups if **A** has injectives. This is for instance the case if **A** is a Grotendieck category (corollary 2.9.5). In these cases $\mathbf{A_e}$ is a genuine additive category and $\mathrm{Ext}^n(-,-)$ are additive bifunctors which duly land in **Ab**.

**3.1.25** We finally make some observations designed to pave the way to the next section.

Suppose $\dot{E} \in \mathscr{E} \in \mathrm{Ext}(C, A)$ and $s: Y \to X$. Then the following diagram is commutative:

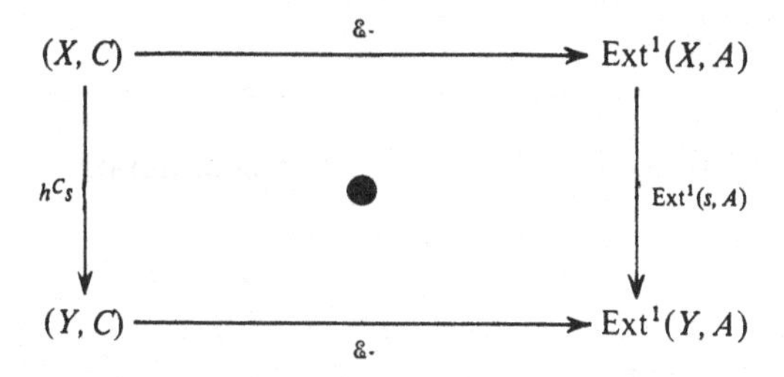


Or, more generally ($n \geq 0$),

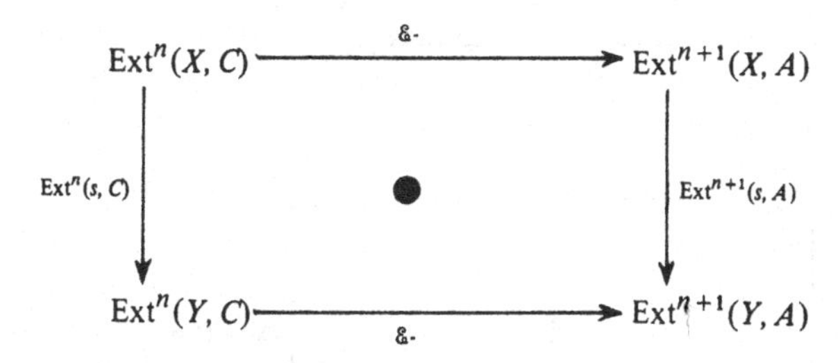


Hence $\mathscr{E}$- is a morphism of contravariant functors:

$$\mathscr{E}\text{-}: \mathrm{Ext}^n(-, C) \to \mathrm{Ext}^{n+1}(-, A).$$

Dually, for $s: X \to Y$,

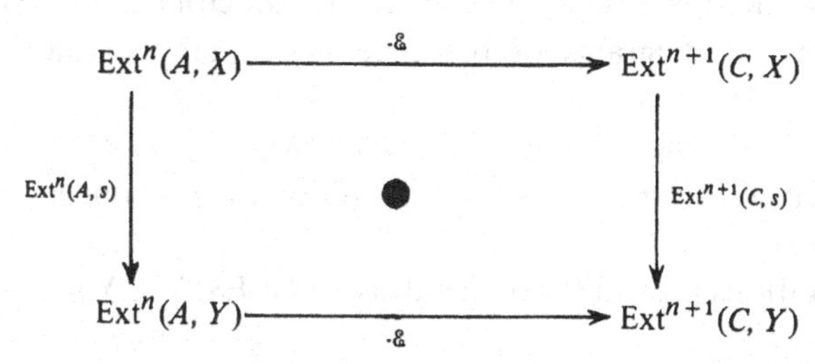


Here -$\mathscr{E}$ is a morphism of covariant functors:

$$\text{-}\mathscr{E}: \mathrm{Ext}^n(A, -) \to \mathrm{Ext}^{n+1}(C, -).$$

Note that here, too, a more accurate notation should distinguish between

$$\mathscr{E}\text{-}: \operatorname{Ext}^n(X, C) \to \operatorname{Ext}^{n+1}(X, A)$$

and

$$\mathscr{E}\text{-}: \operatorname{Ext}^n(Y, C) \to \operatorname{Ext}^{n+1}(Y, A).$$

Suppose, in addition, $\dot{E}' \in \mathscr{E}' \in \operatorname{Ext}(C', A')$ with

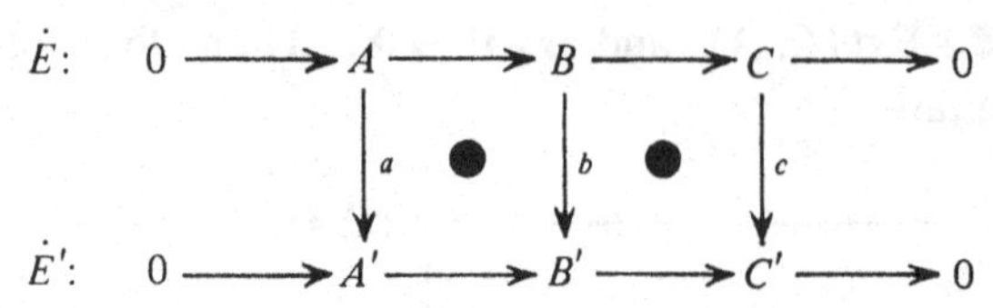


Then, according to 3.1.6, the following diagram is commutative:

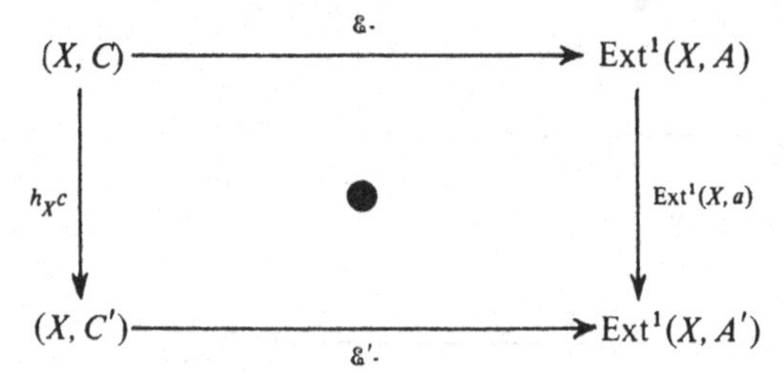


Or, more generally,

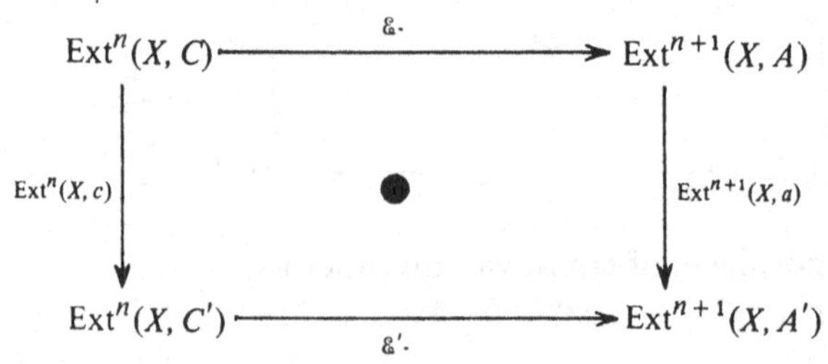


So we see that there is a relation between the functors $\operatorname{Ext}^n(X, \text{-})$ and $\operatorname{Ext}^{n+1}(X, \text{-})$, while diagrams of the type above play a role in this relation.

The sequence of functors $\operatorname{Ext}^0(X, \text{-}), \operatorname{Ext}^1(X, \text{-}), \ldots$ is called a *connected sequence* of functors. A precise definition of this concept will be given in the next section.

Similar conditions prevail for the contravariant $\operatorname{Ext}^n(\text{-}, Y)$.

### 3.2 Connected sequences and satellites

Throughout this section **A** will be an abelian and **B** an additive category.

**3.2.1 Definition** A *connected right sequence* of additive functors from **A** to **B** is a system consisting of:

(i) a sequence of covariant additive functors from **A** to **B**: $F^0, F^1, F^2, \ldots$;

(ii) for every short exact sequence $\dot{A}: 0 \to A' \to A \to A'' \to 0$ in **A** a sequence of morphisms $\delta^n(\dot{A}): F^nA'' \to F^{n+1}A'$ in such a way that the following conditions are satisfied:

(*a*) for all $n \geqslant 0$ the sequence

$$F^nA \to F^nA'' \xrightarrow{\delta^n(\dot{A})} F^{n+1}A' \to F^{n+1}A$$

is a null sequence;

(*b*) for the diagram in **A**

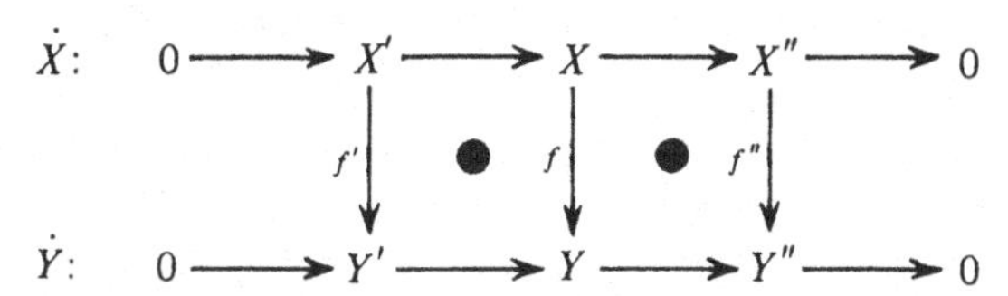

in which both rows are exact the following square is commutative for all $n \geqslant 0$.

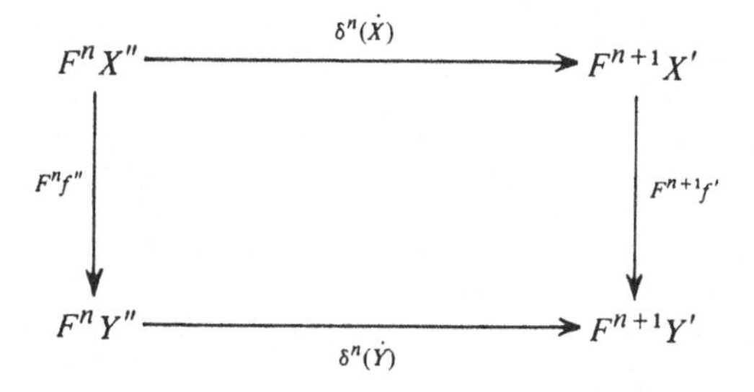

Note that this implies $\delta^n(\dot{X}) = \delta^n(\dot{Y})$ whenever $\dot{X}$ and $\dot{Y}$ represent the same class $\mathscr{A} \in \operatorname{Ext}(A'', A')$.

We denote such a connected right sequence of additive functors by $(F^n; \delta^n)_{n \geqslant 0}$. The sequence

$$0 \to F^0A' \to F^0A \to F^0A'' \xrightarrow{\delta^0(\dot{A})} F^1A' \to F^1A$$

$$\to F^1A'' \xrightarrow{\delta^1(\dot{A})} F^2A' \to \ldots$$

is a complex and natural with respect to the short exact sequence $\dot{A}$. If it is exact for all short exact sequences $\dot{A}$ we speak of an *exact connected sequence* of functors.

Dually a *connected left sequence* of additive functors from **A** to **B** is a sequence $(F_n; \delta_n)_{n\geq 0}$ such that the sequence $(F_n^\circ; \delta_n^\circ)_{n\geq 0}$ is a connected right sequence of additive functors from $\mathbf{A}^\circ$ to $\mathbf{B}^\circ$. Here $F_n^\circ = \Delta F_n \Delta$ and $\delta_n^\circ(\dot{A}) = \Delta\delta_n(\dot{A})\Delta$ (compare 1.2.4) so that $F_n^\circ: \mathbf{A}^\circ \to \mathbf{B}^\circ$ are again covariant functors.

Also contravariant functors may appear as left or right connected sequences. It is customary to consider a contravariant functor from **A** to **B** as a (covariant) functor from $\mathbf{A}^\circ$ to **B**.

The reader may formulate for himself directly (i) a left (or right) connected sequence of contravariant additive functors from **A** to **B**, and (ii) an exact left (or right) connected sequence of contravariant additive functors from **A** to **B**.

In the sequel we will consider right connected sequences of covariant functors unless otherwise stated, and usually refer to them just as connected sequences.

Now let $(F^n; \delta^n)_{n\geq 0}$ and $(G^n; \varepsilon^n)_{n\geq 0}$ be two connected sequences of functors from **A** to **B**. A sequence of morphisms $(\eta^n: F^n \to G^n)_{n\geq 0}$ is called a *morphism from* $(F^n; \delta^n)_{n\geq 0}$ *to* $(G^n; \varepsilon^n)_{n\geq 0}$ provided for every short exact sequence $\dot{A}: 0 \to A' \to A \to A'' \to 0$ in **A** and for every $n \geq 0$ the following square commutes:

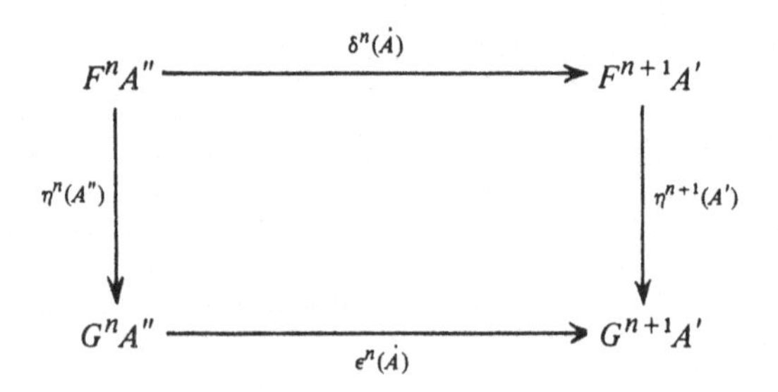


In this way the connected sequences $(F^n; \delta^n)_{n\geq 0}$ from **A** to **B** almost form a category: as in all categories of functors, the question is whether or not the class of morphisms between two connected sequences is a set.

**Exercise** (D. A. Buchsbaum). Prove that in the definition of a connected right sequence $(F^n; \delta^n)_{n\geq 0}$ it is enough to assume condition $(b)$. In other words, the naturality condition $(b)$ already implies the null sequence of $(a)$. Hint: use 3.1.17(iv) for $\mathrm{Ext}^1$.

**3.2.2 Definition** A connected sequence $(F^n; \delta^n)_{n\geq 0}$ is called *universal* provided for each connected sequence $(G^n; \delta^n)_{n\geq 0}$ and morphism of functors $\eta: F^0 \to G^0$ there is a unique morphism

$(\eta^n: F^n \to G^n)_{n\geq 0}$ with $\eta^0 = \eta$; i.e. $\eta$ can be uniquely extended to a morphism from $(F^n; \delta^n)_{n\geq 0}$ to $(G^n; \varepsilon^n)_{n\geq 0}$.

It follows easily that if both $(F^n; \delta^n)_{n\geq 0}$ and $(G^n; \varepsilon^n)_{n\geq 0}$ are universal connected sequences and if $\eta^0: F^0 \to G^0$ is an isomorphism, then the sequences are isomorphic.

Let $F: \mathbf{A} \to \mathbf{B}$ be an additive functor. If there is a universal connected sequence $(F^n; \delta^n)_{n\geq 0}$ with $F^0 = F$ this sequence is, according to the above, unique up to isomorphisms. Such a connected sequence is called a *sequence of right satellites of F.* Again we define left satellites for $F: \mathbf{A} \to \mathbf{B}$ by looking at the right satellites of $F^\circ: \mathbf{A}^\circ \to \mathbf{B}^\circ$.

In order to construct such a sequence of satellites for a given functor $F: \mathbf{A} \to \mathbf{B}$ (with some extra conditions imposed on the categories $\mathbf{A}$ and $\mathbf{B}$) we proceed as follows.

Define a category $\mathbf{A}^e$ in the following way. Objects are the pairs $\ulcorner A, m \urcorner$ with $A \in \mathbf{A}$ and $m$ a nonnegative integer. Morphisms are defined by $(\ulcorner A, m \urcorner, \ulcorner B, n \urcorner)_{\mathbf{A}^e} = \mathrm{Ext}^{n-m}(A, B)$ for $n \geq m$ and 0 for $n < m$. Similar remarks to those made in 3.1.19 apply here. Just as in 3.1.20 one proves that $\mathbf{A}^e$ is an additive category and using 3.1.24 that $\mathbf{A}^e$ is a genuine category if $\mathbf{A}$ has projectives or injectives. Note however that $\mathbf{A}^e$ is not the dual category of $\mathbf{A}_e$.

Next we define covariant additive functors $I^n: \mathbf{A} \to \mathbf{A}^e$ by $I^n A = \ulcorner A, n \urcorner$ and $I^n f = f$ for $A \in \mathbf{A}$ and $f \in (A, B)_{\mathbf{A}}$. Let $\mathscr{E} \in \mathrm{Ext}(A, B)$. If we consider $\mathscr{E}$ as an element of $(\ulcorner A, n \urcorner, \ulcorner B, n+1 \urcorner)_{\mathbf{A}^e}$ we will denote this by $\mathscr{E}(n)$.

Let $F^e: \mathbf{A}^e \to \mathbf{B}$ be an additive functor. Then $F^e I^n: \mathbf{A} \to \mathbf{B}$ is an additive functor which we will call $F^n$. Taking $\delta^n(\dot{E}) = F^e \mathscr{E}(n)$ we obtain a connected sequence of additive functors $(F^n; \delta^n)_{n\geq 0}$ from $\mathbf{A}$ to $\mathbf{B}$ as the reader may check for himself (use 3.1.6 and the previous exercise).

Suppose conversely that we have been given a connected sequence of additive functors $(F^n; \delta^n)_{n\geq 0}$ from $\mathbf{A}$ to $\mathbf{B}$. We will show that there is an additive functor $F^e: \mathbf{A}^e \to \mathbf{B}$ such that $F^n = F^e I^n$ with $\delta^n(\dot{E}) = F^e \mathscr{E}(n)$ for $\dot{E} \in \mathscr{E}$. The functor $F^e: \mathbf{A}^e \to \mathbf{B}$ is defined by $F^e \ulcorner A, n \urcorner = F^n A$ for objects and $F^e f = F^n f$ for morphisms of the type $\ulcorner A, n \urcorner \to \ulcorner B, n \urcorner$ in $\mathbf{A}^e$ (which is just a morphism $f: A \to B$ in $\mathbf{A}$). For morphisms in $(\ulcorner A, n \urcorner, \ulcorner B, n+1 \urcorner)_{\mathbf{A}^e} = \mathrm{Ext}(A, B)$ let $\mathscr{E} \in \mathrm{Ext}(A, B)$, $\dot{E} \in \mathscr{E}$ and define $F\mathscr{E}(n) = \delta^n(\dot{E})$ (we have already seen that $\delta^n(\dot{E})$ does not depend on the choice of the representative $\dot{E} \in \mathscr{E}$). Now for morphisms in $(\ulcorner A, m \urcorner, \ulcorner B, m+n \urcorner)_{\mathbf{A}^e} = \mathrm{Ext}^n(A, B)$ (assuming $n \geq 1$) let $\mathscr{E}^n \in \mathrm{Ext}^n(A, B)$. Consider $\mathscr{E}^n$ as the Yoneda composition of $n$ extensions each of length one. More precisely let $\dot{E}^n \in \mathscr{E}^n$ be

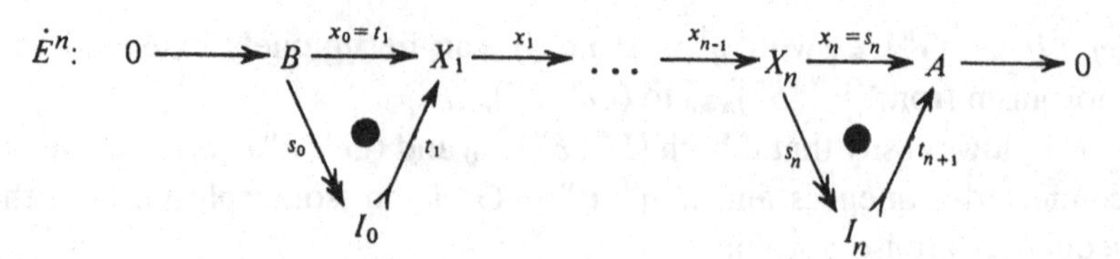


Then $\dot{E}^n = \dot{E}_1\dot{E}_2 \dots \dot{E}_n$ with

$$\dot{E}_i: 0 \to I_{i-1} \xrightarrow{t_i} X_i \xrightarrow{s_i} I_i \to 0 \qquad (1 \leq i \leq n),$$

with $I_0 = B$ and $I_n = A$. Consider $\mathscr{E}_i$ as an element of

$$(\ulcorner I_i, m+n-i \urcorner, \ulcorner I_{i-1}, m+n+1-i \urcorner)_{\mathbf{A}^\bullet} \qquad (1 \leq i \leq n).$$

Then for $\mathscr{E}_i$ the morphism $F\mathscr{E}_i$ has already been defined. So we define $F\mathscr{E}$ by $F\mathscr{E}_1 \circ \dots \circ F\mathscr{E}_n$ and we show that this definition is independent of the chosen representative of $\mathscr{E}^n$. Let therefore $\dot{E}'^n \in \mathscr{E}^n$, directly equivalent with $\dot{E}^n$

$$\begin{array}{lccccccccccccc}
\dot{E}^n: & 0 & \longrightarrow & B & \xrightarrow{x_0} & X_1 & \xrightarrow{x_1} & \dots & \xrightarrow{x_{n-1}} & X_n & \xrightarrow{x_n} & A & \longrightarrow & 0 \\
 & & & \| & & \downarrow p_1 & & & & \downarrow p_n & & \| & & \\
\dot{E}'^n: & 0 & \longrightarrow & B & \xrightarrow[x'_0]{} & X'_1 & \xrightarrow[x'_1]{} & \dots & \xrightarrow[x'_{n-1}]{} & X'_n & \xrightarrow[x'_n]{} & A & \longrightarrow & 0
\end{array}$$

Consider

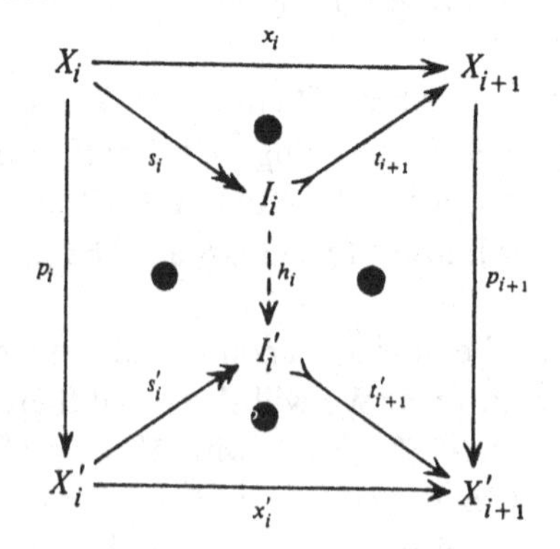


(see lemma 2.5.10).

So we have

$$\begin{array}{ccccccccc}
0 & \longrightarrow & I_{i-1} & \xrightarrow{t_i} & X_i & \xrightarrow{s_i} & I_i & \longrightarrow & 0 \\
 & & \downarrow h_{i-1} & & \downarrow p_i & & \downarrow h_i & & \\
0 & \longrightarrow & I'_{i-1} & \xrightarrow[t'_i]{} & X'_i & \xrightarrow[s'_i]{} & I'_i & \longrightarrow & 0
\end{array}$$

$(1\le i\le n)$, with $h_0=1_B$ and $h_n=1_A$. These diagrams yield the following commutative diagram:

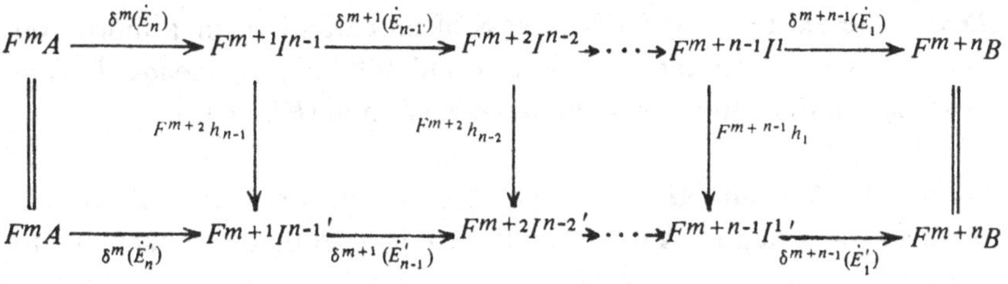

$$\begin{array}{ccccccccc} F^mA & \xrightarrow{\delta^m(\dot E_n)} & F^{m+1}I^{n-1} & \xrightarrow{\delta^{m+1}(\dot E_{n-1})} & F^{m+2}I^{n-2} & \to\dots\to & F^{m+n-1}I^1 & \xrightarrow{\delta^{m+n-1}(\dot E_1)} & F^{m+n}B \\ \| & & \downarrow{\scriptstyle F^{m+2}h_{n-1}} & & \downarrow{\scriptstyle F^{m+2}h_{n-2}} & & \downarrow{\scriptstyle F^{m+n-1}h_1} & & \| \\ F^mA & \xrightarrow[\delta^m(\dot E'_n)]{} & F^{m+1}I^{n-1\prime} & \xrightarrow[\delta^{m+1}(\dot E'_{n-1})]{} & F^{m+2}I^{n-2\prime} & \to\dots\to & F^{m+n-1}I^{1\prime} & \xrightarrow[\delta^{m+n-1}(\dot E'_1)]{} & F^{m+n}B \end{array}$$

From this diagram it follows that

$$\delta^{m+n-1}(\dot E_1)\dots\delta^{m+1}(\dot E_{n-1})\delta^m(\dot E_n)$$
$$=\delta^{m+n-1}(\dot E'_1)\dots\delta^{m+1}(\dot E'_{n-1})\delta^m(\dot E'_n).$$

Thus $F^e\mathscr{E}$ is well defined, since equivalence is just a finite chain of direct equivalences.

For $m<n$ we have $(\ulcorner A,m\urcorner,\ulcorner B,n\urcorner)_{\mathbf{A}^e}=0$. In this case we define $F^e0=0$.

This completes the definition of $F^e$ which can easily be checked to be a functor. It is also clear that $F^n=F^eI^n$.

We finally observe that $F^e$ is an additive functor. By 2.7.2 it is enough to show that $F^e$ preserves every biproduct $\ulcorner\ulcorner A\oplus A,n\urcorner,p_1,p_2,q_1,q_2\urcorner$. But this biproduct is $I^n$ applied to the biproduct $\ulcorner A\oplus A,p_1,p_2,q_1,q_2\urcorner$ in $\mathbf{A}$. Additivity of $F^n=F^eI^n$ then shows that $F^e$ preserves the biproducts $\ulcorner A\oplus A,n\urcorner$.

If $\eta^e: F^e\to G^e$ is a morphism between additive functors from $\mathbf{A}^e$ to $\mathbf{B}$ then $\eta^n=\eta^eI^n: F^eI^n\to G^eI^n$ defines a morphism $(\eta^n)_{n\ge0}: (F^n;\delta^n)_{n\ge0}\to(G^n;\varepsilon^n)_{n\ge0}$ between connected sequences of functors from $\mathbf{A}$ to $\mathbf{B}$, as is easily seen. It follows readily that if $\eta^e$ and $\zeta^e: F^e\to G^e$ yield the same morphisms $(\eta^n)_{n\ge0}=(\zeta^n)_{n\ge0}$ then $\eta^e=\zeta^e$. Indeed, $\eta^e\ulcorner A,n\urcorner=\eta^eI^nA=\eta^nA=\zeta^nA=\zeta^e\ulcorner A,n\urcorner$. Conversely, given such a morphism $(\eta^n)_{n\ge0}$ we can define a morphism $\eta^e: F^e\to G^e$. Applying theorem 1.7.11(iii) we obtain

**3.2.3 Proposition** There is an equivalence between the 'category' of additive functors from $\mathbf{A}^e$ to $\mathbf{B}$ and the 'category' of connected sequences of additive functors from $\mathbf{A}$ to $\mathbf{B}$.

**3.2.4 Corollary** A connected sequence of additive functors $(F^n;\delta^n)_{n\ge0}$ from $\mathbf{A}$ to $\mathbf{B}$ is a sequence of satellites of $F^0=F$ if and only if for every additive functor $G^e:\mathbf{A}^e\to\mathbf{B}$ the mapping from $(F^e,G^e)_{\mathbf{A}^e}$

to $(F, G)_{\mathbf{A}}$, where $G = G^{e}I$, given by $\eta^{e} \to \eta^{e}I^{0}$ is an isomorphism of abelian groups.

*Proof.* The existence and uniqueness of an extension for a morphism $F \to G$ to one between $(F^{n}; \delta^{n})_{n \geq 0}$ and $(G^{n}; \varepsilon^{n})_{n \geq 0}$ means, by the proposition, the finding of a unique morphism in $(F^{e}, G^{e})_{\mathbf{A}^{e}}$.

**3.2.5 Theorem** For every additive functor $F: \mathbf{A} \to \mathbf{B}$ from a small abelian category $\mathbf{A}$ to an additive category $\mathbf{B}$ that is right complete there exists a sequence of satellites.

*Proof.* The functor $T: (\mathbf{A}^{e}, \mathbf{B}) \to (\mathbf{A}, \mathbf{B})$ defined by $TG^{e} = G^{e}I^{0}$ has a left adjoint $S$ (Kan extension along $I^{0} = I$: see theorem 1.8.6; since $\mathbf{A}$ is small $\mathbf{A}^{e}$ is also small). since $I^{0}: \mathbf{A} \to \mathbf{A}^{e}$ is a full embedding we may assume that for $F \in (\mathbf{A}, \mathbf{B})$ the equality $(SF)I^{0} = F$ holds (see remark after theorem 1.8.6). We will now show that if $F: \mathbf{A} \to \mathbf{B}$ is an additive functor, the Kan extension of $F$ along $I^{0}$, by which we mean $SF$, is also additive. To see this we go over the construction as given in theorem 1.8.6 once more and translate everything in terms of $\mathbf{A}$ and $\mathbf{A}^{e}$. Suppose $A^{e}$ and $A^{e'}$ are objects of $\mathbf{A}^{e}$, $f_1$ and $f_2$ morphisms from $A^{e}$ to $A^{e'}$. We have to show that $SF(f_1 + f_2) = SFf_1 + SFf_2$.

For $SFf_1$ we have

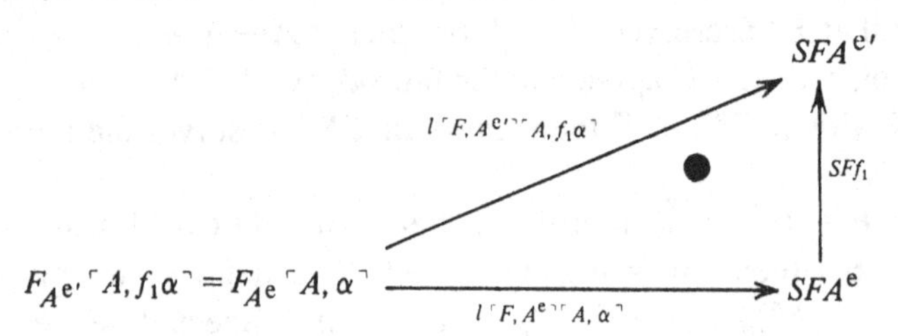


for all $\ulcorner A, \alpha \urcorner$ with $\alpha: IA = \ulcorner A, 0 \urcorner \to A^{e}$. Similarly for $SFf_2$ and $SF(f_1 + f_2)$.

Furthermore we have the following three commuting diagrams in the category $\mathbf{A}^{e}$ (see diagrams on top of page 165).

This yields in the category $\mathbf{H}_{A^{e'}}$ the morphisms

$$\begin{pmatrix}1\\1\end{pmatrix}: \ulcorner A, f_1\alpha + f_2\alpha \urcorner \to \ulcorner A \oplus A, (f_1\alpha \ f_2\alpha) \urcorner$$

$$\begin{pmatrix}1\\0\end{pmatrix}: \ulcorner A, f_1\alpha \urcorner \to \ulcorner A \oplus A, (f_1\alpha \ f_2\alpha) \urcorner$$

$$\begin{pmatrix}0\\1\end{pmatrix}: \ulcorner A, f_2\alpha \urcorner \to \ulcorner A \oplus A, (f_1\alpha \ f_2\alpha) \urcorner$$

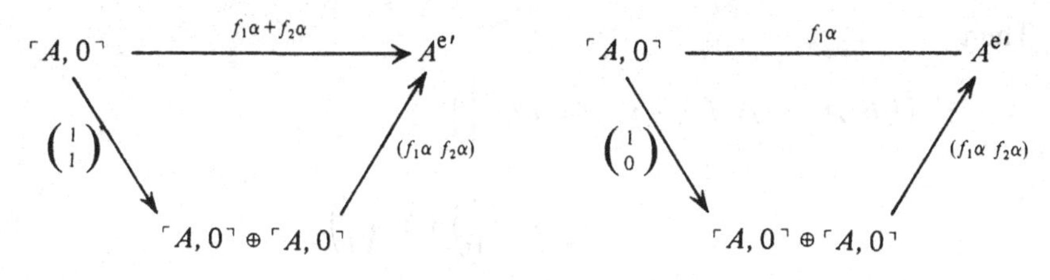


$$\begin{array}{ccc} \ulcorner A, 0 \urcorner & \xrightarrow{f_2\alpha} & A^{e'} \\ {\scriptstyle \binom{0}{1}} \searrow & & \nearrow {\scriptstyle (f_1\alpha\ f_2\alpha)} \\ & \ulcorner A, 0 \urcorner \oplus \ulcorner A, 0 \urcorner & \end{array}$$

Note that $I\binom{1}{1} = \binom{1}{1}$, $I\binom{1}{0} = \binom{1}{0}$ and $I\binom{0}{1} = \binom{0}{1}$. So we obtain

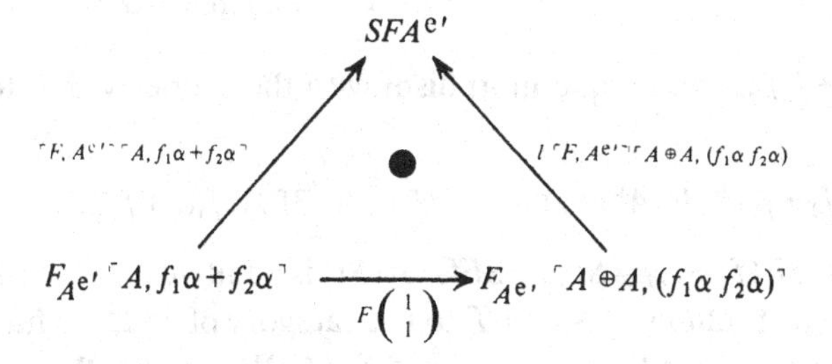


and

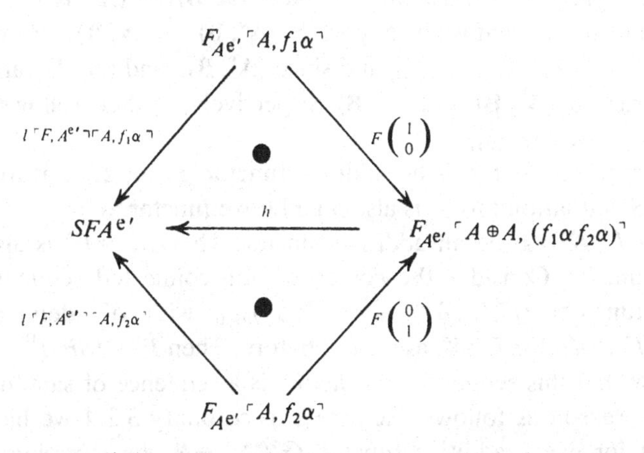


Denote $l \ulcorner F, A^{e'} \urcorner \ulcorner A \oplus A, (f_1\alpha\ f_2\alpha) \urcorner$ by $h$. Since $F$ is additive we have

$$F\binom{1}{1} = F\binom{1}{0} + F\binom{0}{1}.$$

Then

$$l\ulcorner F, A^{e'}\urcorner\ulcorner A, f_1\alpha + f_2\alpha\urcorner = hF\binom{1}{1}$$

$$= hF\binom{1}{0} + hF\binom{0}{1}$$

$$= l\ulcorner F, A^{e'}\urcorner\ulcorner A, f_1\alpha\urcorner + l\ulcorner F, A^{e'}\urcorner\ulcorner A, f_2\alpha\urcorner.$$

Consequently we have

$$\begin{aligned}((SF)f_1 + (SF)f_2)l\ulcorner F, A^{e}\urcorner\ulcorner A, \alpha\urcorner &= (SF)(f_1)l\ulcorner F, A^{e}\urcorner\ulcorner A, \alpha\urcorner + (SF)(f_2)l\ulcorner F, A^{e}\urcorner\ulcorner A, \alpha\urcorner \\ &= l\ulcorner F, A^{e'}\urcorner\ulcorner A, f_1\alpha\urcorner + l\ulcorner F, A^{e'}\urcorner\ulcorner A, f_2\alpha\urcorner \\ &= l\ulcorner F, A^{e'}\urcorner\ulcorner A, f_1\alpha + f_2\alpha\urcorner.\end{aligned}$$

Since $SF(f_1+f_2)$ is the unique morphism with the property that for all $\ulcorner A, \alpha\urcorner$

$$SF(f_1+f_2)l\ulcorner F, A^{e}\urcorner\ulcorner A, \alpha\urcorner = l\ulcorner F, A^{e'}\urcorner\ulcorner A, f_1\alpha + f_2\alpha\urcorner$$

it follows that $SF(f_1+f_2) = SFf_1 + SFf_2$. So $SF$ is additive if $F$ is additive. Denoting the restrictions of $S$ and $T$ to the category of additive functors by the same symbols we thus have proved $S: (\mathbf{A}, \mathbf{B})_+ \to (\mathbf{A}^{e}, \mathbf{B})_+$. Since $I$ is additive it is clear that we also have $T: (\mathbf{A}^{e}, \mathbf{B})_+ \to (\mathbf{A}, \mathbf{B})_+$. From the fact that $(S\text{-}, \text{-})_{(\mathbf{A}^{e},\mathbf{B})} \simeq (\text{-}, T\text{-})_{(\mathbf{A},\mathbf{B})}$ and since $(\mathbf{A}^{e}, \mathbf{B})_+$ and $(\mathbf{A}, \mathbf{B})_+$ are full subcategories of $(\mathbf{A}^{e}, \mathbf{B})$ and $(\mathbf{A}, \mathbf{B})$ respectively, it then follows that $(S\text{-}, \text{-})_{(\mathbf{A}^{e},\mathbf{B})_+} \simeq (\text{-}, T\text{-})_{(\mathbf{A},\mathbf{B})_+}$.

It is easy to see that $T$ is an additive functor. From 2.7.3 it follows then that $S$, left adjoint to $T$, is also an additive functor.

Now let $F: \mathbf{A} \to \mathbf{B}$ be an additive functor. Then $SF = F^{e}$ is also an additive functor. Consider the corresponding connected sequence of additive functors $(F^{e}I^n; \delta^n)_{n\geq 0} = (F^n; \delta^n)_{n\geq 0}$ with $\delta^n$ defined by $\delta^n(\dot{E}) = SF(\mathscr{E}(n))$ for $\dot{E} \in \mathscr{E}$, as we did before. Then $F^0 = (SF)I^0 = F$.

To show that this sequence of functors is a sequence of satellites of $F^0 = F$ we reason as follows. Referring to corollary 3.2.4 we have to show that for every additive functor $G^{e}: \mathbf{A}^{e} \to \mathbf{B}$ the mapping from $(F^{e}, G^{e})_{(\mathbf{A}^{e},\mathbf{B})_+}$ to $(F^{e}I^0, G^{e}I^0)_{(\mathbf{A},\mathbf{B})_+}$ given by $\eta^{e} \mapsto \eta^{e}I^0$ is an isomorphism of abelian groups. But $(SF, G^{e})_{(\mathbf{A}^{e},\mathbf{B})_+} \rightrightarrows (F, TG^{e})_{(\mathbf{A},\mathbf{B})_+}$ is given by $\eta^{e} \mapsto \eta^{e}I^0$ according to the remark concluding 1.8.6.

*Proof.* Suppose **C** has injectives and let $C$ be an object of **C**. Then there is an injective object $Q^0$ in **C** with a monomorphism $C \rightarrowtail Q^0$. Suppose therefore that an exact sequence $0 \to C \to Q^0 \to \ldots \to Q^n$ has already been constructed with injective objects $Q^i$ $(i \leq n)$ and let $Q^n \to R$ be the cokernel of the last morphism. Then choose an injective object $Q^{n+1}$ and a monomorphism $R \rightarrowtail Q^{n+1}$. Attaching the composition $Q^n \twoheadrightarrow R \rightarrowtail Q^{n+1}$ to the tail of the existing sequence we see that $0 \to C \to Q^0 \to \ldots \to Q^n \to Q^{n+1}$ is exact with injective objects $Q^i$ $(i \leq n+1)$. Thus $C$ has an injective resolution by induction. The converse is trivial.

We shall often denote such a resolution by $\ulcorner C, Q^+ \urcorner$. It should be distinguished from the complex

$$Q^+ : 0 \to Q^0 \to Q^1 \to Q^2 \to \ldots$$

In the sequel we will often be dealing with the category $\mathbf{C}^+$, the full subcategory of $\hat{\mathbf{C}}$ consisting of all right complexes, or $\mathbf{C}^-$ of left complexes.

**Definition** Let **C** be an exact and additive category. Two morphisms $\hat{f}$ and $\hat{g}: \hat{C} \to \hat{D}$ in $\hat{\mathbf{C}}$ are called *chain homotopic* provided there exists a sequence of morphisms $(s^{i+1}: C^{i+1} \to D^i)_{i \in \mathbf{Z}}$

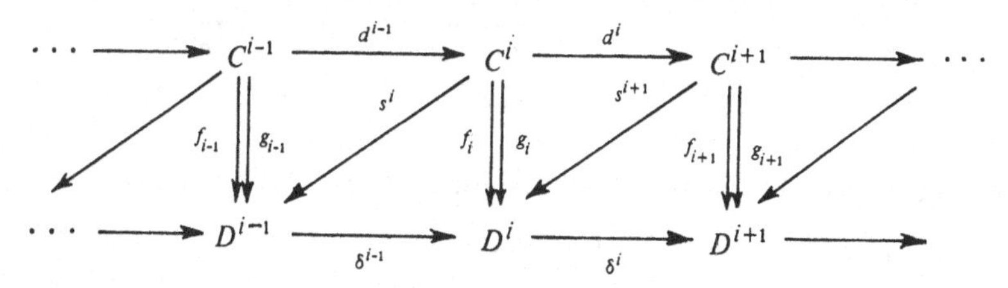


such that $f^i - g^i = s^{i+1}d^i + \delta^{i-1}s^i$ for all $i \in \mathbb{Z}$.

This yields an equivalence relation between the morphisms from $\hat{C}$ to $\hat{D}$ and it is easy to see that if $\hat{f} \sim \hat{f}'$ and $\hat{g} \sim \hat{g}'$ then $(\hat{f} + \hat{g}) \sim (\hat{f}' + \hat{g}')$. Therefore $\hat{f} \sim \hat{g}$ is equivalent to $\hat{f} - \hat{g} \sim \hat{0}$. If $T: \mathbf{C} \to \mathbf{B}$ is an additive functor and $\hat{f} \sim \hat{g}: \hat{C} \to \hat{D}$, then $\hat{T}\hat{f} \sim \hat{T}\hat{g}: \hat{T}\hat{C} \to \hat{T}\hat{D}$ in $\hat{\mathbf{B}}$, since additive functors obviously preserve chain homotopies.

If $\hat{1}_{\hat{C}} \sim \hat{0}_{\hat{C}}: \hat{C} \to \hat{C}$ through a chain homotopy $(s_i)_{i \in \mathbf{Z}}$, the latter is called a *contracting homotopy* and the complex $\hat{C}$ is called *contractible.* All this applies equally well to the full subcategory $\mathbf{C}^+ \subset \hat{\mathbf{C}}$, or $\mathbf{C}^-$ for that matter.

**Exercises** $(a)$ If $\hat{f} \sim \hat{g}: \hat{C} \to \hat{D}$ and $\hat{h} \sim \hat{k}: \hat{D} \to \hat{E}$, then $\hat{h}\hat{f} \sim \hat{k}\hat{f} \sim \hat{k}\hat{g}: \hat{C} \to \hat{E}$.

(*b*) A contractible complex is exact, but not conversely.

**3.3.3 Theorem** Let **C** be an exact and additive category. Suppose $0 \to C \xrightarrow{d^{-1}} C^0 \xrightarrow{d^0} C^1 \to \ldots$ is an exact complex under $C$ and $0 \to D \xrightarrow{\delta^{-1}} Q^0 \xrightarrow{\delta^0} Q^1 \to \ldots$ is an injective complex under $D$. Let $f: C \to D$ be a given morphism in **C**. Then there exists a morphism $f^+: C^+ \to Q^+$ which extends $f$, i.e. with $f^{i+1}d^i = \delta^i f^i$, $i \geqslant -1$, where $f^{-1} = f$. If $g^+$ is another extension of $f$ the morphisms $f^+$ and $g^+$ are chain homotopic. Furthermore $\ulcorner f, f^{+}\urcorner$ and $\ulcorner f, g^{+}\urcorner$ are chain homotopic.

Caution. One should be aware of the distinction between the complexes $C^+$ and $\ulcorner C, C^{+}\urcorner$ and the morphisms out of them ($f^+$ and $\ulcorner f, f^{+}\urcorner$ respectively).

*Proof.* (i) We first write down the epic–monic factorization of both complexes:

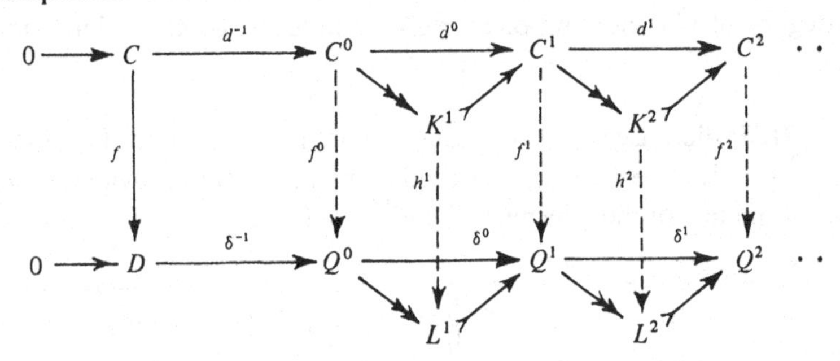

Putting $C = K^0$ and $D = L^0$ we have for $i \geqslant 0$ the short exact sequences

$$0 \to K^i \to C^i \to K^{i+1} \to 0$$

and

$$0 \to L^i \to Q^i \to L^{i+1} \to 0.$$

Now suppose the morphism $h^i: K^i \to L^i$ is given:

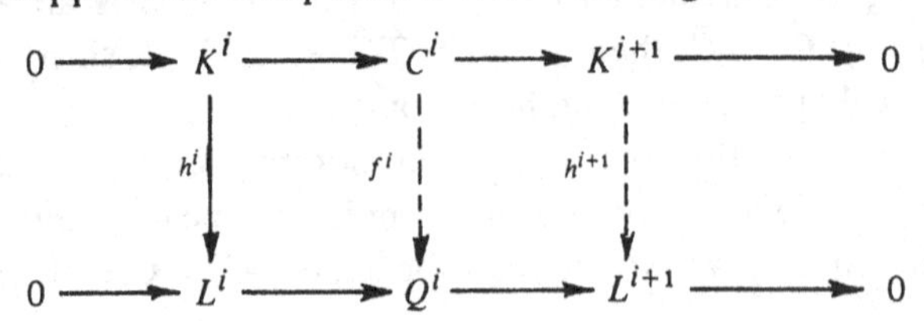

Due to the injectivity of the object $Q^i$ there exists a morphism $f^i: C^i \to Q^i$ making the left square commutative. Since both sequences are exact there exists a unique $h^{i+1}: K^{i+1} \to L^{i+1}$ making the right square commutative. The desired extension of $f$ is constructed by induction in this way, starting with $h^0 = f$. It is clear that the squares are commutative.

(ii) If $f^+$ and $g^+$ are two extensions of $f: C \to D$ the morphism $k^+ = f^+ - g^+$ is an extension of $0: C \to D$. It is therefore sufficient to show that an extension $k^+: C^+ \to Q^+$ of $0: C \to D$ is always chain homotopic to $0^+: C^+ \to Q^+$, i.e. $\ulcorner 0, k^+ \urcorner$ is chain homotopic to $\ulcorner 0, 0^+ \urcorner$. In order to prove this we first establish a lemma.

(iii) Suppose that in the following diagram the upper row is exact and the object $Q^n$ is injective. Furthermore suppose $\delta^{n-1}\delta^{n-2} = 0$, $\delta^{n-1}k^{n-1} = k^n d^{n-1}$ and $k^{n-1} = s^n d^{n-1} + \delta^{n-2}s^{n-1}$. Then there is a morphism $s^{n+1}: F^{n+1} \to Q^n$ with $k^n = s^{n+1}d^n + \delta^{n-1}s^n$:

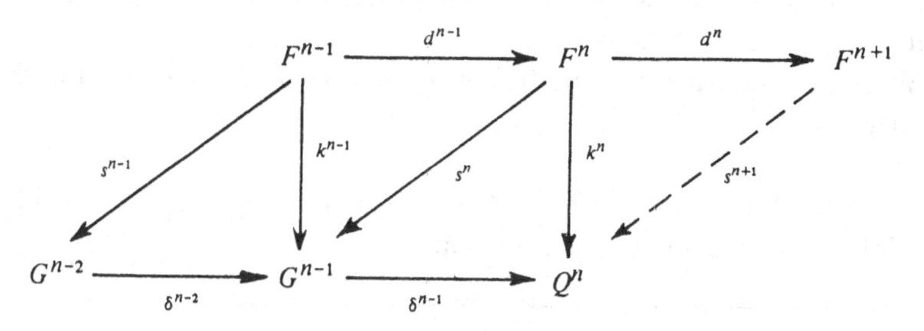


*Proof.* Put $l = k^n - \delta^{n-1}s^n: F^n \to Q^n$. Then

$$
\begin{aligned}
ld^{n-1} &= k^n d^{n-1} - \delta^{n-1}s^n d^{n-1} \\
&= \delta^{n-1}(k^{n-1} - s^n d^{n-1}) \\
&= \delta^{n-1}\delta^{n-2}s^{n-1} \\
&= 0: F^{n-1} \to Q^n.
\end{aligned}
$$

Since $F^{n-1} \to F^n \to F^{n+1}$ is exact and since we have the epic–monic factorization $F^n \twoheadrightarrow K \rightarrowtail F^{n+1}$, the row in this diagram is exact:

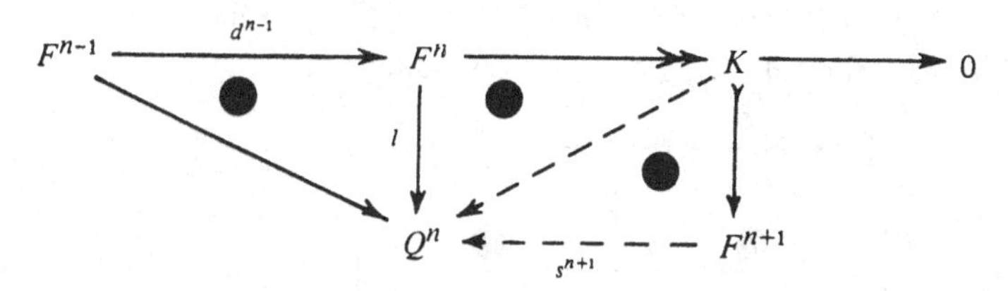


The fact that $K$ is the cokernel of $d^{n-1}$ and $ld^{n-1} = 0$ implies that $l$ factors uniquely through $K$. From $K \rightarrowtail F^{n+1}$ follows the existence of a morphism $s^{n+1}: F^{n+1} \to Q^n$ with $s^{n+1}d^n = l = k^n - \delta^{n-1}s^n$, which proves the lemma.

(iv) The following diagram satisfies the conditions of the lemma:

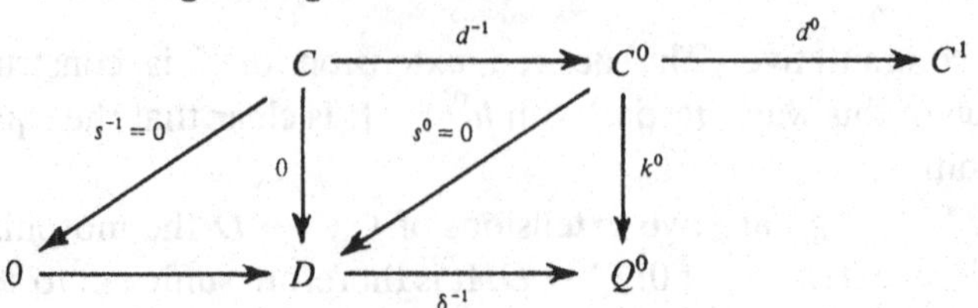


Thus $s^1: C^1 \to Q^0$ exists such that $k^0 = s^1 d^0 + \delta^{-1} s^0$ (starting point for the induction). Then we construct, using the lemma, inductively $s^{i+1}: C^{i+1} \to Q^i$ for all $i \geqslant 0$, satisfying $k^i = s^{i+1} d^i + \delta^{i-1} s^i$ for $i \geqslant 0$. This means that the morphisms $k^+$ and $0^+: C^+ \to Q^+$ and also the morphisms $\ulcorner 0, k^{+}\urcorner$ and $\ulcorner 0, 0^{+}\urcorner: \ulcorner C, C^{+}\urcorner \to \ulcorner D, Q^{+}\urcorner$ are chain homotopic (see definition).

We now come to the homology of complexes and their morphisms. Let **C** be an exact category.

We leave the proof of the following lemma as an exercise for the reader.

**3.3.4 Lemma** Let $C' \overset{d'}{\to} C \overset{d}{\to} C''$ be a null sequence. Consider the following epic–monic factorizations:

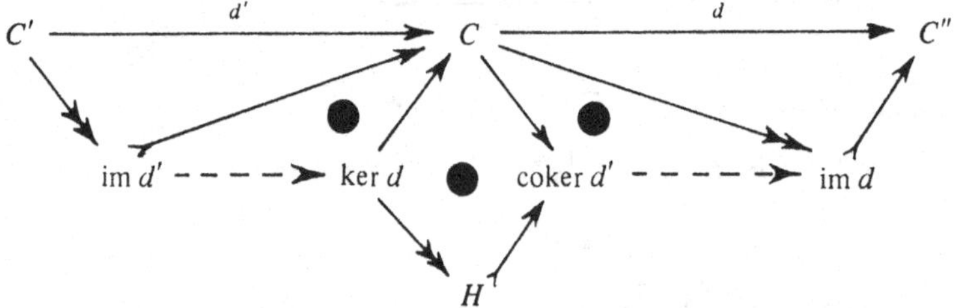


Then the following hold.

(*a*) There are unique morphisms $\operatorname{im} d' \to \ker d$ and $\operatorname{coker} d' \to \operatorname{im} d$ making the triangles commutative.

(*b*) The sequences $0 \to \operatorname{im} d' \to \ker d \to H \to 0$ and $0 \to H \to \operatorname{coker} d' \to \operatorname{im} d \to 0$ are exact.

(*c*) $H$ equals 0 if and only if $C' \to C \to C''$ is exact.

Consider the following commutative diagram

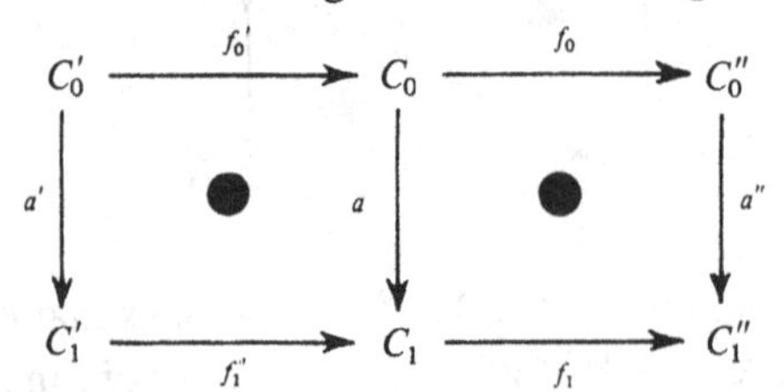


in which both rows are null sequences. Apply the construction of the lemma to both of these null sequences. Let us call the lemma-diagram obtained by using the top or bottom row $I_0$ and $I_1$ respectively. Between the corresponding objects of diagram $I_0$ and diagram $I_1$ one finds uniquely determined morphisms making everything commute. So one finds a unique $h: H_0 \to H_1$.

Now let $\hat{C} = (C^i)_{i\in\mathbf{Z}}$ be a complex. For $n \in \mathbb{Z}$ we write $C' = C^{n-1}$, $C = C^n$ and $C'' = C^{n+1}$. Then we construct the object $H$ as in our last lemma. We call this object $H$ the $n^{\text{th}}$ *homology of the complex* $(C^i)_{i\in\mathbf{Z}}$ and we denote it by $H^n(C^i)_{i\in\mathbf{Z}}$ or by $H^n\hat{C}$. For a morphism $(f^i)_{i\in\mathbf{Z}}: (C^i)_{i\in\mathbf{Z}} \to (D^i)_{i\in\mathbf{Z}}$ also denoted by $\hat{f}: \hat{C} \to \hat{D}$ we have $H^n\hat{f}: H^n\hat{C} \to H^n\hat{D}$. One verifies readily that each $H^n$ is a functor from $\hat{\mathbf{C}}$ to $\mathbf{C}$. If $\mathbf{C}$ is additive, and hence $\hat{\mathbf{C}}$ also, it follows immediately that $H^n: \hat{\mathbf{C}} \to \mathbf{C}$ is an additive functor. Most of the time we will be dealing with the subcategory of right complexes $\mathbf{C}^+$.

**3.3.5 Lemma** Let $\hat{C}$ be a complex in the exact category $\mathbf{C}$. Denote by $\ker d^n\hat{C}$ and $\operatorname{coker} d^{n-1}\hat{C}$ the objects $\ker d$ and $\operatorname{coker} d'$ from lemma 3.3.4 in the case that $C = C^n$. Then the following sequence is exact in $\mathbf{C}$:

$$0 \to H^n\hat{C} \to \operatorname{coker} d^{n-1}\hat{C} \to \ker d^n\hat{C} \to H^{n+1}\hat{C} \to 0.$$

*Proof.* Join the exact sequences in 3.3.4($b$) for $C = C^n$ and $C = C^{n+1}$.

Let $\hat{C}: 0 \to \hat{C}' \to \hat{C} \to \hat{C}'' \to 0$ be a short exact sequence of complexes in $\hat{\mathbf{C}}$. Then we may construct a connecting morphism $\delta^n(\hat{C}): H^n\hat{C}'' \to H^{n+1}\hat{C}'$ in the following way. First consider the following diagram

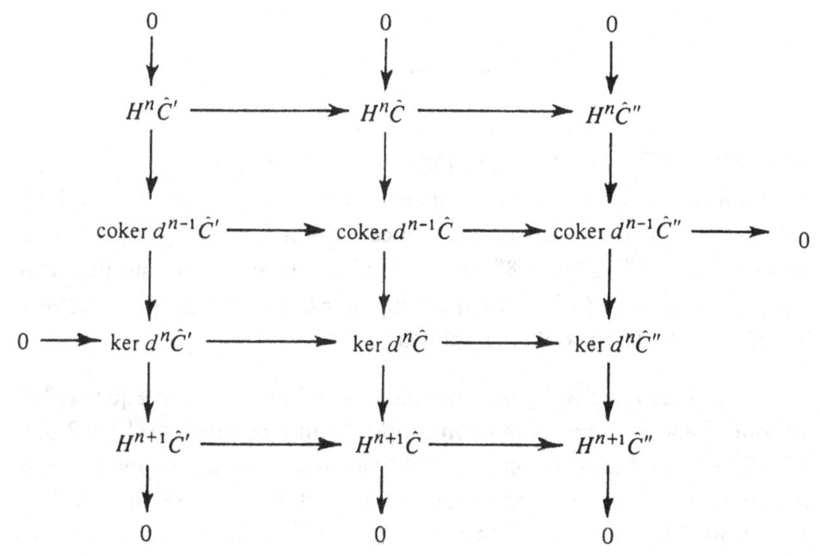

in which the two middle rows are exact (corollary to lemma 2.6.9, the ker–coker sequence).

Then the existence of the morphism $\delta^n(\dot{\hat{C}})\colon H^n\hat{C}'' \to H^{n+1}\hat{C}'$ follows from the snake-lemma (see 2.6.9). Since all our constructions are natural it is easy to verify condition (ii)(*b*) of 3.2.1. Thus we have

**3.3.6 Theorem** The sequence of functors $(H^n;\delta^n)_{n\geq 0}$ is an exact connected right sequence of functors from $\mathbf{C}^+$ to $\mathbf{C}$.

**3.3.7 Theorem** Let $\mathbf{C}$ be an exact and additive category and let $C^+$ and $D^+$ be objects in $\mathbf{C}^+$. Suppose $f^+, g^+\colon C^+ \to D^+$ are chain homotopic. Then $H^nf^+ = H^ng^+\colon H^nC^+ \to H^nD^+$.

*Proof.* Since $H^n$ is an additive functor and since $h^+ = f^+ - g^+\colon C^+ \to D^+$ is chain homotopic with $0^+\colon C^+ \to D^+$ it is sufficient to prove that $H^nh^+ = 0^+\colon H^nC^+ \to H^nD^+$. Let us suppose that the homotopy of $h^+$ with $0^+$ is achieved by $(s^i\colon C^i \to D^{i-1})_{i\geq 0}$, assuming that $D_{-1} = 0$.

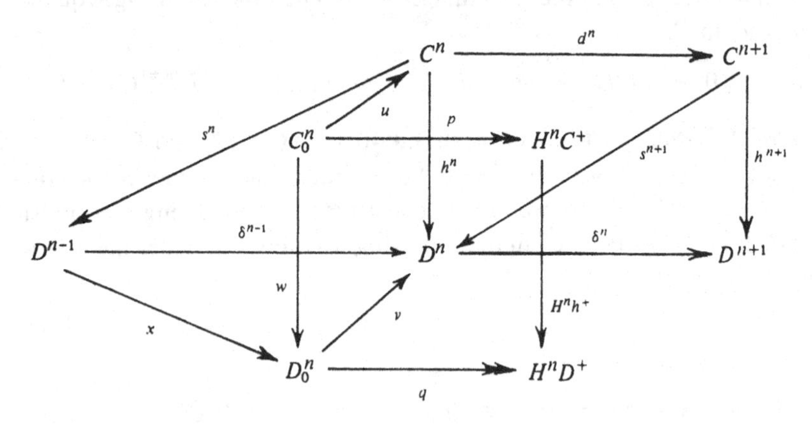


Let $u\colon C_0^n \to C^n = \ker d^n$ and $v\colon D_0^n \to D^n = \ker \delta^n$. Let $w\colon C_0^n \to D_0^n$ be the unique morphism such that $vw = h^nu$, and $x$ be the unique factorization of $\delta^{n-1}$ through $v$. Since $h^n = s^{n+1}d^n + \delta^{n-1}s^n$ we find $vw = (s^{n+1}d^n + \delta^{n-1}s^n)u = \delta^{n-1}s^nu = vxs^nu$. Since $v$ is monic, this implies $w = xs^nu$ and $H^nh^+p = qw = qxs^nu = 0$. Since $p$ is epic it follows that $H^nh^+ = 0$. The proof also applies to $\hat{\mathbf{C}}$.

**Exercise** Show by an example in $\mathbf{Ab}$ that $H^nf^+$ can equal $H^ng^+$ without $f^+$ and $g^+$ being chain homotopic. Compare exercise (*b*) in 3.3.2.

Now let $\mathbf{C}$ and $\mathbf{B}$ be exact and additive categories and suppose $\mathbf{C}$ has injectives. Let $T\colon \mathbf{C} \to \mathbf{B}$ be an additive functor. We take an injective resolution $\ulcorner C, Q^{+}\urcorner$ of $C \in \mathbf{C}$: $0 \to C \to Q^0 \to Q^1 \to \ldots$ With this sequence we associate the null sequence $TQ^+: 0 \to TQ^0 \to TQ^1 \to TQ^2 \to \ldots$ of which we consider the homology $H^nTQ^+$. We want to associate with $T: \mathbf{C} \to \mathbf{B}$ a sequence of functors $H^n: \mathbf{C} \to \mathbf{B}$.

To this purpose we consider the category $\ulcorner \mathbf{C}, \mathbf{Q}^+ \urcorner$ whose objects are the injective resolutions $\ulcorner C, Q^+ \urcorner: 0 \to C \to Q^0 \to Q^1 \to \ldots$ of $C \in \mathbf{C}$ and whose morphisms are $\ulcorner f, f^+ \urcorner: \ulcorner X, Q^+ \urcorner \to \ulcorner Y, R^+ \urcorner$ with $f: X \to Y$ in $\mathbf{C}$ and $f^+ = (f^n: Q^n \to R^n)_{n \geq 0}$ in $\mathbf{C}^+$. By choosing for every object $C \in \mathbf{C}$ an injective resolution $\ulcorner C, Q^+ \urcorner$ (which is possible by assumption) and by choosing for every $f: X \to Y$ in $\mathbf{C}$ an extension $\ulcorner f, f^+ \urcorner: \ulcorner X, Q^+ \urcorner \to \ulcorner Y, R^+ \urcorner$ (made possible by theorem 3.3.3) we obtain an additive functor $J: \mathbf{C} \to \ulcorner \mathbf{C}, \mathrm{Q}^+ \urcorner$. With the aid of $T: \mathbf{C} \to \mathbf{B}$ we now construct additive functors $F^n: \ulcorner \mathbf{C}, \mathbf{Q}^+ \urcorner \to \mathbf{B}$ for $n \geq 0$. For any object $\ulcorner C, Q^+ \urcorner$ in $\ulcorner \mathbf{C}, \mathbf{Q}^+ \urcorner$ we define $F^n \ulcorner C, Q^+ \urcorner = H^nTQ^+$. One can easily check that $F^n$ is an additive functor from $\ulcorner \mathbf{C}, \mathbf{Q}^+ \urcorner$ to $\mathbf{B}$. The point of theorems 3.3.3 and 3.3.7 is that our procedure is unambiguous.

**Exercise** Prove that the functor $F^nJ: \mathbf{C} \to \mathbf{B}$ is unique up to isomorphism; i.e. if $J': \mathbf{C} \to \ulcorner \mathbf{C}, \mathbf{Q}^+ \urcorner$ is obtained in the manner described above, but possibly by making other choices for the injective resolutions and extensions of morphisms $f: X \to Y$ in $\mathbf{C}$ to the chosen injective resolutions, the resulting functors $F^nJ'$ are isomorphic to the functors $F^nJ$.

A more sophisticated approach is to modify the category $\ulcorner \mathbf{C}, \mathbf{Q}^+ \urcorner$ by taking as objects injective resolutions of objects in $\mathbf{C}$ and as morphisms homotopy classes of morphisms extending the same morphism $f: X \to Y$ in $\mathbf{C}$. One then obtains a category equivalent to the category $\mathbf{C}$; see 1.7.11.

Still another possibility is to take as morphisms for the category $\ulcorner \mathbf{C}, \mathbf{Q}^+ \urcorner$ just the morphisms $f: X \to Y$ in $\mathbf{C}$. Two different injective resolutions for the same object $C \in \mathbf{C}$ would then be isomorphic in this category which also is equivalent to $\mathbf{C}$.

Whichever variant one chooses, this justifies the following important definition.

**3.3.8 Definition** Let $\mathbf{C}$ and $\mathbf{B}$ be exact and additive categories and suppose $\mathbf{C}$ has injectives. Let $T: \mathbf{C} \to \mathbf{B}$ be an additive functor. The functors $R^nT = F^nJ: \mathbf{C} \to \mathbf{B}$ for $n \geq 0$, for $F^n$: $\ulcorner \mathbf{C}, \mathbf{Q}^+ \urcorner \to \mathbf{B}$ and $J: \mathbf{C} \to \ulcorner \mathbf{C}, \mathbf{Q}^+ \urcorner$ as defined above, are called the *right derived functors of* $T$. These functors are additive. Note that it is apparently enough to know the values of $T$ on injective objects of $\mathbf{C}$ and morphisms between them in order to define and study its derived functors $R^nT$.

Remarks. Again suppose that **C** and **B** are exact and additive categories and let **C** have injectives. Suppose $T: \mathbf{C} \to \mathbf{B}$ is additive. The following assertions hold.

(i) If $T$ is left exact then $R^0T \simeq T$.

(ii) If $T$ is exact then $R^nT = 0$ for $n \geqslant 1$.

(iii) If $Q$ is injective then $(R^0T)Q \simeq TQ$ and $(R^nT)Q = 0$ for $n \geqslant 1$.

Now suppose **C** and **B** are again exact and additive categories while **C** has projectives. Suppose $T: \mathbf{C} \to \mathbf{B}$ is additive. Consider the dual categories $\mathbf{C}^\circ$ and $\mathbf{B}^\circ$ and the functor $T^\circ: \mathbf{C}^\circ \to \mathbf{B}^\circ$. We may now construct the functors $R^nT^\circ: \mathbf{C}^\circ \to \mathbf{B}^\circ$ and then $(R^nT^\circ)^\circ: \mathbf{C} \to \mathbf{B}$. Thus we define $L_nT: \mathbf{C} \to \mathbf{B}$ by $L_nT = (R^nT^\circ)^\circ$ for $n \geqslant 0$ and we call these functors $L_nT$ the *left derived functors of T*. The construction of $L_nTC$ amounts to the following. First consider any projective resolution of $C: \ldots \to P_2 \to P_1 \to P_0 \to C \to 0$. Then consider the null sequence $TP_2 \to TP_1 \to TP_0 \to 0$ and for each triple $TP_{n+1} \to TP_n \to TP_{n-1}$ determine the homology according to 3.3.4.

Further, if $T: \mathbf{C} \to \mathbf{B}$ is a contravariant functor then we consider $T: \mathbf{C}^\circ \to \mathbf{B}$, being a covariant functor. A projective resolution in **C** then yields an injective resolution in $\mathbf{C}^\circ$. Thus if **C** has projectives we may construct the right derived functors $R^nT: \mathbf{C}^\circ \to \mathbf{B}$. In the same way we may construct the left derived functors $L_nT: \mathbf{C}^\circ \to \mathbf{B}$ provided **C** has injectives. The reader may adapt (i) to (iii) to these cases.

## 3.4 Satellites and derived functors

**3.4.1 Definition** An additive functor $F: \mathbf{C} \to \mathbf{B}$ between exact and additive categories is called *effaceable* provided for every object $C \in \mathbf{C}$ there exists a monomorphism $m: C \rightarrowtail C'$ in **C** such that $Fm = 0$. The morphism $m$ is called an *effacement* of the object $C$ for $F$.

**3.4.2 Proposition** Suppose $F: \mathbf{C} \to \mathbf{B}$ is an additive functor. Then the following hold.

(i) If $F$ is effaceable then $FQ = 0$ for every injective $Q$ in **C**.

(ii) If **C** has injectives and $FQ = 0$ for every injective $Q$ in **C** then $F$ is effaceable.

*Proof.* (i) Suppose $F$ is effaceable and $Q$ is injective in **C**. Then there is a monomorphism $m: Q \rightarrowtail D$ in **C** with $Fm = 0$. Due to injectivity there is a morphism $n: D \to Q$ with $nm = 1_Q$. Hence $1_{FQ} = F(nm) = F_n \circ F_m = 0$ so $FQ = 0$.

(ii) Suppose **C** has injectives and suppose $FQ = 0$ for every injective $Q$

in **C**. For any given $C \in \mathbf{C}$, there is a monomorphism $m: C \rightarrowtail Q$ with $Q$ injective. Now $FQ = 0$; hence $Fm = 0$. So $m$ is an effacement of $C$.

**Corollary** Let **C** and **B** be exact and additive categories while **C** has injectives. Suppose $T: \mathbf{C} \to \mathbf{B}$ is additive. Then $R^nT: \mathbf{C} \to \mathbf{B}$ is effaceable for $n \geq 1$ (see 3.3.8 remark (iii)).

**3.4.3 Theorem** Let **A** be abelian and **B** exact and additive. Suppose $(S^n; \delta^n)_{n \geq 0}$ is an exact connected sequence of functors from **A** to **B** with $S^n$ effaceable for $n \geq 1$. Then $(S^n; \delta^n)_{n \geq 0}$ is a universal connected sequence of functors.

*Proof.* Before proving this result we want to remark that the requirement that $(S^n; \delta^n)_{n \geq 0}$ is an exact connected sequence is more than we need. As one will notice from the proof it suffices that for every short exact sequence $0 \to A \to B \to C \to 0$ the complex

$$0 \to S^0A \to S^0B \to S^0C \xrightarrow{\delta^0} S^1A \to S^1B \to S^1C \xrightarrow{\delta^1} S^2A \to \dots$$

be exact at $S^iC$ and $S^{i+1}A$ for all $i \geq 0$. We first establish three lemmas.

**Lemma (i)** Let $0 \to A \to B' \to C' \to 0$ and $0 \to A \to B'' \to C'' \to 0$ be exact in **A**. Then there is a commutative diagram in **A** (in which the last column is a sum diagram) with the middle row exact:

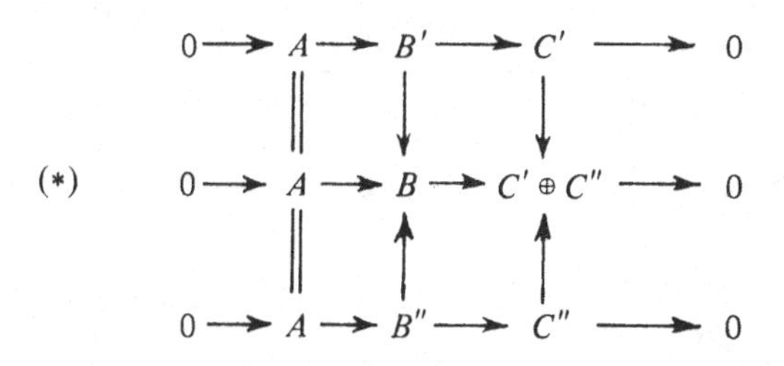

*Proof.* If we call the top extension $\dot{E}'$ and the bottom one $\dot{E}''$, the middle extension is just $(1, 1)(\dot{E}' \oplus \dot{E}'')$. Compare 3.1.19.

**Lemma (ii)** If the following commutative diagram has exact rows

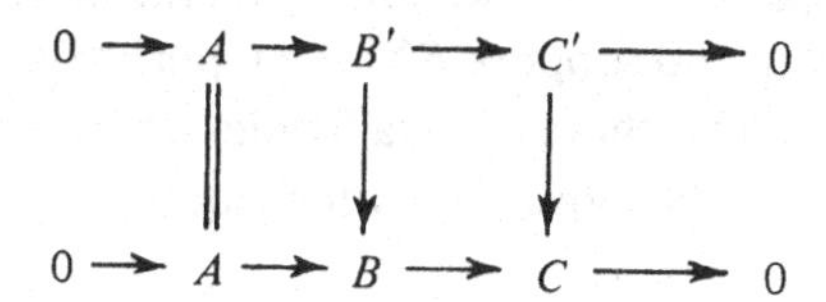

and if $(S^n; \delta^n)_{n\geq 0}$ and $(T^n; \varepsilon^n)_{n\geq 0}$ are connected sequences of functors from **A** to **B**, if furthermore for the morphism of functors $\phi^k: S^k \to T^k$ the following square in **B** is commutative for $k \geq 0$

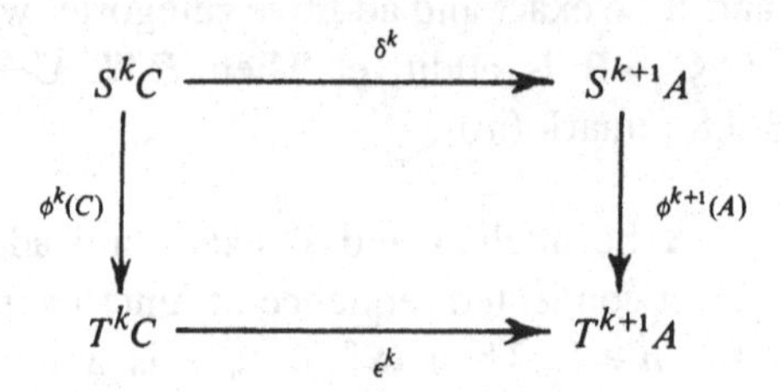

then the next square is also commutative.

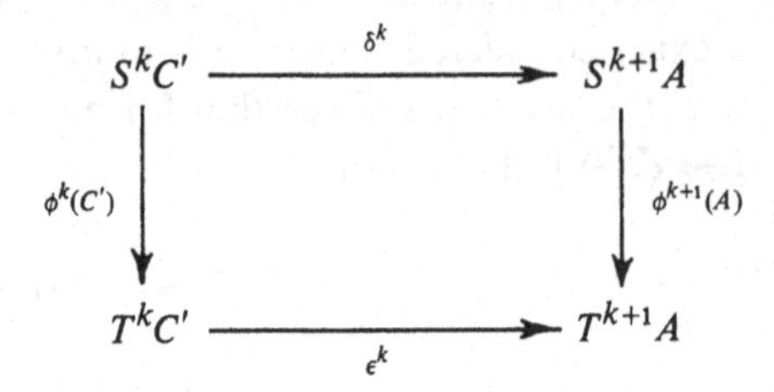

*Proof.* The assertion follows directly from the commutativity of the following diagram:

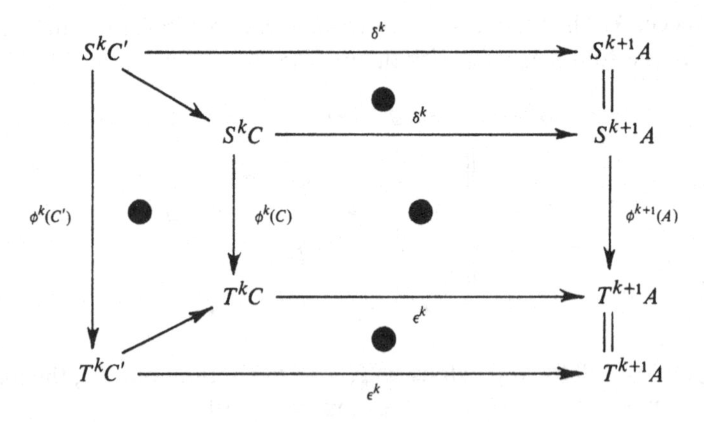

**Lemma (iii)** Suppose $0 \to A \xrightarrow{q} B \to C \to 0$ is short exact in **A** and $(S^n; \delta^n)_{n\geq 0}$ and $(T^n; \varepsilon^n)_{n\geq 0}$ are connected sequences of functors from **A** to **B** such that $(S^n; \delta^n)_{n\geq 0}$ is exact. Additionally let the morphism of functors $\phi^k: S^k \to T^k$ be given and suppose $S^{k+1}q = 0$ for $k \geq 0$. Then $\delta^k: S^kC \to S^{k+1}A$ is epic and there is a unique morphism $\phi^{k+1}(A): S^{k+1}A \to T^{k+1}A$ such that the following square is commutative:

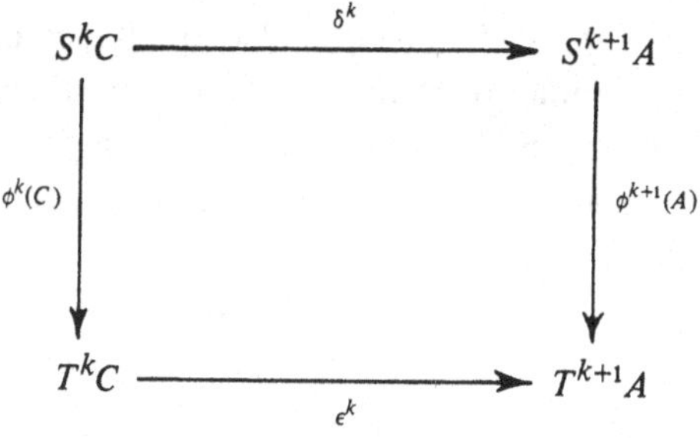


*Proof.* Since $(S^n; \delta^n)_{n \geq 0}$ is exact and since $S^{k+1}q = 0$ the following sequence is exact: $S^kB \to S^kC \xrightarrow{\delta^k} S^{k+1}A \xrightarrow{0} S^{k+1}B$ (see remark at the beginning of the proof).

It follows that $\delta^k: S^kC \to S^{k+1}A$ is epic and hence equals coker $(S^kB \to S^kC)$. This implies that in the following diagram there is a unique morphism $\phi^{k+1}(A)$ making the diagram commute:

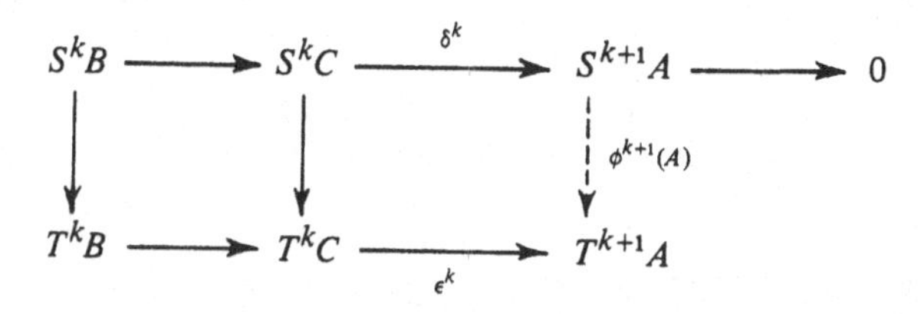


**Proof of theorem 3.4.3** Assume we have already constructed a morphism of functors $\phi^k: S^k \to T^k$. For every object $A \in \mathbf{A}$ there is an effacement $q': A \rightarrowtail B'$ for $S^{k+1}$ (i.e. $S^{k+1}q' = 0$). Consider the short exact sequence $0 \to A \xrightarrow{q'} B' \to C' \to 0$ to which lemma (iii) is applicable. So there is a unique $\phi_0^{k+1}(A): S^{k+1}A \to T^{k+1}A$ such that the following square commutes:

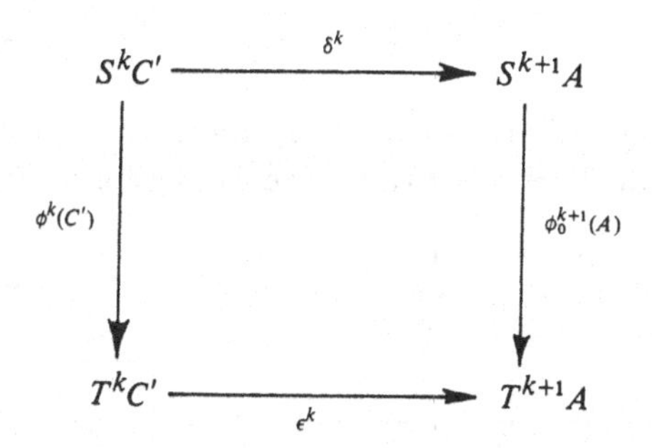


We have to show that $\phi_0^{k+1}(A)$ fulfils the requirements for an arbitrary short exact sequence starting with $A$. Let $0 \to A \to B'' \to C'' \to 0$ be any such short exact sequence. Consider the diagram (*) in lemma (i). Denote the morphisms $A \to B'$ and $A \to B$ by $q'$ and $q$ respectively. Since

$S^{k+1}q' = 0$ we also have $S^{k+1}q = 0$. Hence $q$ is also an effacement of $A$ for $S^{k+1}$. So we may apply lemma (iii) to the middle row of diagram (*). This yields the existence and uniqueness of a morphism $\phi^{k+1}(A)\colon S^{k+1}A \to T^{k+1}A$ such that the following diagram commutes:

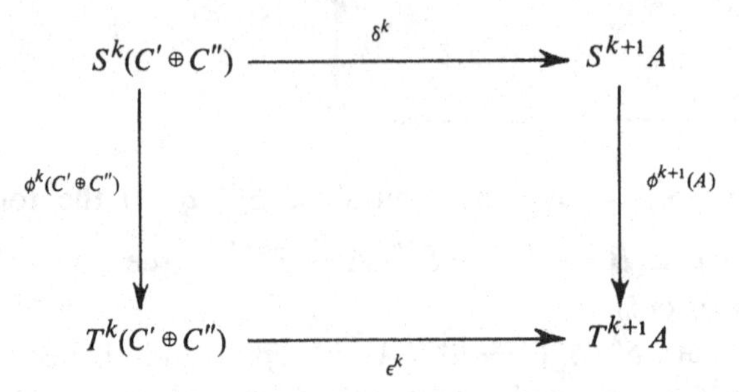


Using lemma (ii) for the upper half of diagram (*) one finds $\phi^{k+1}(A) = \phi_0^{k+1}(A)$. Using lemma (ii) for the lower half of diagram (*) one finds

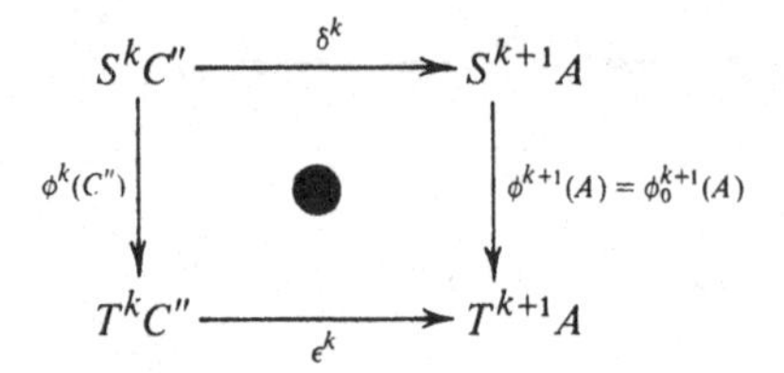


Thus $\phi^{k+1}(A)$ fulfils all the requirements provided we can also show that it is natural in $A$. Let therefore $g$ be any morphism from $A$ to $D$ in $\mathbf{A}$. Take any effacement $q\colon A \rightarrowtail B$ for $S^{k+1}$ and construct the cocartesian square

$$\begin{array}{ccc} A & \overset{q}{\rightarrowtail} & B \\ {\scriptstyle g}\downarrow & & \downarrow \\ D & \longrightarrow & E \end{array}$$

in which we know that $D \to E$ is monic. Then there exists a unique commutative diagram with short exact rows (see exercise after 2.4.1°):

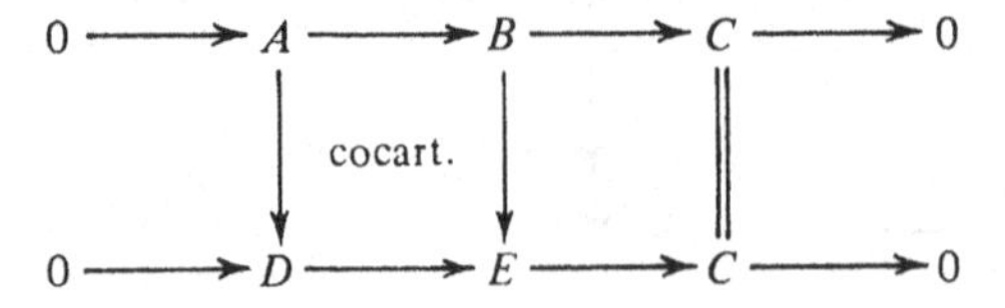


Then consider the diagram on top of page 183.

The triangles I and II are both commutative since $(S^n; \delta^n)_{n \geq 0}$ and $(T^n; \varepsilon^n)_{n \geq 0}$ are connected sequences of functors. The squares III and IV are commutative due to the construction of $\phi^{k+1}(A)$ and $\phi^{k+1}(D)$. From

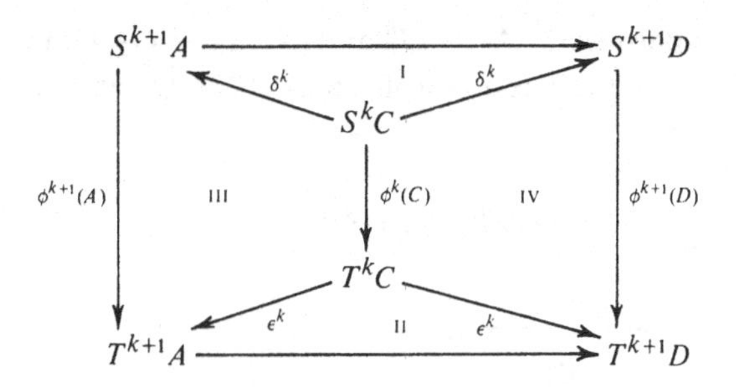


lemma (iii) it follows that $\delta^k: S^kC \to S^{k+1}A$ is epic ($A \rightarrowtail B$ is an effacement for $S^{k+1}$). This yields the commutativity of the big square and so $\phi^{k+1}: S^{k+1} \to T^{k+1}$ is a morphism of functors.

The proof of theorem 3.4.3 is now completed by induction, starting with a given morphism of functors $\phi^0: S^0 \to T^0$.

Our next goal is to prove the existence of an exact sequence of right satellites for a given left exact functor. We first develop the necessary preliminary lemmas. Let $\mathbf{A}$ be an abelian category with injectives and let $\mathbf{B}$ be an exact and additive category. Finally let $\dot{\mathbf{A}}$ be the category whose objects are the short exact sequences $\dot{A}: 0 \to A_0 \to A_1 \to A_2 \to 0$ in $\mathbf{A}$ and whose morphisms are

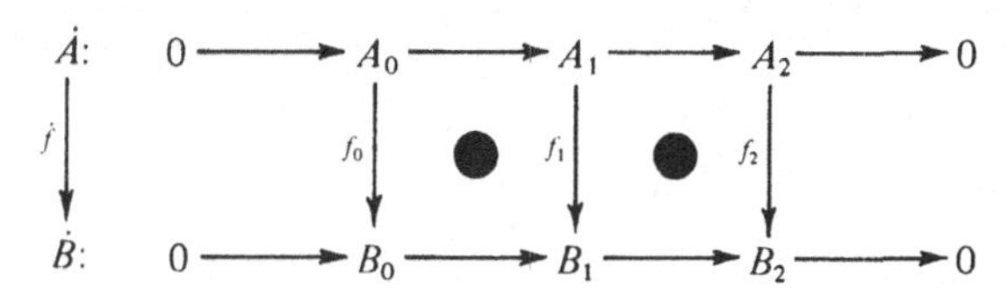


**Exercise** Prove that $\dot{f}: \dot{A} \to \dot{B}$ is monic (epic) in $\dot{\mathbf{A}}$ if and only if $f_1: A_1 \to B_1$ is monic (epic) in $\mathbf{A}$. If $f_1: A_1 \to B_1$ is an isomorphism then $\dot{f}: \dot{A} \to \dot{B}$ is a bimorphism. Prove that $\dot{\mathbf{A}}$ is not balanced (definition 2.5.3), and that $\dot{\mathbf{A}}$ is additive but not exact (unless $\mathbf{A}$ is trivial).

A sequence $0 \to \dot{A} \to \dot{Q}^0 \to \dot{Q}^1 \to \dot{Q}^2 \to \ldots$ in $\dot{\mathbf{A}}$ is called an *injective resolution of the object* $\dot{A}$ *in* $\dot{\mathbf{A}}$ if the sequences $0 \to A_i \to Q_i^0 \to Q_i^1 \to Q_i^2 \to \ldots$ are injective resolutions in $\mathbf{A}$ ($i = 0, 1, 2$). We will use the notation $\ulcorner \dot{A}, \dot{Q}^+ \urcorner$ for this resolution. It follows directly from the injectivity of the $Q_0^n$ that the exact sequences $0 \to Q_0^n \to Q_1^n \to Q_2^n \to 0$ split for $n \geq 0$.

**3.4.4 Lemma** Every object $\dot{A}: 0 \to A_0 \to A_1 \to A_2 \to 0$ in $\dot{\mathbf{A}}$ has an injective resolution.

*Proof.* Consider the object $\dot{A}^n: 0 \to A_0^n \xrightarrow{a_{01}} A_1^n \xrightarrow{a_{12}} A_2^n \to 0$ in $\dot{\mathbf{A}}$.

There are monomorphisms $a_0^n: A_0^n \to Q_0^n$ and $a_2^n: A_2^n \to Q_2^n$ with $Q_0^n$ and $Q_2^n$ injective in $\mathbf{A}$. Then there is a morphism $k: A_1^n \to Q_0^n$ with $k \circ a_{01} = a_0^n$. Define

$$a_1^n = \begin{pmatrix} k \\ a_2^n \circ a_{12} \end{pmatrix}: A_n^1 \to Q_0^n \oplus Q_2^n = Q_1^n:$$

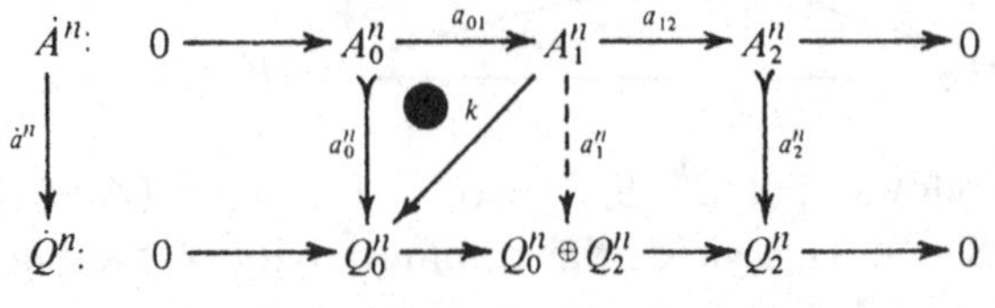

$$\begin{array}{ccccccccccc}
\dot{A}^n: & 0 & \longrightarrow & A_0^n & \xrightarrow{a_{01}} & A_1^n & \xrightarrow{a_{12}} & A_2^n & \longrightarrow & 0 \\
\downarrow{\scriptstyle \dot{a}^n} & & & \downarrow{\scriptstyle a_0^n} & \swarrow{\scriptstyle k} & \downarrow{\scriptstyle a_1^n} & & \downarrow{\scriptstyle a_2^n} & & \\
\dot{Q}^n: & 0 & \longrightarrow & Q_0^n & \longrightarrow & Q_0^n \oplus Q_2^n & \longrightarrow & Q_2^n & \longrightarrow & 0
\end{array}$$

Verify that both squares commute. Since the kernels of the morphisms $a_i^n$ form an exact sequence, it follows that $a_1^n$ is a monomorphism.

Then we may construct, with the aid of the nine-lemma (see 2.6.6), a short exact sequence

$$\dot{A}^{n+1}: 0 \to A_0^{n+1} \to A_1^{n+1} \to A_2^{n+1} \to 0$$

with $\dot{b}^n: \dot{Q}^n \to \dot{A}^{n+1}$ in $\dot{\mathbf{A}}$ (take cokernels) for which

$$0 \to A_i^n \overset{a_i^n}{\to} Q_i^n \overset{b_i^n}{\to} A_i^{n+1} \to 0$$

is short exact as well ($i = 0, 1, 2$).

Finally the injective resolution

$$0 \to \dot{A} \xrightarrow{\dot{d}^{-1}} \dot{Q}^0 \overset{\dot{d}^0}{\to} \dot{Q}^1 \overset{\dot{d}^1}{\to} \dot{Q}^2 \to \ldots$$

we are looking for is constructed by induction, starting with $\dot{A}^0 = \dot{A}$ and putting $\dot{a}^0 = \dot{d}^{-1}: \dot{A}^0 \to \dot{Q}^0$ and $\dot{a}^{n+1}\dot{b}^n: \dot{Q}^n \to \dot{Q}^{n+1}$ for $n \geq 0$.

**3.4.5 Lemma** If the objects $\dot{A}$ and $\dot{B}$ of $\dot{\mathbf{A}}$ have injective resolutions $\ulcorner \dot{A}, \dot{Q}^+ \urcorner$ and $\ulcorner \dot{B}, \dot{R}^+ \urcorner$ then, for every $\dot{f}: \dot{A} \to \dot{B}$, there is an extension $(\dot{f}^n: \dot{Q}^n \to \dot{R}^n)_{n \geq 0}$ i.e. $\ulcorner \dot{f}, \dot{f}+ \urcorner: \ulcorner \dot{A}, \dot{Q}^+ \urcorner \to \ulcorner \dot{B}, \dot{R}^+ \urcorner$.

*Proof.* We first factor the injective resolution $\ulcorner \dot{A}, \dot{Q}^+ \urcorner$ in epic–monic fashion as follows, reversing our procedure in the previous lemma (see diagram on top of page 185).

It follows easily (2.6.6, nine-lemma) that all sequences $0 \to A_0^n \to A_1^n \to A_2^n \to 0$ are short exact (for $n \geq 0$) and hence are elements of $\dot{\mathbf{A}}$. The same holds for the injective resolution $\ulcorner B, R^+ \urcorner$. Then we have the following situation in $\dot{\mathbf{A}}$ (see diagram in centre of page 185).

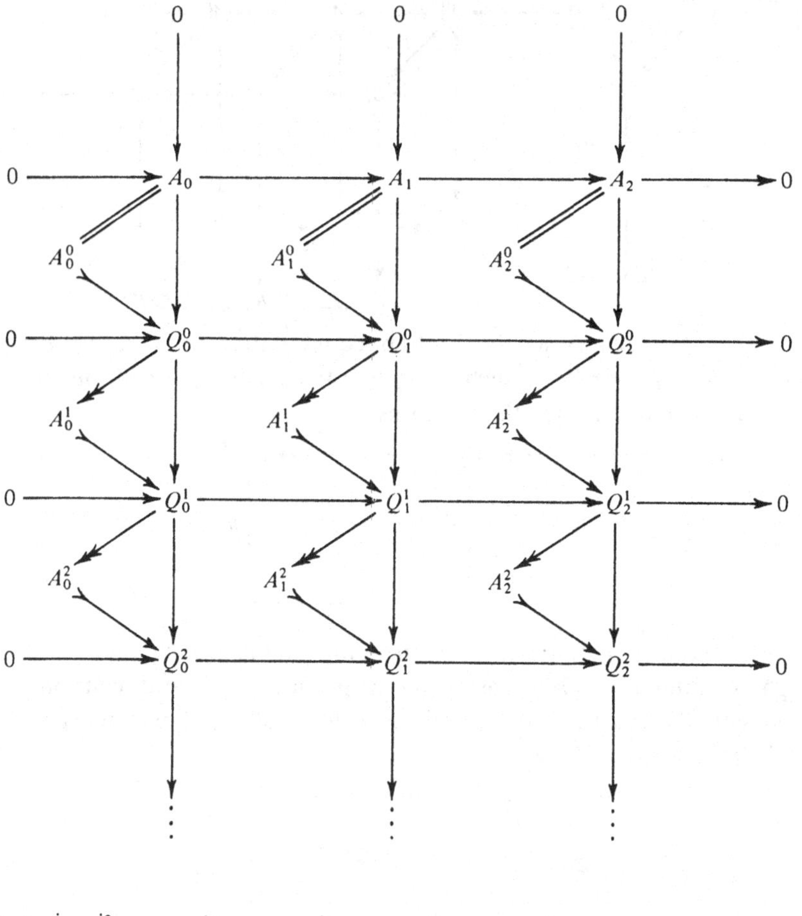


$$
\begin{array}{ccccccccccccccc}
0 \to \dot{A} = \dot{A}^0 & \longrightarrow & \dot{Q}^0 & \longrightarrow & \dot{A}^1 & \longrightarrow & \dot{Q}^1 & \longrightarrow & \dot{A}^2 & \longrightarrow & \dot{Q}^2 & \longrightarrow & \cdots \\
\dot{f}\downarrow \quad \downarrow \dot{g}^0 & & \downarrow \dot{f}^0 & & \downarrow \dot{g}^1 & & \downarrow \dot{f}^1 & & \downarrow \dot{g}^2 & & \downarrow \dot{f}^2 & & \\
0 \to \dot{B} = \dot{B}^0 & \longrightarrow & \dot{R}^0 & \longrightarrow & \dot{B}^1 & \longrightarrow & \dot{R}^1 & \longrightarrow & \dot{B}^2 & \longrightarrow & \dot{R}^2 & \longrightarrow & \cdots
\end{array}
$$

where we are looking for the morphisms $f^i$ and $g^i$ $(i \geq 0)$. We will do this by induction. Suppose $g^n: \dot{A}^n \to \dot{B}^n$ has been constructed. We then have the following situation (recall that the exact sequences $\dot{Q}^n$ and $\dot{R}^n$ are split): (See diagram on top of page 186).
Due to the injectivity of $R_0^n$ and $R_2^n$ there are morphisms $f_0^n: Q_0^n \to R_0^n$ and $f_2^n: Q_2^n \to R_2^n$ making the vertical squares on both sides commutative.

Then one may verify that the morphism

$$f_1^n = f_0^n \oplus f_2^n : Q_1^n = Q_0^n \oplus Q_2^n \ \to\ R_0^n \oplus R_2^n = R_1^n$$

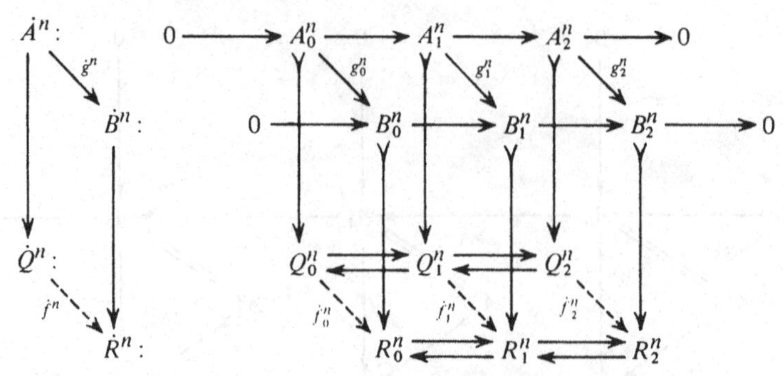

makes both squares at the bottom commutative so that we have obtained $\dot{f}^n: \dot{Q}^n \to \dot{R}^n$ in such a way that the single square at the left is commutative in $\dot{\mathbf{A}}$. Then the diagrams

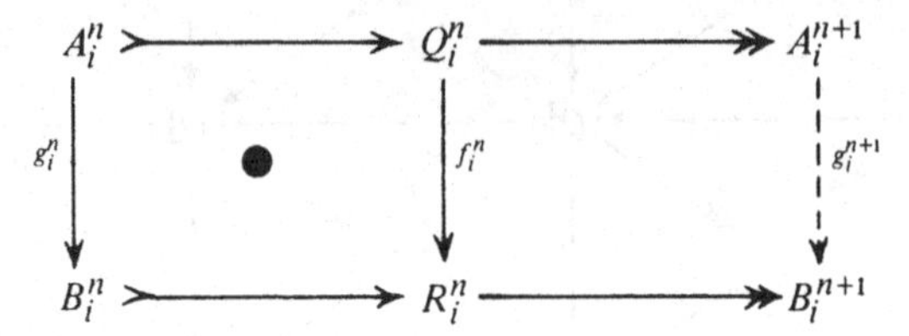

yield unique morphisms $g_i^{n+1}$ $(i = 0, 1, 2)$ making the squares on the right commutative. By chasing the appropriate (three-dimensional!) diagram, it is not hard to ascertain that one really obtains a morphism $\dot{g}^{n+1}: \dot{A}^{n+1} \to \dot{B}^{n+1}$ in $\dot{\mathbf{A}}$:

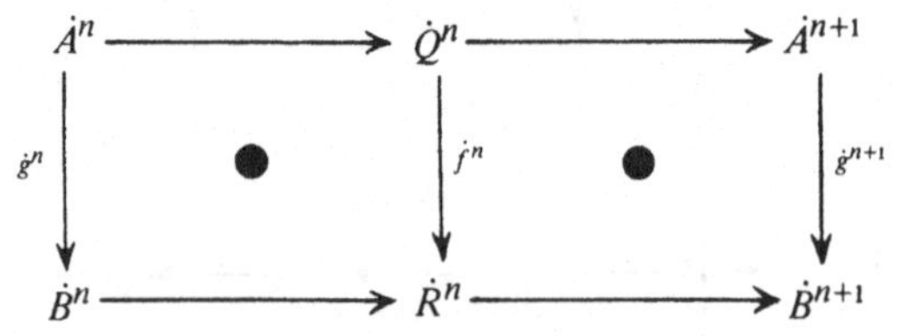

In the proof of this, one uses that the $Q_i^n \to A_i^{n+1}$ are epic.

The extension $\dot{f}^n: (\dot{Q}^n \to \dot{R}^n)_{n \geq 0}$ is now constructed by induction starting with $g^0 = f$.

Let $\mathbf{A}$ be an abelian category with injectives and $\mathbf{B}$ an exact and additive category. We denote by $\ulcorner \dot{\mathbf{A}}, \dot{\mathbf{Q}}^+ \urcorner$ the category of all injective resolutions of objects in $\dot{\mathbf{A}}$ (short exact sequences in $\mathbf{A}$). In this category the morphsims are $\ulcorner \dot{f}, \dot{f}^+ \urcorner: \ulcorner \dot{X}, \dot{Q}^+ \urcorner \to \ulcorner \dot{Y}, \dot{R}^+ \urcorner$ where $f: \dot{X} \to \dot{Y}$ and $f^+ = (\dot{f}^n: \dot{Q}^n \to \dot{R}^n)_{n \geq 0}$. By choosing an injective resolution for every object $\dot{A}$ in $\dot{\mathbf{A}}$ (which is possible by lemma 3.4.4) we find a functor $\dot{J}: \dot{\mathbf{A}} \to \ulcorner \dot{\mathbf{A}}, \dot{\mathbf{Q}}^+ \urcorner$ since any morphism $\dot{f}: \dot{X} \to \dot{Y}$ has an extension $\ulcorner \dot{f}, \dot{f}^+ \urcorner: \ulcorner \dot{X}, \dot{Q}^+ \urcorner \to \ulcorner \dot{Y}, \dot{R}^+ \urcorner$ by lemma 3.4.5. To prove the following theorem we will construct additive functors $\dot{F}^n: \ulcorner \dot{\mathbf{A}}, \dot{\mathbf{Q}}^+ \urcorner \to \dot{\mathbf{B}}$ to the category of all short exact sequences in $\mathbf{B}$.

**3.4.6 Theorem** Let **A** be an abelian category with injectives and **B** an exact and additive category. Suppose $T: \mathbf{A} \to \mathbf{B}$ is an additive functor. Then there is a sequence of morphisms $(\delta^n)_{n \geq 0}$ such that for every short exact sequence $\dot{A}: 0 \to A' \to A \to A'' \to 0$ in **A** the morphism $\delta^n: R^nTA'' \to R^{n+1}TA'$ makes $(R^nT; \delta^n)_{n \geq 0}$ an exact and universal connected sequence of additive functors from **A** to **B**. If, additionally, $T$ is left exact, then $(R^nT; \delta^n)_{n \geq 0}$ are right satellites of $T$.

*Proof*. Consider $\dot{J}\dot{A} = \ulcorner \dot{A}, \dot{Q}^+ \urcorner$ for a short exact sequence $\dot{A}$ in **A**. Since for all $n \geq 0$ the sequences $0 \to Q'^n \to Q^n \to Q''^n \to 0$ are split exact (due to injectivity) and since $T$ is additive and hence preserves biproducts, the sequences $0 \to TQ'^n \to TQ^n \to TQ''^n \to 0$ are also split exact. So the sequence $0 \to TQ'^+ \to TQ^+ \to TQ''^+ \to 0$ is again short exact in the category $\mathbf{B}^+$ consisting of right complexes in **B**. This sequence then yields the objects $(R^nT)A'$, $(R^nT)A$ and $(R^nT)A''$ in **B** (see definition 3.3.8) for $n \geq 0$: they are just the homology objects of the right complexes $TQ'^+$, $TQ^+$ and $TQ''^+$ respectively. Finally the sequence of morphisms $(\delta^n)_{n \geq 0}$ exists and yields an exact connected right sequence of functors from **A** to **B** by theorem 3.3.6. We are therefore (except for the last assertion of the theorem) left with two things to prove:

(i) that the sequence $(\delta^n)_{n \geq 0}$ does not depend on the chosen injective resolution;

(ii) that the obtained sequence $(R^nT; \delta^n)_{n \geq 0}$ is universal.

ad (i). We prove a more general result. Let $\ulcorner \dot{X}, \dot{Q}^+ \urcorner$ and $\ulcorner \dot{Y}, \dot{R}^+ \urcorner$ be two injective resolutions of short exact sequences in **A** denoted by $\dot{X}$ and $\dot{Y}$. Let $\dot{f}: \dot{X} \to \dot{Y}$ be a morphism in $\dot{\mathbf{A}}$. Then it follows from lemma 3.4.5 that there is an extension $(\dot{f}^n: \dot{Q}^n \to \dot{R}^n)_{n \geq 0}$ of $\dot{f}$. Since the connecting morphisms $\delta^n$ are natural we have

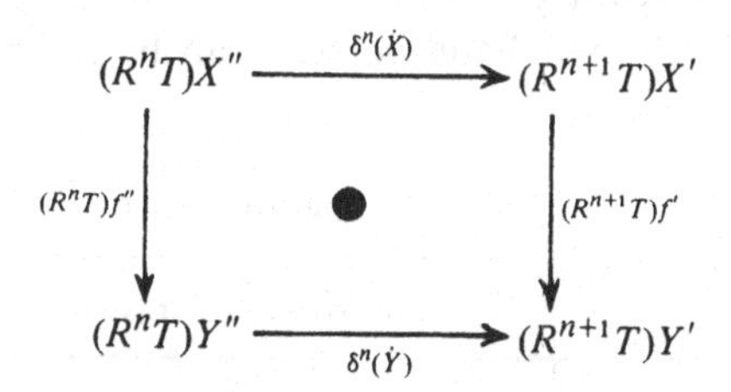


Each $(R^nT)\dot{f}: (R^nT)\dot{X} \to (R^nT)\dot{Y}$ becomes a morphism in the category $\dot{\mathbf{B}}$. Let $\dot{F}^n$ associate with a morphism $\ulcorner \dot{f}, \dot{f}^+ \urcorner: \ulcorner \dot{X}, \dot{Q}^+ \urcorner \to \ulcorner \dot{Y}, \dot{R}^+ \urcorner$ in $\ulcorner \dot{\mathbf{A}}, \dot{\mathbf{Q}}^+ \urcorner$, the category of injective resolutions of objects in $\dot{\mathbf{A}}$, the morphism $(R^nT)\dot{f}$ in $\dot{\mathbf{B}}$. This association does not depend on the chosen extension (since different extensions of the same morphism are homotopic). Furthermore $\dot{F}^n$ is an additive functor from $\ulcorner \dot{\mathbf{A}}^+, \dot{\mathbf{Q}}^{+\urcorner}$ to $\dot{\mathbf{B}}$.

Now if $\ulcorner \dot{A}, \dot{Q}^{+\urcorner}$ and $\ulcorner \dot{A}, \dot{R}^{+\urcorner}$ are two injective resolutions of the same short exact sequence $\dot{A}$ in $\mathbf{A}$ then the following diagram is commutative for all $n \geqslant 0$, where the $\delta s$ stem from $\dot{Q}^+, \dot{R}^+$ respectively.

$$\begin{array}{ccc} (R^nT)A'' & \xrightarrow{\delta^n} & (R^{n+1}T)A' \\ {\scriptstyle (R^nT)1_{A''} = 1_{(R^nT)A''}} \downarrow & & \downarrow {\scriptstyle (R^{n+1}T)1_{A'} = 1_{(R^{n+1}T)A'}} \\ (R^nT)A'' & \xrightarrow[\delta^n]{} & (R^{n+1}T)A' \end{array}$$

This means that the dependence of the sequence $(\delta^n)_{n \geqslant 0}$ on the injective resolution for the short exact sequence $\dot{A}$ was only apparent.

ad (ii). The sequence $(R^nT; \delta^n)_{n \geqslant 0}$ is an exact sequence of additive functors from $\mathbf{A}$ to $\mathbf{B}$ such that $R^nT$ is effaceable for $n \geqslant 1$ by the corollary to proposition 3.4.2. But then $(R^nT; \delta^n)_{n \geqslant 0}$ is universal by 3.4.3.

If $T$ is left exact we have $R^0T \simeq T$ (see remarks following 3.3.8). Hence the sequence $(R^nT; \delta^n)_{n \geqslant 0}$ is a sequence of right satellites of $T$. This finishes the proof.

We leave to the reader the formulation of the theorem for left satellites, which follows by applying the above to $T^\circ: \mathbf{A}^\circ \to \mathbf{B}^\circ$.

We finally make some comments. The last theorem proves the existence of a sequence of right satellites for a functor $T$ under certain conditions. A more general result in this direction is due to Grothendieck [18, 2.2].

In section 3.2 we proved the existence of a sequence of right satellites for a functor $T$ under different conditions.

Summarizing we get the following comparison. As we remarked before, the smallness of $\mathbf{A}$ in the left hand version can be somewhat relaxed.

| | theorem 3.2.5 | theorem 3.4.6 |
|---|---|---|
| $T$ | additive | additive and left exact |
| $\mathbf{A}$ | small abelian | abelian with injectives |
| $\mathbf{B}$ | additive and right complete | exact and additive |
| then | $T$ has a sequence of right satellites | $T$ has an exact sequence of right satellites |

**3.4.7 Theorem** Let **A** be an abelian and **B** an exact category. Suppose $(F^n; \delta^n)_{n\geq 0}$ is an exact connected right sequence of functors from **A** to **B** (so $F^0$ is left exact). Call an object $A \in \mathbf{A}$ *acyclic* with respect to this sequence if $F^nA = 0$ for all $n \geq 1$ and call a long exact sequence $0 \to A \to A^0 \to A^1 \to A^2 \to \ldots$ an *acyclic resolution of* $A$ if each $A^i$ is acyclic ($i \geq 0$). Then for such an acyclic resolution $\ulcorner A, A^+ \urcorner$ we have $F^nA = H^n(F^0A^+)$.

*Proof.* Let $K^n \rightarrowtail A^n$ be the kernel of $A^n \to A^{n+1}$. Then, owing to the exactness of the long sequence, $0 \to K^n \to A^n \to K^{n+1} \to 0$ is short exact for $n \geq 0$ if we put $K^0 = A$. For each $n \geq 0$ we now have in **B** the long exact sequence

$$L^n: 0 \to F^0K^n \to F^0A^n \to F^0K^{n+1} \xrightarrow{\delta 0} F^1K^n$$

$$\to F^1A^n \to F^1K^{n+1} \xrightarrow{\delta 1} F^2K^n \to \ldots$$

Since $F^iA^n = 0$ for $i \geq 1$ the sequences $0 \to F^iK^{n+1} \to F^{i+1}K^n \to 0$ are exact for all $i \geq 1$ which means $F^iK^{n+1} \simeq F^{i+1}K^n$ for all $i \geq 1$. So we have $F^nA \simeq F^1K^{n-1}$ $(F^1K^{n-1} \simeq F^2K^{n-2} \ldots \simeq F^nK^0 = F^nA)$. To determine the homology $H^n(F^0A^+)$ we first remark that since $F^0$ is left exact it follows from the exactness of the sequence $0 \to K^n \to A^n \to A^{n+1}$ that $F^0K^n \to F^0A^n = \ker(F^0A^n \to F^0A^{n+1})$. Then the homology $H^n(F^0A^+)$ is found from the diagram

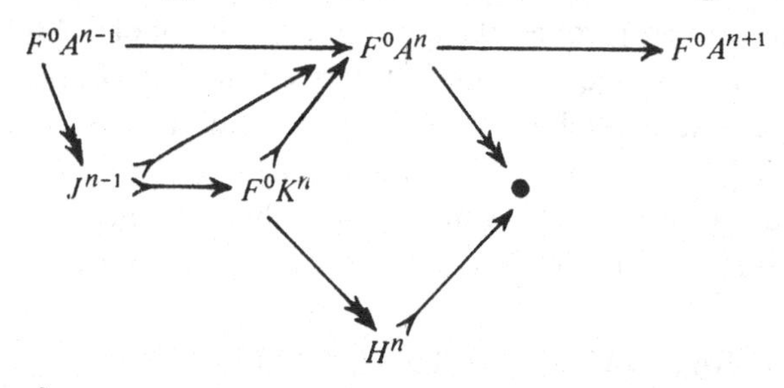


where $F^0K^n \to H^n = \operatorname{coker}(J^{n-1} \to F^0K^n) = \operatorname{coker}(F^0A^{n-1} \to F^0K^n)$. From the long exact sequence $L^{n-1}$ we take the exact piece

$$F^0A^{n-1} \to F^0K^n \xrightarrow{\delta} F^1K^{n-1} \to 0.$$

But then $F^0K^n \to F^1K^{n-1} = \operatorname{coker}(F^0A^{n-1} \to F^0K^n)$. Comparison yields $H^n \simeq F^1K^{n-1}$. Thus we find $F^nA \simeq H^n(F^0A^+)$.

Remark. The above applies in particular to the right derived functors $R^nT$ of a left exact functor $T: \mathbf{A} \to \mathbf{B}$ satisfying the conditions of theorem 3.4.6. According to 3.3.8 remark (iii), injectives are certainly acyclic with respect to the sequence $R^nT$ but the converse is not necessarily true. We shall use this result extensively in chapter 4 when dealing with sheaf cohomology.

**3.4.8 Examples** (*a*) Let $\mathbf{A}$ be an abelian category with injectives, and $A \in \mathbf{A}$. The additive functor $h_A: \mathbf{A} \to \mathbf{Ab}$ being left exact, in virtue of theorem 3.4.6 the sequence of derived functors $(R^n h_A; \delta^n)_{n\geq 0}$ is an exact sequence of right satellites of $h_A$. When $\mathbf{A}$ is small (or at least not too large) we saw in 3.2.6(*a*) that $(\text{Ext}^n\ (A, \text{-}); \mathscr{E}\text{-})_{n\geq 0}$ is a sequence of right satellites of $h_A$. Therefore $(R^n h_A; \delta^n) \simeq (\text{Ext}^n\ (A, \text{-}); \mathscr{E}\text{-})$. This gives a different proof that the sequence of Exts is exact. However, the proof in 3.1.23 has the advantage that we did not postulate the existence of injectives.

Now suppose that $\mathbf{A}$ has projectives, so there are injective resolutions in $\mathbf{A}^\circ$. The right derived functors $R^n h^B: \mathbf{A}^\circ \to \mathbf{Ab}$ exist and we see that $(R^n h^B; \delta^n)_{n\geq 0} \simeq (\text{Ext}^n\ (\text{-}, B); \text{-}\mathscr{E})$. This gives us the formula $R^n h_A(B) \simeq R^n h^B(A)_{n\geq 0}$, since both identify with $\text{Ext}^n\ (A, B)$. This again illustrates the exceptionally pleasant properties of the basic bifunctor $(\text{-}, \text{-}) = H: \mathbf{A}^\circ \times \mathbf{A} \to \mathbf{Ab}$.

(*b*) This last property is not automatically satisfied by every bifunctor. We shall prove it for the bifunctor $\text{-} \otimes_R \text{-}: \mathbf{M}_R \times {}_R\mathbf{M} \to \mathbf{Ab}$ for any given ring $R$. Since we shall only deal with a fixed $R$, we shall drop it from the notation of the tensor product and its derived functors. The category ${}_R\mathbf{M}$ having projectives, the appropriate version of 3.4.6 tells us that, with $M \in \mathbf{M}_R$, the sequence of left derived functors $L_n(M \otimes \text{-})$ is also a sequence of left satellites of $(M \otimes \text{-})$. The latter were christened $\text{Tor}_n\ (M, \text{-})$ in 3.2.6(*c*). Since $M \otimes \text{-}$ is right exact, theorem 3.4.6 proves the promised result that with each short exact sequence $0 \to N' \to N \to N'' \to 0$ in ${}_R\mathbf{M}$ is associated a long exact sequence

$$\begin{aligned}
\ldots &\to \text{Tor}_{n+1}\ (M, N'') \xrightarrow{\delta} \text{Tor}_n\ (M, N') \\
&\to \text{Tor}_n\ (M, N) \to \text{Tor}_n\ (M, N'') \\
\ldots &\to \text{Tor}_1\ (M, N'') \xrightarrow{\delta} M \otimes N' \\
&\to M \otimes N \to M \otimes N' \to 0.
\end{aligned} \tag{1}$$

Next observe that the tensor product, as a left adjoint, is right continuous. Tensoring with a free module is therefore an exact functor, and so is tensoring with a direct summand of a free module, i.e. a projective module. Take a short exact sequence $\dot{M} = 0 \to M' \to M \to M'' \to 0$ in $\mathbf{M}_R$. If $N \in {}_R\mathbf{M}$ has $\ulcorner P^{\cdot}, N \urcorner$ for a projective resolution, the complex $\dot{M} \otimes P^{\cdot}$ has short exact rows $0 \to M' \otimes P_n \to M \otimes P_n \to M'' \otimes P_n \to 0$. Hence, taking homology, we find a second long exact sequence

$$\begin{aligned} \ldots &\to \mathrm{Tor}_{n+1}(M'', N) \xrightarrow{\delta} \mathrm{Tor}_n(M', N) \\ &\to \mathrm{Tor}_n(M, N) \to \mathrm{Tor}_n(M'', N) \\ \ldots &\to \mathrm{Tor}_1(M'', N) \xrightarrow{\delta} M' \otimes N \\ &\to M \otimes N \to M' \otimes N \to 0. \end{aligned} \tag{2}$$

For distinction, call the left derived functors of $- \otimes N$ by the name $\hat{\mathrm{Tor}}_n(-, N)$. The sequences (1) and (2) are of course also exact for $\hat{\mathrm{Tor}}$. We shall prove that

$$\mathrm{Tor}_n(M, N) \simeq \hat{\mathrm{Tor}}_n(M, N)$$

for all $M \in \mathbf{M}_R$, $N \in {}_R\mathbf{M}$, first tackling the case $n = 1$. Let $P \twoheadrightarrow M$ in $\mathbf{M}_R$, $P' \twoheadrightarrow N$ in ${}_R\mathbf{M}$ with $P$ and $P'$ projective, and call the respective kernels $K$ and $K'$. Consider the following commuting diagram with exact rows and columns:

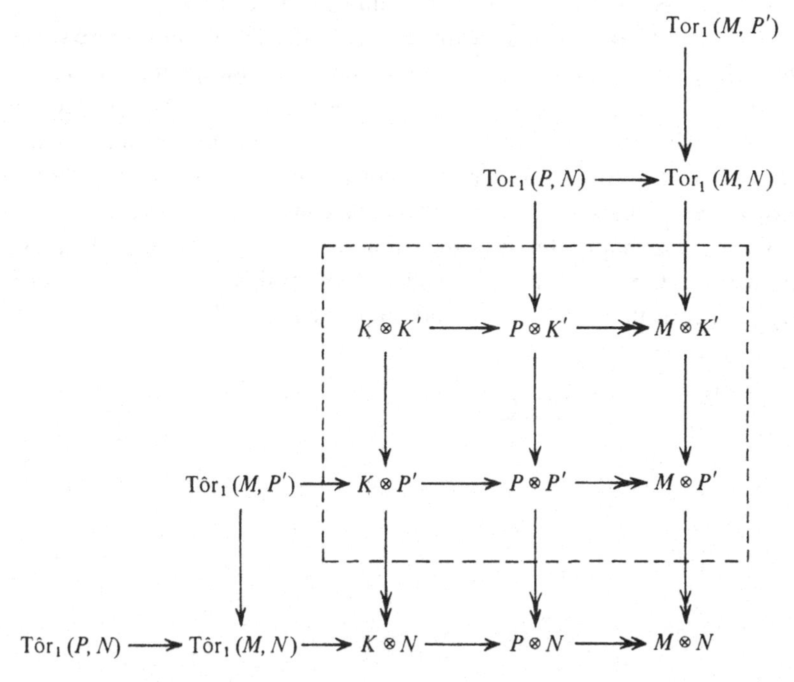


Now $\mathrm{Tor}_1(P, N) = 0 = \mathrm{Tor}_1(M, P')$ because $P \otimes -$ is exact while $P'$ is $\mathrm{Tor}_n(M, -)$ acyclic, $n \geqslant 1$; see the remarks after 3.3.8. Similarly

$\hat{\mathrm{Tor}}_1(M, P') = 0 = \hat{\mathrm{Tor}}_1(P, N)$. Applying the snake-lemma to the bracketed rectangle, we obtain an exact sequence

$$0 \to \mathrm{Tor}_1(M, N) \to K \otimes N \to P \otimes N \to M \otimes N \to 0.$$

But the bottom row of the diagram shows the same exact sequence, with $\mathrm{Tor}_1(M, N)$ replaced by $\hat{\mathrm{Tor}}_1(M, N)$. Hence $\mathrm{Tor}_1(M, N) \simeq \hat{\mathrm{Tor}}(M, N)$.

To prove the same for arbitrary $n$, we use the method already featuring in the proof of 3.4.7, which the French call 'décalage'. Let

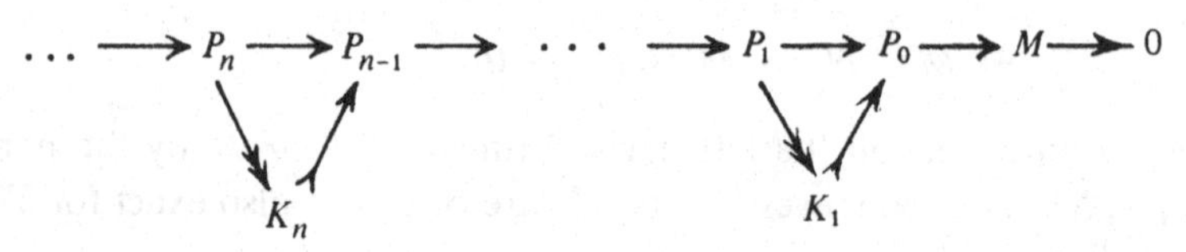


be a projective resolution in $\mathbf{M}_R$, factored in epic–monic fashion. By applying the sequence (2) for Tor to the short exact sequences $0 \to K_i \to P_i \to K_{i-1} \to 0$, $1 \leq i \leq n-1$, with $K_0 = M$, we find that $\mathrm{Tor}_n(M, N) \simeq \mathrm{Tor}_{n-1}(K_1, N) \simeq \ldots \simeq \mathrm{Tor}_1(K_{n-1}, N)$. The latter is isomorphic to $\hat{\mathrm{Tor}}_1(K_{n-1}, N)$ by the above. Now applying the analogue of sequence (1) for $\hat{\mathrm{Tor}}$ to the same short exact sequences, we find that $\hat{\mathrm{Tor}}_1(K_{n-1}, N) \simeq \hat{\mathrm{Tor}}_n(M, N)$, thus gaining our objective.

Modules which are acyclic with respect to the Tor sequence are called *flat*. They are important in algebra and algebraic geometry. As an exercise, the reader may prove that in **Ab** the flat objects are precisely the torsionfree abelian groups, which explains the name torsion functors for Tor. Since $\mathbb{Q}$ is torsionfree but not projective, we see that though projective modules are obviously flat, the converse is not true.

To end this chapter let us remark that the uses of homological algebra are manifold, but will not be touched upon in this book. The reader is urged to consult [9], [25], [30] and, in particular, [20].

# 4

# Sheaves and their cohomology

## 4.1 Introduction

Sheaves and their cohomology are an important tool in such diverse mathematical disciplines as algebraic topology, theory of analytic functions, algebraic geometry and others. Though there are differences in their use in these different areas, they have an underlying pattern in common which is best expressed in categorical language. In particular adjoint functors facilitate the understanding of rather complex situations. Historically, in fact, a good deal of category theory and homological algebra was developed in order to treat the cohomology theory of sheaves which is considerably more intricate than that of modules; cf. [18].

In this chapter we shall present the elements of the theory which will provide good illustrational material for some of the work in previous chapters. It is outside the scope of this book to treat applications of sheaves and their cohomology. This would require detailed knowledge of each particular field. Treatises entirely devoted to sheaves are [6], [17], [42].

Recently a far-reaching categorical generalization of sheaf theory has been studied. A definitive account of these so-called topoi [43] still remains to be written.

In the following sections we will assign to every topological space $X$ the category $\mathbf{P}(X)$ of presheaves on $X$. The category $\mathbf{F}(X)$ of sheaves on $X$ is a full subcategory of $\mathbf{P}(X)$. The embedding $\mathbf{F}(X) \to \mathbf{P}(X)$ has a left adjoint which is called the sheafification functor. This functor is a reflection since the embedding is full and faithful.

A continuous map $f: X \to Y$ induces functors $\tilde{f}_*$, $f_*$, $\tilde{f}^*$ and $f^*$ for which the following diagram commutes.

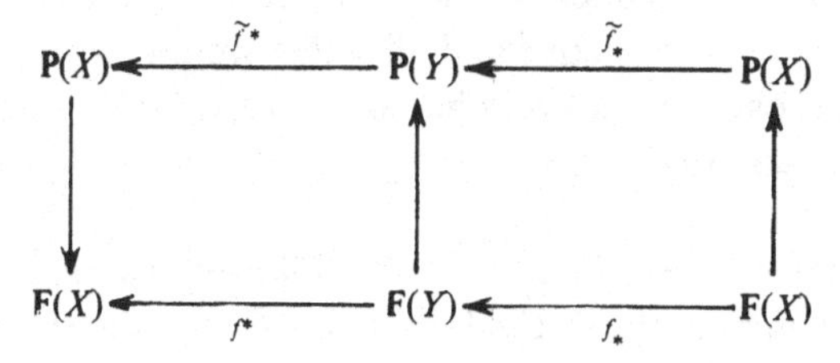

The functors at the top and bottom of the left square are left adjoint to those of the right square.

Sections 4.2 to 4.7 deal with sheaves of sets. In 4.8 and 4.9 the theory is shown to apply also to sheaves of modules. In that case the categories **P**($X$) and **F**($X$) are Grothendieck and the functors **P**($X$)$\to$**F**($X$), $\tilde{f}_*$, $\tilde{f}^*$ and $f^*$ are exact, while **F**($X$)$\to$**P**($X$) and $f_*$ are left exact. In section 4.10 we do homological algebra for sheaves of modules. This gives an important application of chapter 3. The right satellites of **F**($X$)$\to$**P**($X$) are discussed, the cohomology presheaf functors, and the right satellites of $\Gamma(X, \text{-})$: **F**($X$)$\to {}_R$**M**, the cohomology module functors, and in particular the behaviour of these under restriction of the sheaf to locally closed subspaces of $X$.

The category **F**($X$) has injectives, but a construction of those is not given since sections 4.11 and 4.12 treat other more general constructions of the cohomology presheaves instead, based on theorem 3.4.7 concerning acyclic objects.

## 4.2 Concrete sheaves

**4.2.1 Definitions** A map $\pi: F \to X$ between two topological spaces $F$ and $X$ is called a *local homeomorphism* provided that for all $f \in F$ there is an open neighbourhood $O_f$ of $f$ and an open neighbourhood $U_x$ of $x = \pi f$ such that $\pi | O_f: O_f \to U_x$ is a homeomorphism (bijective and continuous in both directions). Since locally its inverse is continuous, $\pi$ is also an open mapping.

A *concrete sheaf* is a triple $\ulcorner F, \pi, X \urcorner$ where $F$ and $X$ are topological spaces and $\pi: F \to X$ is a local homeomorphism. The topological space $F$ is called the *total space*, $\pi$ is called the *projection* and $X$ is called the *base space*. For any $x \in X$ the set $F_x = \overset{-1}{\pi} x$ is called the *stalk of* $x$. For any open $U \subset X$ a continuous map $s: U \to F$ such that $\pi s = \text{id}_U$ is called a *section on* $U$. A section is therefore a local inverse of the projection. The set of all sections on $U$ is denoted by $\Gamma(U, F)$.

For readers familiar with covering spaces [39, p. 62] it is instructive to compare both notions. A covering space is certainly a concrete sheaf. But it is easy to see that a concrete sheaf $\ulcorner F, \pi, X \urcorner$ need not be a covering space, even if $\pi$ is surjective and all stalks are finite.

Note that in the French literature what we have called a concrete sheaf is referred to as an *espace étalé*.

**4.2.2 Examples** Check in each of the following cases that the triple $\ulcorner F, \pi, X \urcorner$ is a concrete sheaf. Find the stalks $F_x$ and the sections $s$

on $U$ (in particular for those $U$ that are connected and for $U = X$).

($a$) $\pi = 1_{\mathbb{R}}: \mathbb{R} \to \mathbb{R}$.

($b$) $\pi: (0, 1) \hookrightarrow \mathbb{R}$.

($c$) $\pi: X \times D \to X$, with $D$ any discrete topological space. This is called the *constant sheaf on X* with stalks all equal to $D$.

($d$) $\pi: \mathbb{R} \to \mathbb{R}/\mathbb{Z}$.

($e$) $X = \mathbb{R}$ and $F = \{\ulcorner x, 0 \urcorner\}_{x<0} \cup \{\ulcorner x, 1 \urcorner\}_{x \geqslant 0} \cup \{\ulcorner x, -1 \urcorner\}_{x \geqslant 0}$ with $\pi: F \to X$ defined by $\ulcorner x, a \urcorner \mapsto x$. Embedded in $\mathbb{R}^2$ the total space $F$ is like a fork with two loose prongs. Define $B(\ulcorner x, a \urcorner, \varepsilon) = \{\ulcorner y, b \urcorner \in F \mid |x - y| < \varepsilon$ and $|a - b| \leqslant 1\}$. Endow $F$ with the topology defined by these open balls (which is non Hausdorff!).

($f$) Generalize example ($e$) in the following way: start from example ($c$); choose an open $X_0 \subset X$ and define in $X \times D$ the equivalence relation given by $\ulcorner x, a \urcorner \sim \ulcorner y, b \urcorner$ if and only if $x = y$ and either $a = b$ or $x \in X_0$. Find the natural topology for $F = (X \times D)/\sim$ and define $\pi: F \to X$ in the obvious way.

In each of the following cases explain why $\ulcorner F, \pi, X \urcorner$ is not a sheaf.

($g$) $\pi$ is the identity mapping of $\mathbb{R}$ (with the usual topology) onto the real line with the discrete topology.

($h$) $\pi: \mathbb{R} \to \mathbb{R}^2$ defined by $x \mapsto \ulcorner x, 0 \urcorner$.

($i$) Example ($e$) with the condition $x < 0$ in the definition of $F$ changed into $x > 0$.

($j$) Example ($f$), without the condition that $X_0$ should be open.

### 4.2.3 Some properties of $\ulcorner F, \pi, X \urcorner$

($a$) $F = \bigcup_x F_x$.

($b$) Every stalk has the discrete topology (if not, $\pi$ would not be locally injective).

($c$) If $X$ is a $T_1$-space, so is $F$. For suppose $f_1$ and $f_2$ are two elements of $F$ not in the same stalk. Since then $\pi f_1 \neq \pi f_2$ there is an open subset $U$ of $X$ containing $\pi f_1$ but not containing $\pi f_2$. Then $\pi^{-1} U$ is an open subset of $F$ containing $f_1$ but not $f_2$. For $T_1$-space, see 4.12.

($d$) If $X$ is a discrete topological space, so is $F$. To see this let $f \in F$ and let $\pi f = x \in X$. Furthermore assume that $O_f$ is an open subset of $F$ containing $f$ such that $\pi O_f$ and $O_f$ are homeomorphic under $\pi$. Then $\{f\} = F_x \cap O_f$. Since both $F_x$ and $O_f$ are open $\{f\}$ is also open.

($e$) The inverse implications of ($c$) and ($d$) hold if $\pi: F \to X$ is surjective.

($f$) If $s: U \to F$ is a section and if $V \subset U$ ($V$ also open in $X$) then the restriction of $s$ to $V$ (notation $s|V$) is a section.

($g$) Sections that coincide in some point coincide on an open set containing that point.

*Proof.* Let $s_1: U_1 \to F$ and $s_2: U_2 \to F$ be sections which coincide for $x \in U_1 \cap U_2$. The point $s_1x = s_2x$ is contained in an open set $O$ such that $\pi|O$ is injective. Consider the open set $V = \overset{-1}{s_1}O \cap \overset{-1}{s_2}O$ that contains $x$. For $y \in V$ one has $s_1y \in O$ and $s_2y \in O$; furthermore $\pi s_1 y = y = \pi s_2 y$; as $\pi|O$ is injective it follows that $s_1y = s_2y$, which means that $s_1|V = s_2|V$.

($h$) For an open set $O$ of $F$ such that $\pi|O$ is injective, the map $s: \pi O \to F$, defined by $sx$ being the only element of $F_x \cap O$, is a section defined on $\pi O$ with $O = s(\pi O)$.

*Proof.* Let $O'$ be open in $F$; then $\overset{-1}{s}O' = \pi(O \cap O')$ which is open in $X$. Hence $s$ is continuous. Also $\pi O$ is open and $\pi \circ s = \text{id}$.

($i$) For any section $s: U \to F$ the set $sU$ is open in $F$ and homeomorphic to $U$; $s = (\pi|sU)^{-1}$.

*Proof.* Let $x \in U$; $sx$ is contained in an open set $O$ such that $\pi|O$ is injective. Then $(\pi|O)^{-1} = s': \pi O \to F$ is a section with $s'x = sx$. Therefore there is an open set $U_0$ in $X$, such that $s|U_0 = s'|U_0$. But $sU_0 = s'U_0 = O \cap \overset{-1}{\pi}U_0$ which is open in $F$, so that $sx \in sU_0 \subset sU$. Hence $sU$ is open in $F$. The other statements are self evident.

($j$) From properties ($g$), ($h$) and ($i$) it follows that the collection

$$\{sU | U \text{ is open in } X \text{ and } s \text{ is a section on } U\}$$

is a base for the open sets of $F$. This means that the topology of the total space $F$ is determined by the sections.

**4.2.4** As a particularly important example we will consider what is called the *sheaf of germs* of continuous functions on a topological space $X$. First consider all real valued continuous functions defined on open subsets of $X$. For every $x \in X$ we define an equivalence relation for these functions. Let $f_i$ ($i = 1, 2$) be two real valued functions defined on the open subsets $U_i$ respectively. Then define $f_1 \sim f_2$ provided there is an open set $U \subset U_1 \cap U_2$ containing $x$ with $f_1|U = f_2|U$. The equivalence class of $f$ at $x$ is denoted by $[f]_x$. It is called the *germ of $f$ at $x$*. The set of all germs at $x$ is denoted by $F_x$. Then we put $F = \bigcup_{x \in X} F_x$ and we define $\pi: F \to X$ by $\pi[f]_x = x$. To make $\ulcorner F, \pi, X \urcorner$ into a concrete sheaf we need a topology for $F$ such that $\pi$ is a local homeomorphism. For this purpose define for any continuous real valued function $f$ defined on the open subset $U \subset X$ the map $s_f: U \to F$ by $s_f x = [f]_x$. Then $\pi \circ s_f = \text{id}$. It is clear that $F = \bigcup s_f U$, where the union is taken over all couples consisting of a continuous real valued function $f$ defined on an open subset $U \subset X$. Since $s_{f_1} U_1 \cap s_{f_2} U_2$ is either void or equal to a set $s_f U$ for some couple $f$ and $U$, one may take the collection $\{s_f U\}$ as a base for a topology of $F$. We leave it to the reader to check that this makes $\ulcorner F, \pi X \urcorner$ into a concrete sheaf. The mappings $s_f: U \to F$ are then sections on $U$ and for any section $s: U \to F$ one may prove that there is a function $f$ defined on $U$ such that $s = s_f$. We note that, in general, the topological space $F$ is not Hausdorff as is shown by the following example. Take $X = \mathbb{R}$ and consider the continuous functions $f_1 \equiv 0$ and $f_2(x) = 0$ for $x < 0$ and $f_2(x) = x$ for $x \geqslant 0$. These two functions have different germs at $x = 0$. The reader should convince himself that it is impossible to separate these two points in the sheaf of germs by disjoint open subsets.

**4.2.5 Definition** Let $X$ be a fixed topological space. By $\mathbf{C}(X)$ we denote the category of concrete sheaves on $X$ where morphisms between $\ulcorner F, \pi, X \urcorner$ and $\ulcorner G, \rho, X \urcorner$ are the continuous maps $\lambda: F \to G$ such that $\rho\lambda = \pi$.

**4.2.6 Lemma** A morphism $\lambda: \ulcorner F, \pi, X \urcorner \to \ulcorner G, \rho, X \urcorner$ in $\mathbf{C}(X)$, considered just as a continuous map $\lambda: F \to G$, is open and locally injective.

*Proof.* The last assertion follows from the facts that $\rho\lambda = \pi$ and $\pi$ is locally injective. To prove that $\lambda$ is an open mapping it is sufficient to show that for every open subset $U$ of $X$ and any section $s$ on $U$ the subset $\lambda(sU)$ is open in $G$ (these subsets $sU$ form a base for the topology of $F$). But $\lambda s$ is a section on $U$ for the sheaf $G$ since $\rho\lambda s = \pi s = \text{id}$. Therefore $\lambda(sU)$ is open in $G$.

It follows from the lemma that $\ulcorner F, \lambda, G \urcorner$ is a concrete sheaf on $G$.

It also follows from the lemma that if $\lambda$ is bijective it is an isomorphism.

**Exercise** Prove that the concrete sheaf of 4.2.2($e$) is isomorphic with the concrete sheaf of all germs of continuous functions on $\mathbb{R}$ that are restrictions of the function $f_1$ and $f_2$ described in the last part of 4.2.4.

## 4.3 Presheaves

**4.3.1 Definition** Let $X$ be a topological space. Let $\mathbf{O}(X)$ be the category whose objects are the open sets of $X$ and whose morphisms are the inclusion maps $i^{U_1}_{U_2}$: $U_1 \to U_2$ for open subsets $U_1$ and $U_2$ of $X$ such that $U_1 \subset U_2$. A *presheaf* is a contravariant functor $P$: $\mathbf{O}(X) \to \mathbf{Sets}$.

The morphisms $P(i^{U_1}_{U_2})$: $PU_2 \to PU_1$ will be denoted by $P^{U_2}_{U_1}$ and will be called the *restriction morphisms.*

**Examples** (*a*) We may consider a concrete sheaf $\ulcorner F, \pi, X \urcorner$ as a presheaf. For each open $U$ in $X$ let $\Gamma(U, X)$ be the set of sections $s$: $U \to F$ and for each pair of open sets $U_1 \subset U_2$ let the restriction morphism $\operatorname{res}^{U_2}_{U_1}$: $\Gamma(U_2, F) \to \Gamma(U_1, F)$ be the map which assigns to any $s \in \Gamma(U_2, F)$ its restriction $s|U_1 \in \Gamma(U_1, F)$. Note that $\Gamma(\varnothing, F)$ is a one point set.

(*b*) Let $PU$ be the set of all continuous real valued functions on the open subset $U$ of $X$. Again take the restrictions as the restriction morphisms.

(*c*) Let $X$ be a finite set with the discrete topology. Let $|U|$ denote the number of elements in the open subset $U$ of $X$. Define $PU = \{i \in \mathbb{Z} | 1 \leq i \leq |U|\}$ for any nonempty open subset $U$ of $X$ and define $P\varnothing = \{0\}$. Define $P^{U_1}_{U_2}$: $PU_1 \to PU_2$ for $U_2 \subset U_1$ by $P^{U_1}_{U_2}(i) = \min\{i, |U_2|\}$.

(*d*) $X$ as above. Define $QU = \{i \in \mathbb{Z} | 0 \leq i \leq |U| - 1\}$ for $U \neq \varnothing$ and $Q\varnothing = \{0\}$. For $U_2 \subset U_1$ define $Q^{U_1}_{U_2}$: $QU_1 \to QU_2$ by $Q^{U_1}_{U_2}(i) = \max\{0, |U_2| - i\}$.

Remark. The functors $P$ and $Q$: $\mathbf{O}^\circ(X) \to \mathbf{Sets}$ from the last two examples are isomorphic.

**4.3.2** Since $\mathbf{O}(X)$ is a small category, we may consider the category $\mathbf{P}(X) = (\mathbf{O}^\circ(X), \mathbf{Sets})$ of all presheaves on $X$ (see 1.3.3). We recall that the objects of this category are the presheaves $P$ on $X$, where the morphisms are the morphisms or natural transformations between these presheaves. This means that for $U$ and $V$ open in $X$ with $V \subset U$ a morphism $\nu$: $P \to Q$ gives rise to the following commutative diagram:

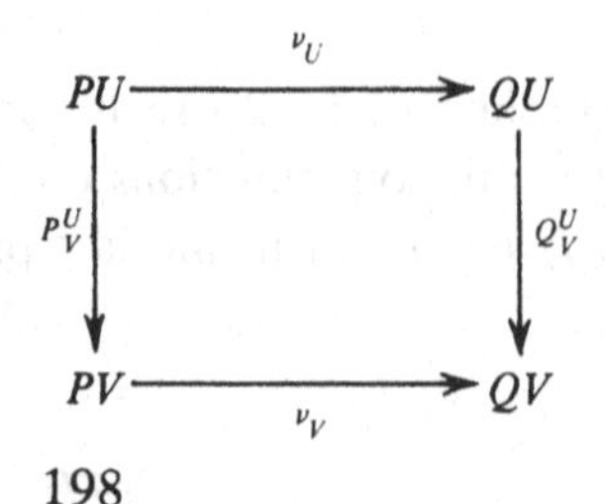

Here we write $\nu_U$ and $\nu_V$ rather than $\nu(U)$ and $\nu(V)$ conforming to standard practice.

We may now define a functor $T: \mathbf{C}(X) \to \mathbf{P}(X)$ in the following way. To any concrete sheaf in $\mathbf{C}(X)$ we assign the presheaf described in 4.3.1$(a)$ and for a morphism $f: F_1 \to F_2$ of concrete sheaves we define $\nu_U: \Gamma(U, F_1) \to \Gamma(U, F_2)$ by $\nu_U(s) = f \circ s$. This is seen to yield a morphism of functors (presheaves on $X$) $\nu: \Gamma(\text{-}, F_1) \to \Gamma(\text{-}, F_2)$. So $T$, defined by putting $TF = \Gamma(\text{-}, F)$, is a functor from the category $\mathbf{C}(X)$ of concrete sheaves on $X$ to the category $\mathbf{P}(X)$ of the presheaves on $X$. We could prove now that this functor $T$ is fully faithful. But this fact will become apparent later on.

**4.3.3** Thus far the concept of stalk has not been introduced in connection with presheaves. We will proceed to do so. Let $P$ be a presheaf on $X$. The open sets containing $x \in X$ constitute a direct set when ordered according to $U \leqslant V$ if and only if $V \subset U$. Consequently the system $\{P_V^U: PU \to PV\}$ with $x \in V \subset U$ is a directed system and so the direct limit $\varinjlim (PU)_{U \ni x}$ exists (see 1.6.2$(a)$).

**Definition** The *stalk in* $x \in X$ of a presheaf $P$ on $X$ is $P_x = \varinjlim (PU)_{U \ni x}$.

Now for $s_1 \in PU_1$ and $s_2 \in PU_2$, with $x \in U_1 \cap U_2$, we put $s_1 \underset{x}{\sim} s_2$ provided there is an open set $V \subset U_1 \cap U_2$ containing $x$ with the property $P_V^{U_1} s_1 = P_V^{U_2} s_2$. This yields an equivalence relation. We denote the equivalence class of an element $s$ (section defined on an open set $U$ containing $x$) at $x$ by $[s]_x$. The set of all $[s]_x$ for a fixed $x$ is then $\varinjlim (PU)_{U \ni x}$. The map $P_x^V: PV \to \varinjlim (PU)_{U \ni x}$ for $x \in V$ is given by $P_x^V s = [s]_x$.

Let $F$ be a concrete sheaf. We want to recover the stalks of $F$ from the presheaf $TF = \Gamma(\text{-}, F)$. For any open set $U$ containing $x$ the mapping $\text{res}_x^U: \Gamma(U, F) \to F_x$, given by $\text{res}_x^U s = sx$, has the property that for any open set $V$ such that $x \in V \subset U$ the condition $\text{res}_x^U = \text{res}_x^V \circ \text{res}_V^U$ holds. Therefore there is a unique morphism $\text{res}_x: \varinjlim \Gamma(U, F)_{U \ni x} \to F_x$ such that the following diagram commutes for every open set $V$ containing $x$. (See diagram on top of page 200.)

The map $\text{res}_x$ is injective (see 4.2.3$(g)$) and also surjective (see 4.2.3$(h)$). Hence $\text{res}_x$ is a bijection between the sets $\varinjlim \Gamma(U, F)_{U \ni x}$ and $F_x$. Thus we have proved that for a concrete sheaf $F$ the map $\text{res}_x: (TF)_x \to F_x$ yields an isomorphism between the stalks of $TF$ and the stalks of $F$. The fact that $F_x$ is considered here as a

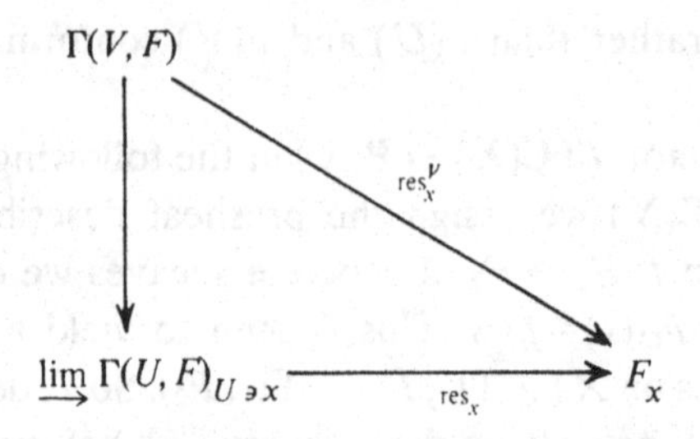


set rather than a topological space is justified by the fact that its topology is discrete (see 4.2.3(*b*)). We see that the functor $T: \mathbf{C}(X) \to \mathbf{P}(X)$ transforms stalks into stalks.

For a morphism $\nu: P \to Q$ between two presheaves in $\mathbf{P}(X)$, we have for every $x \in X$ a unique map $\nu_x: P_x \to Q_x$ with the property $\nu_x \circ P_x^U = Q_x^U \circ \nu_U$; in diagram

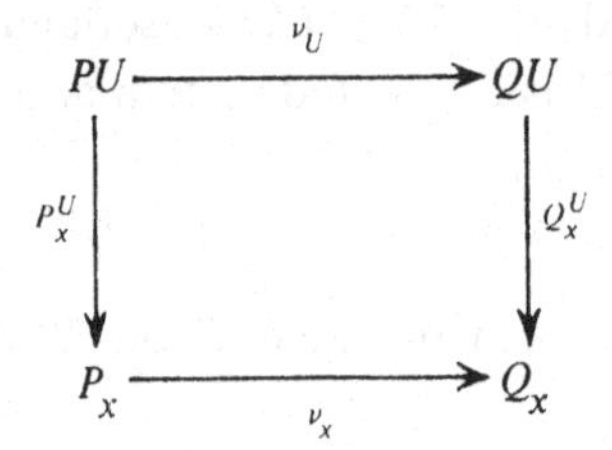


The map $\nu_x$ is given by $\nu_x[s]_x = [\nu_U s]_x$ for $x \in U$; $\nu_x$ is just $\varinjlim (\nu_U)_{U \ni x}$.

**4.3.4** Let us call a concrete sheaf $\ulcorner F, \pi, X \urcorner$ trivial provided $\pi: F \to X$ is a homeomorphism, and a presheaf $P$ on $X$ trivial provided every $PU$ contains exactly one element and almost trivial provided every $P_x$ contains exactly one element. A trivial presheaf is almost trivial but the converse need not be true. See examples (*c*) and (*d*) in 4.3.1.

For a concrete sheaf $F$ on $X$ the following three properties are equivalent.

(i) $TF$ is almost trivial.
(ii) $F$ is trivial.
(iii) $TF$ is trivial.

**Exercises** (*a*) Conclude from the above that for the presheaf $P$ on $X$ in example (*c*) of 4.3.1 there is no concrete sheaf $F$ on $X$ such that $P \simeq TF$.

(*b*) Determine the stalks of the presheaf in example (*b*) of 4.3.1. Compare with 4.2.4.

## 4.4 The sheafification of presheaves

Let $P$ be a presheaf on $X$. We want to construct a concrete sheaf $\ulcorner F, \pi, X \urcorner$ which, in a sense to be made precise later, is very much like $P$. We want the stalks of $F$ to be the stalks of $P$. So we define the set $F$ to be the disjoint union of the stalks of $P$:

$$F = \{[s]_x | x \in U, \ U \text{ open in } X \text{ and } s \in PU\}.$$

Furthermore we define $\pi: F \to X$ by $\pi[s]_x = x$. So we already have $\pi^{-1}x = P_x$. It is our goal to construct $F$ in such a way that the element $s \in PU$ will be found among the sections $U \to F$. Therefore we define, for $s \in PU$, the mapping $\tilde{s}: U \to F$ by $\tilde{s}x = [s]_x$. Then we already have $\pi \circ \tilde{s} = \text{id}$. In order to obtain $\ulcorner F, \pi, X \urcorner$ as a concrete sheaf having at least as sections on $U$ these mappings $\tilde{s} \in \Gamma(U, F)$, we should make sure (see 4.2.3(*i*)) that the sets $\tilde{s}U = \{[s]_x | x \in U\}$ are all open in $F$.

**4.4.1 Lemma** The following assertions are true.

(i) For $s_i \in PU_i$ $(i = 1, 2)$ there is an open set $U \subset U_1 \cap U_2$ such that $\tilde{s}_1 U_1 \cap \tilde{s}_2 U_2 = \tilde{s}U$ for $s = P^{U_i}_U s_i$ $(i = 1, 2)$.

(ii) $F = \bigcup\{\tilde{s}U | U \text{ open in } X; s \in PU\}$.

(iii) The collection $\{\tilde{s}U | U \text{ open in } X; s \in PU\}$ is a base for a topology in $F$.

(iv) The mapping $\pi: F \to X$ is continuous, open and locally injective for the topology defined by the base in (iii). Hence $\ulcorner F, \pi, X \urcorner$ is a concrete sheaf.

(v) For $x \in X$ it follows that $F_x = P_x$.

(vi) For $x \in PU$ it follows that $\tilde{s} \in \Gamma(U, F)$.

(vii) Define $\Phi_U(P): PU \to \Gamma(U, F)$ by $s \mapsto \tilde{s}$. Then we have $\Phi(P): P \to TF$ in $\mathbf{P}(X)$.

*Proof.* (i) The following holds:

$$\begin{aligned}\tilde{s}_1 U_1 \cap \tilde{s}_2 U_2 &= \{[s_1]_x | x \in U_1 \cap U_2 \quad \text{and} \quad [s_1]_x = [s_2]_x\} \\ &= \{[s_1]_x | x \in U\}\end{aligned}$$

with $U = \{x \in U_1 \cap U_2 | [s_1]_x = [s_2]_x\}$. For $x \in U$, therefore, $s_1 \underset{x}{\sim} s_2$. Hence $x$ has an open neighbourhood $W$ with $P^{U_1}_W s_1 = P^{U_2}_W s_2$. It follows that $U$ is open since it is a neighbourhood for all of its points. If $s = P^{U_1}_U s_1$, we have $\tilde{s}_1 U_1 \cap \tilde{s}_2 U_2 = \{[s]_x | x \in U\} = U$. Note that $U$ might be void.

(ii) is trivial.

(iii) follows from (i) and (ii).

(iv) Let $U$ be open in $X$. Then

$$\pi^{-1}U = \{[s]_x | x \in U \text{ and } x \text{ has an open neighbourhood } V \text{ such that } s \in PV\}$$
$$= \{[s]_x | x \in V \subset U \text{ with } s \in PV\}$$
$$= \bigcup\{sV | V \subset U, s \in PV\}.$$

Hence $\pi^{-1}U$ is open in $F$ and thus $\pi: F \to X$ is continuous. The fact that $\pi$ is an open mapping and locally injective is evident from the way $\pi$ operates on the base elements for the topology of $F$.

(v) is trivial.

(vi) Let $s \in PU$ and consider $\tilde{s}: U \to F$. Take a base element $\tilde{s}_1 U_1$ of the topology of $F$. Then

$$\tilde{s}^{-1}(\tilde{s}_1 U_1) = \{x \in U | \tilde{s}x \in \tilde{s}_1 U_1\} = \pi(\tilde{s}U \cap \tilde{s}_1 U_1).$$

This set is open in $X$, so $\tilde{s}: U \to F$ is continuous. Since also $\pi \circ \tilde{s} = \mathrm{id}$ we have proved $\tilde{s} \in \Gamma(U, V)$.

(vii) For the open sets $U$ and $V$ such that $V \subset U$ we have to verify that $\Phi_V(P) \circ P_V^U = \mathrm{res}_V^U \circ \Phi_U(P)$. Take $s \in PU$. Then we have

$$(\Phi_V(P) \circ P_V^U)s = (x \in V) \mapsto [P_V^U s]_x = (\mathrm{res}_V^U \circ \Phi_U(P))s.$$

Hence $\Phi(P): P \to TF$ in $\mathbf{P}(X)$.

**4.4.2** Let $S$ assign to any presheaf $P$ on $X$ the concrete sheaf $\ulcorner F, \pi, X \urcorner$ described above. We want to show that $S$ defines a functor, i.e. assigns to each morphism $\nu: P \to Q$ of presheaves on $X$ a morphism $SP \to SQ$ of concrete sheaves on $X$. Let $\ulcorner F, \pi, X \urcorner = SP$ and $\ulcorner G, \rho, X \urcorner = SQ$. Consider the mapping $f: F \to G$ defined by $(f|F_x: F_x \to G_x) = (\nu_x: P_x \to Q_x)$ (see the end of 4.3.3). We claim that $f: F \to G$ thus defined is a morphism $SP \to SQ$ in $\mathbf{C}(X)$. For in the first place it is clear that $\rho \circ f = \pi: [s]_x \mapsto x$. So it is sufficient to show that $f$ is continuous. Take a base element $\tilde{s}_1 U$ for the topology of $G$ ($U$ open in $X$ and $s_1 \in QU$). Then

$$f^{-1}(\tilde{s}_1 U) = \{[s]_x | x \in U \text{ and } x \text{ has an open neighbourhood } V \text{ with } s \in PV \text{ such that } s_1 \underset{x}{\sim} \nu_V s\}.$$

For open $V \subset X$ and $s \in PV$ we define $W(V, s) = \{x \in U \cap V | s_1 \underset{x}{\sim} \nu_V s\}$. It turns out that these sets $W(V, s)$ are open in $X$. This implies that $\{[s]_x | x \in W(V, s)\}$ is open in $F$. Then $f^{-1}(\tilde{s}_1 U) = \bigcup_{V,s} \{[s]_x | x \in W(V, s)\}$ and so their union is open in $F$. Thus $f: F \to G$ is continuous.

We make $S$ into a functor from $\mathbf{P}(X)$ to $\mathbf{C}(X)$ by assigning to $\nu: P \to Q$ in $\mathbf{P}(X)$ the morphism $f: SP \to SQ$ just defined.

Furthermore the morphisms $\Phi(P)$ of 4.4.1(vii) give rise to a morphism of functors $\Phi: I_{\mathbf{P}(X)} \to TS$. To prove this we have to show the commutativity of the following diagram for any morphism $\nu: P \to Q$ in $\mathbf{P}(X)$:

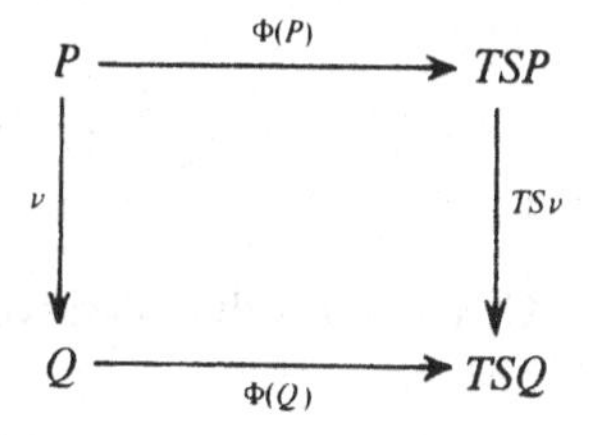


Take an open set $U \subset X$ and $s \in PU$. Then

$$\Phi_U(P)s = (x \in U \mapsto [s]_x) = \tilde{s} \in \Gamma(U, SP).$$

Furthermore $S\nu: SP \to SQ$ operates as $[s]_x \mapsto [\nu_V s]_x$ for $x \in V$ and $s \in PV$. So

$$((TS\nu)_U \circ \Phi_U(P))s = ((x \in U) \mapsto [\nu_U s]_x) = (\Phi_U(Q) \circ \nu_U)s.$$

The functor $S: \mathbf{P}(X) \to \mathbf{C}(X)$ is called the *sheafification functor* and the morphism of functors $\Phi: I_{\mathbf{P}(X)} \to TS$ is called the *sheafification morphism*.

**4.4.3 Proposition** There exists an isomorphism of functors $\Psi: ST \to I_{\mathbf{C}(X)}$.

*Proof.* Let $F$ be a concrete sheaf on $X$. Then $TF = \Gamma(-, F)$ is a presheaf on $X$. The points of the total space $F'$ of $STF$ are $[s]_x$, with $x \in X$ and $s \in \Gamma(U, F)$ for $U$ open and containing $x$. Consider the mapping $f: F' \to F$ given by $[s]_x \mapsto sx$. This mapping is injective due to 4.2.3($g$) and open according to 4.2.3($i$). From the injectivity it follows that $f^{-1}(sU) = \tilde{s}U$ for $s \in \Gamma(U, F)$. This means that $f$ is continuous by 4.2.3($j$). It is clear that $f$ is also surjective. So $f: F' \to F$ is an isomorphism in **Top**. Since for the projections $\pi$ of $F$ and $\pi'$ of $F'$ evidently $\pi \circ f = \pi'$ and $\pi' \circ f^{-1} = \pi$, the isomorphism $f$ indeed yields an isomorphism $\Psi(F): STF \xrightarrow{\sim} F$ in the category $\mathbf{C}(X)$. It remains to be proved that the diagram on top of page 204 commutes.

In this diagram $f$ and $g$ operate according to $[s]_x \mapsto sx$. The morphism $h'$ operates as $[s]_x \mapsto [h \circ s]_x$. Therefore $h \circ f$ and $g \circ h'$ both act as $[s]_x \mapsto (h \circ s)x$.

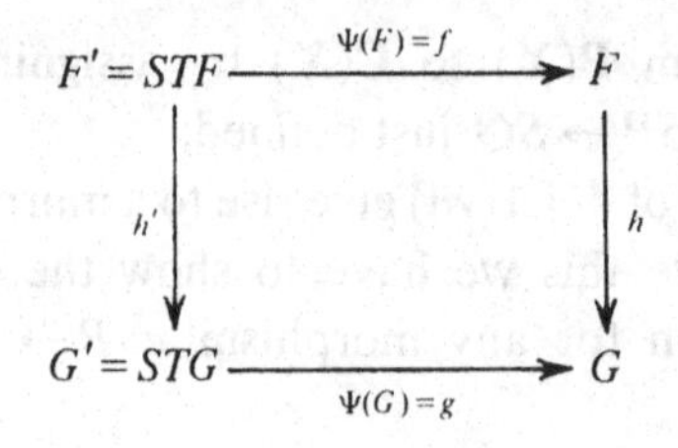


**4.4.4 Proposition**

$$\mathbf{P}(X) \underset{T}{\overset{S}{\rightleftarrows}} \mathbf{C}(X)$$

is an adjoint situation where $T$ embeds $\mathbf{C}(X)$ as a full reflective subcategory of $\mathbf{P}(X)$.

*Proof.* In the first place for the morphisms of functors $\Phi\colon I_{\mathbf{P}(X)} \to TS$ and $\Psi\colon ST \to I_{\mathbf{C}(X)}$ the following hold.

(i) For $P \in \mathbf{P}(X)$, $SP \xrightarrow{S\Phi(P)} STSP \xrightarrow{\Psi(SP)} SP = SP \xrightarrow{1_{SP}} SP$.

(ii) For $F \in \mathbf{C}(X)$, $TF \xrightarrow{\Phi(TF)} TSTF \xrightarrow{T\Psi(F)} TF = TF \xrightarrow{1_{TF}} TF$.

Since we already know that $\Psi(SP)$ and $\Psi(F)$ (and hence also $T\Psi(F)$) are isomorphisms, it remains to be shown that $S\Phi(P)$ and $\Phi(TF)$ are their respective inverses. Analysing the first sequence for $x$ in an open set $U \subset X$ and for $s \in PU$ we find $[s]_x \mapsto [\tilde{s}]_x \mapsto \tilde{s}x = [s]_x$. In the second sequence we find for $s \in \Gamma(U, F) = (TF)U$ that $s \mapsto \tilde{s} \mapsto f \circ \tilde{s}$ if we denote $\Psi(F)$ by $f$. To show that $s = f \circ \tilde{s}$, take $x \in U$. Then we have $(f \circ \tilde{s})x = f(\tilde{s}x) = f([s]_x) = sx$.

Inferring from the fact that $\Psi$ is an isomorphism that $T$ embeds the category of concrete sheaves fully into the category of presheaves (see theorem 1.7.8) we have proved our proposition.

**Exercises** (*a*) Apply the functor $S$ to the presheaves $P$ and $Q$ of examples (*c*) and (*d*) in 4.3.1. Also determine $\Phi(P)$ and $\Phi(Q)$.

(*b*) A presheaf $P$ on $X$ is almost trivial if and only if $SP$ is a trivial concrete sheaf (see 4.3.4).

(*c*) Apply the functor $S$ to the presheaf of germs of continuous functions (example (*b*) in 4.3.1) and compare with 4.2.4.

(*d*) Let $D$ be a set. We define the presheaf $\tilde{D}$ on $X$ by $\tilde{D}U = D$ for all open $U \subset X$ and $\tilde{D}^U_V = 1_D$ for open $V \subset U$; $\tilde{D}$ is called the *constant presheaf on $X$ with stalk $D$*. Prove that $S\tilde{D}$ is isomorphic with the concrete sheaf $X \times D \to X$ of example (*c*) of 4.2.2. The concrete sheaf $\hat{D} = S\tilde{D}$ is therefore a constant sheaf on $X$ with stalk $D$.

## 4.5 Sheaves

**4.5.1 Definition** A *sheaf on X* is a presheaf $P$ on $X$ such that there is a concrete sheaf $F$ on $X$ with the property that $P \simeq TF$ in $\mathbf{P}(X)$. We denote this full subcategory of $\mathbf{P}(X)$ by $\mathbf{F}(X)$. The embedding $\mathbf{F}(X) \to \mathbf{P}(X)$ will be denoted by $T'$. The functor $T$ will be factored as $T'J$ where $J: \mathbf{C}(X) \to \mathbf{F}(X)$. Since $T$ is a faithful embedding it follows that $J$ is an equivalence (see 1.7.11). Write $S' = JS$. Put together in a diagram we have

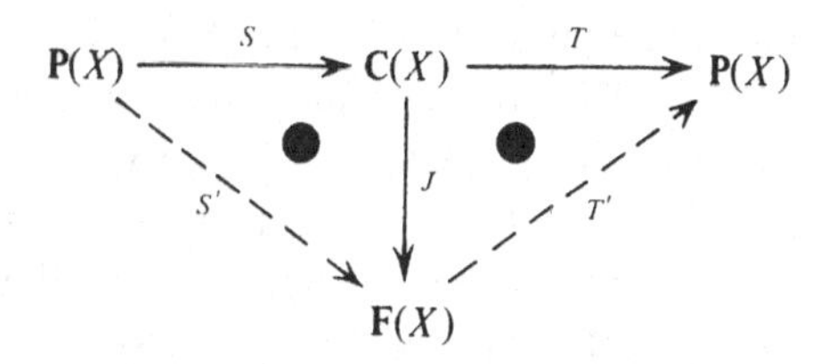


Now $\mathbf{F}(X)$ is a full reflective subcategory of $\mathbf{P}(X)$ and the functor $S'$ is left adjoint to $T'$. It is left to the reader to make the adjunctions explicit.

In the following we shall develop criteria for a presheaf to be a sheaf.

**4.5.2 Definition** An open covering $\{U_\alpha\}_\alpha$ of an open set $U \subset X$ is said to *separate* $s$ and $t \in PU$ for $P \in \mathbf{P}(X)$ provided there is an index $\alpha$ such that $P^U_{U_\alpha}s \neq P^U_{U_\beta}t$. A collection $\{s_\alpha \in PU_\alpha\}$ is called *compatible* provided for every pair of indices $\alpha$ and $\beta$ the equality $P^{U_\alpha}_{U_{\alpha\beta}}s_\alpha = P^{U_\beta}_{U_{\alpha\beta}}s_\beta$ holds, where $U_{\alpha\beta} = U_\alpha \cap U_\beta$.

**4.5.3 Theorem** A presheaf $P$ on $X$ is a sheaf on $X$ if and only if it satisfies the following two conditions.

S(i) For every open set $U \subset X$ and every pair $s, t \in PU$ with $s \neq t$, every open covering $\{U_\alpha\}_\alpha$ of $U$ separates $s$ and $t$.

S(ii) For every open set $U \subset X$ and every open covering $\{U_\alpha\}_\alpha$ of $U$ and every compatible system $\{s_\alpha \in PU_\alpha\}_\alpha$, there is an element $s \in PU$ with $s_\alpha = P^U_{U_\alpha}s$ for all $\alpha$.

*Proof.* We will give the proof in three steps, each step being a separate lemma.

**Lemma (a)** A presheaf $P$ on $X$ is a sheaf on $X$ if and only if $P$ and $TSP$ are isomorphic.

*Proof.* If $P$ and $TSP$ are isomorphic $P$ is a sheaf by definition. Suppose conversely that $P$ is a sheaf. Then there is a concrete sheaf $F \in \mathbf{C}(X)$ and an isomorphism in $\mathbf{P}(X)$, say $f: P \rightrightarrows TF$. Then $TSTf: TSP \rightrightarrows TSTF$ while $T\Psi(F): TSTF \rightrightarrows TF$. Compose with $f^{-1}$ to obtain $TSP \rightrightarrows P$.

**Lemma (b)** For $P \in \mathbf{P}(X)$ and $U$ open in $X$ the map $\Phi_U(P): PU \to \Gamma(U, SP)$ is injective if and only if $P$ satisfies condition S(i) (such a presheaf will be called a *monic presheaf*).

*Proof.* Let $\Phi_U(P)$ be injective. Take $s_1$ and $s_2 \in PU$ with $s_1 \neq s_2$. Suppose $\{U_\alpha\}_\alpha$ is an open covering of $U$. Then $\Phi_U(P)s_i = \tilde{s}_i$ with $\tilde{s}_i: x \to [s_i]_x$ $(i = 1, 2)$. As $\Phi_U(P)$ is injective, so that $\Phi_U(P)s_1 \neq \Phi_U(P)s_2$, there must be a point $x \in U$ with $[s_1]_x \neq [s_2]_x$. Thus there is an index $\alpha$ with $x \in U_\alpha$ and $P^U_{U_\alpha}s_1 \neq P^U_{U_\alpha}s_2$. This means that $\{U_\alpha\}_\alpha$ separates $s_1$ and $s_2$.

Now suppose $\Phi_U(P)$ is not injective. Then there are $s_1$ and $s_2$ in $PU$ with $s_1 \neq s_2$ and $\Phi_U(P)s_1 = \Phi_U(P)s_2$. Hence $[s_1]_x = [s_2]_x$ for all $x \in U$. Every $x$ therefore has an open neighbourhood $U_x$ with $P^U_{U_x}s_1 = P^U_{U_x}s_2$. This open covering $\{U_x\}_{x \in U}$ does not separate $s_1$ and $s_2$.

**Lemma (c)** Suppose $P \in P(X)$ is a monic presheaf. Then $P$ is a sheaf if and only if $P$ satisfies S(ii).

*Proof.* Let $\Phi_U(P)$ be surjective, $\{U_\alpha\}_\alpha$ an open covering of $U$ and $\{s_\alpha \in PU_\alpha\}$ a compatible set. For every index $\alpha$ we have $\Phi_{U_\alpha}(P)s_\alpha = \tilde{s}_\alpha$ with $\tilde{s}_\alpha: U_\alpha \to [s_\alpha] \subset F$ where $F$ denotes the total space of $SP$. Due to the compatibility of the collection $\{s_\alpha \in PU_\alpha\}$ it follows that $\tilde{s}_\alpha | U_{\alpha\beta} = \tilde{s}_\beta | U_{\alpha\beta}$, again writing $U_{\alpha\beta} = U_\alpha \cap U_\beta$. Thus one may define $t: U \to F$ by $t | U_\alpha = \tilde{s}_\alpha$. Then $t \in \Gamma(U, SP)$. Since $\Phi_U(P)$ is surjective there exists an elements $s \in PU$ with $\Phi_U(P)s = t$. For this element $s$ we have

$$\Phi_{U_\alpha}(P) \circ P^U_{U_\alpha}s = \mathrm{res}^U_{U_\alpha} \circ \Phi_U(P)s = t | U_\alpha = \tilde{s}_\alpha = \Phi_{U_\alpha}(P)s_\alpha.$$

Since $\Phi_{U_\alpha}(P)$ is injective it follows that $P^U_{U_\alpha}s = s_\alpha$.

Now assume $\Phi_{U_\alpha}(P)$ is not surjective. Then there is an element $t \in \Gamma(U, SP)$ with the property that, for every $s \in PU$, $\Phi_U(P)s \neq t$. For $x \in U$ we have $tx = [t_x]_x$ for $t_x \in PV_x$, $V_x$ an open neighbourhood of $x$ in $U$. Hence $\tilde{t}x = t_x x$. Then $x$ has an open neighbourhood $U_x \subset V_x$ such that $t | U_x = \tilde{t}_x | U_x$. Call $P^{V_x}_{U_x}t_x = s_x \in PU_x$. Then $t | U_x = \tilde{s}_x$, so $\tilde{s}_x | U_{xy} = t | U_{xy} = \tilde{s}_y | U_{xy}$. Since $\Phi_{U_{xy}}(P)$ is injective for every $x$ and $y$, it follows that $\{s_x \in PU_x\}_{x \in U}$ is a compatible set for the open covering $\{U_x\}_{x \in U}$ of $U$. Suppose there was an element $s \in PU$ with $P^U_{U_x}s = s_x$. Then we would have

$$\mathrm{res}^U_{U_x} \circ \Phi_U(P)s = \Phi_{U_x}(P) \circ P^U_{U_x}s = \Phi_{U_x}(P)s_x = \tilde{s}_x = t | U_x$$

for all $x \in U$. This would mean $\Phi_U(P)s = t$, yielding a contradiction.

Let $P$ be a presheaf on $X$, let $U \subset X$ be open and let $\{U_\alpha\}_\alpha$ be an open covering of $U$. Consider the following diagram in **Sets**

$$PU \xrightarrow{d} \prod_\alpha P(U_\alpha) \overset{d_1}{\underset{d_2}{\rightrightarrows}} \prod_{\alpha,\beta} P(U_\alpha \cap U_\beta)$$

where $d$, $d_1$ and $d_2$ are defined by $ds = (P^U_{U_\alpha} s)_\alpha$; $d_1(s_\alpha)_\alpha = (P^{U_\alpha}_{U_{\alpha\beta}} s_\alpha)_{\alpha\beta}$ and $d_2(s_\beta)_\beta = (P^{U_\beta}_{U_{\alpha\beta}} s_\beta)_{\alpha\beta}$. Then it is clear that $d_1 \circ d = d_2 \circ d$. Furthermore S(i) is apparently equivalent with the injectivity of $d$, whereas S(ii) is equivalent with the fact that $d$ maps surjectively onto the equalizer of $d_1$ and $d_2$ for all open $U$ and all open coverings $\{U_\alpha\}_\alpha$. So $P$ is a sheaf if and only if $d$ is an equalizer of $d_1$ and $d_2$.

**4.5.4 Examples** (*a*) Consider for any open $U \subset X$ the presheaf $h^U$: $\mathbf{O}°(X) \to$ **Sets**. According to Yoneda's lemma we have for every presheaf $P$ on $X$ the isomorphism $(h^U, P) \simeq PU$. Denoting the concrete sheaf $\ulcorner U, i, X \urcorner$ by $H^U$, where $i$ is the inclusion $U \to X$, one sees that $h^U = TH^U$. So $h^U$ is even a sheaf for every open $U \subset X$. Expressed more functorially: $h^-$ is a contravariant functor from $\mathbf{O}(X)$ to $\mathbf{P}(X)$ (even from $\mathbf{O}(X)$ to $\mathbf{F}(X)$) and

$$\text{-(-)} \simeq (h^-, \text{-})_{\mathbf{P}(X)}: \mathbf{O}°(X) \times \mathbf{P}(X) \to \textbf{Sets} \quad \text{(see 1.4.2).}$$

Verify that for any concrete sheaf $F$ on $X$, $\Gamma(U, F) = (H^U, F)_{\mathbf{C}(X)}$. Also verify that $h^U$ satisfies the criteria S(i) and S(ii). Prove that $h^X$ is a final object in both $\mathbf{P}(X)$ and $\mathbf{F}(X)$ and that $h^\varnothing$ is an initial object in $\mathbf{F}(X)$ but not in $\mathbf{P}(X)$. Also prove the special case of the Yoneda lemma $PU \simeq (h^U, P)_{\mathbf{P}(X)}$ directly.

(*b*) An example closely connected with the previous one is the following. Let $X$ and $Y$ be two topological spaces. We define a presheaf $P_Y$ on $X$ by $P_Y U = (U, Y)_{\mathbf{Top}}$ for all open $U \subset X$ and for $V \subset U$ we define $P^U_{YV}$: $P_Y U \to P_Y V$ in the obvious way $(f: U \to Y) \mapsto (f|V)$. Thus the presheaf of example (*b*) in 4.3.1 is just $P_{\mathbf{R}}$. Verify that $P_Y$ satisfies the criteria S(i) and S(ii). Also verify that $F$: **Top** $\to \mathbf{F}(X)$ can be made into a functor (provided $X \neq \varnothing$) which is faithful.

(*c*) A finite presheaf (the points of the compass). Let $X$ be the set consisting of the four elements north, east, south and west: $X = \{n, e, s, w\}$. We endow $X$ with a topology whose open sets are $\varnothing$, $N = \{n\}$, $S = \{s\}$ and $A = \{n, s\}$, $E = \{n, e, s\}$, $W = \{n, w, s\}$ and $X = \{n, e, s, w\}$. One easily verifies that this indeed yields a topology for $X$ and that $\mathbf{O}°(X)$ looks like

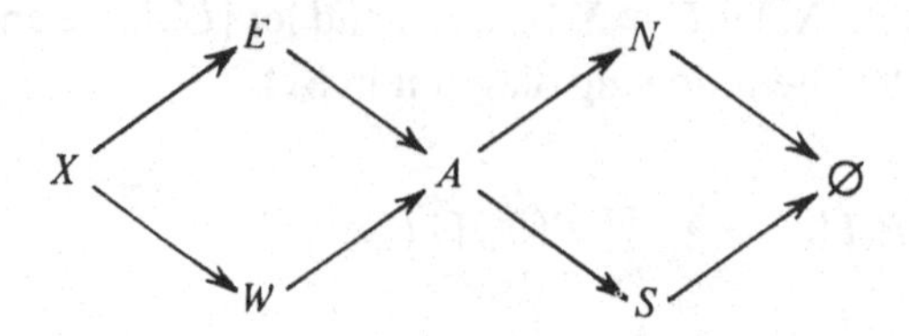


We construct the following presheaf $P$ on $X$: $PU = U$ for $U \neq \emptyset$ and $P\emptyset = \{a\}$. We define $P^U_V$: $PU \to PV$ by $x \mapsto x$ provided $x \in PV$ and if not then $x \mapsto x'$ where $x'$ is the point of the compass adjacent to $x$ going clockwise. One verifies easily that this yields a presheaf. Since the covering $\{N, S\}$ of $A \subset X$ does not separate the elements $n$ and $s$ of $PA$ (theorem 4.5.3) $P$ is not a sheaf. For the stalks of the presheaf $P$ we find $P_n \simeq N$, $P_e \simeq E$, $P_s \simeq S$ and $P_w \simeq W$. Hence the total space of the concrete sheaf $SP$ contains 8 points. Since $P_n$ and $P_s$ both contain only one element there is only one section in $\Gamma(A, SP)$. We see that $\Phi_A(P)$: $PA \to \Gamma(A, SP)$ is not injective so $P$ is not a sheaf. However $\Phi_A(P)$ is surjective. Compute the number of sections in the sets $\Gamma(U, SP)$ for $U \in \{N, E, S, W, \emptyset, X\}$.

Remark. Henceforth we shall more or less identify the categories $\mathbf{F}(X)$ and $\mathbf{C}(X)$. That is, we shall work in $\mathbf{F}(X)$ but whenever it is convenient we shall consider a sheaf $F$ as a concrete sheaf. So then $FU = \Gamma(U, F)$ and we choose our notation according to our point of view.

### 4.6 Change of base space

The functor $S'$: $\mathbf{P}(X) \to \mathbf{F}(X)$ looks like a natural transformation (or morphism) between the 'functors' $\mathbf{P}$ and $\mathbf{F}$: **Top** $\to$ **Cat**, where **Cat** is the pseudo-category (see 1.3.3) whose objects are categories and whose morphisms are functors between these categories. We therefore look for functors $\mathbf{P}(X) \to \mathbf{P}(Y)$ and $\mathbf{F}(X) \to \mathbf{F}(Y)$ induced by a morphism $f: X \to Y$ in **Top**, co- or contravariantly. Both possibilities occur as we shall presently see.

For the time being we stick to a fixed chosen continuous mapping $f: X \to Y$. Open sets in $X$ will be denoted by $U$, those in $Y$ by $V$. Presheaves on $X$ are written $P$, those on $Y$ are called $Q$. Sheaves on $X$ or $Y$ are denoted by $F$ or $G$ respectively. Morphisms in $\mathbf{P}(X)$ and $\mathbf{F}(X)$ will be denoted by $\xi$, those in $\mathbf{P}(Y)$ and $\mathbf{F}(Y)$ by $\eta$.

For our purpose we start to define the functor $\tilde{f}$: $\mathbf{O}^\circ(Y) \to \mathbf{O}^\circ(X)$ by $V \mapsto f^{-1}V$. Then to a presheaf $P$ on $X$ we attach the presheaf $P\tilde{f}$: $\mathbf{O}^\circ(Y) \to$ **Sets** on $Y$. For a morphism $\xi$: $P_1 \to P_2$ in $\mathbf{P}(X)$ we have the morphism $\xi\tilde{f}$: $P_1\tilde{f} \to P_2\tilde{f}$ in $\mathbf{P}(Y)$. Thus the assignment $P \mapsto P\tilde{f}$

apparently defines a functor from $\mathbf{P}(X)$ to $\mathbf{P}(Y)$. We denote this functor by $\tilde{f}_*: \mathbf{P}(X) \to \mathbf{P}(Y)$. Put concretely this amounts to the following. For $P \in \mathbf{P}(X)$ the presheaf $\tilde{f}_* P$ on $Y$ assigns to the open subsets $V_1$ and $V_2$ of $Y$ with $V_1 \supset V_2$ the sets $P(f^{-1}V_1)$ and $P(f^{-1}V_2)$ respectively and furthermore

$$\tilde{f}_* P(V_1 \to V_2) = P^{f^{-1}V_1}_{f^{-1}V_2}: P(f^{-1}V_1) \to P(f^{-1}V_2).$$

For $\xi: P_1 \to P_2$ in $\mathbf{P}(X)$ the morphism $\tilde{f}_*\xi: \tilde{f}_*P_1 \to \tilde{f}_*P_2$ in $\mathbf{P}(Y)$ is defined, for $V$ open in $Y$, as

$$(\tilde{f}_* \xi)_V = \xi_{f^{-1}V}: P_1(f^{-1}V) \to P_2(f^{-1}V).$$

**4.6.1 Proposition** The functor $\tilde{f}_*: \mathbf{P}(X) \to \mathbf{P}(Y)$ preserves sheaves.

*Proof.* Suppose $P \in \mathbf{F}(X)$. We have to show that $\tilde{f}_* P \in \mathbf{F}(Y)$. Let $V$ be open in $Y$ and $(V_\alpha)_\alpha$ be an open covering of $V$ in $Y$. Since $\bigcup_\alpha f^{-1}V_\alpha = f^{-1}V$ and $f^{-1}(V_\alpha \cap V_\beta) = f^{-1}V_\alpha \cap f^{-1}V_\beta$, the morphism $d'$ is the equalizer of $d_1'$ and $d_2'$, hence so is $d$ of $d_1$ and $d_2$ in the diagram

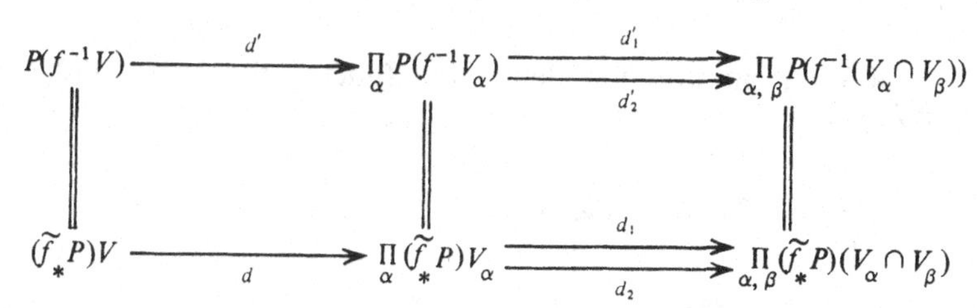


So $\tilde{f}_* P$ is a sheaf on $Y$ by 4.5.3.

According to 4.5.1 we may define $f_* = \tilde{f}_* | \mathbf{F}(X): \mathbf{F}(X) \to \mathbf{F}(Y)$. For a (pre)sheaf $P$ on $X$ we call $\tilde{f}_* P$ the *direct image of $P$ under $f$*. The following diagram commutes:

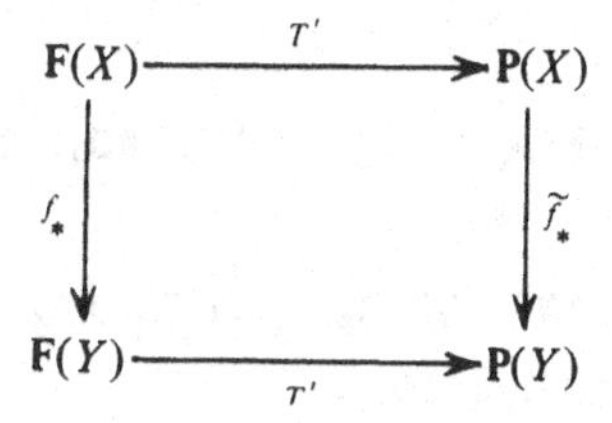


Note that we actually should have written $T'(X)$ in the upper row and $T'(Y)$ in the lower row to distinguish between these two functors.

**4.6.2 Definitions** Let $P$ be a presheaf on $X$. Then we define for any $A \subset X$, $P[A] = \varinjlim PU_{U \supset A}$ and for $A \supset B$,

$P_B^A$: $\varinjlim PU_{U\supset A} \to \varinjlim PU_{U\supset B}$; here $U$ ranges over the open subsets of $X$ containing $A$ and $B$ respectively. For open $U \subset X$ one of course has $PU \simeq P[U]$.

Now we are in a position to define a presheaf $P$ on $X$ for a given presheaf $Q$ on $Y$. Suppose $Q \in \mathbf{P}(Y)$. Then $\bar{f}^*Q \in \mathbf{P}(X)$ is defined by $((\bar{f}^*Q)U = Q[fU]$ and $(\bar{f}^*Q)(U_1 \to U_2) = Q_{fU_2}^{fU_1}$ for $U$, $U_1$ and $U_2$ open in $X$ and $U_1 \supset U_2$.

For a morphism $\eta: Q_1 \to Q_2$ in $\mathbf{P}(Y)$ we define $\bar{f}^*\eta: \bar{f}^*Q_1 \to \bar{f}^*Q_2$ by $(\bar{f}^*\eta)_U = \eta_{fU} = \varinjlim (\eta_V: Q_1V \to Q_2V)_{V\supset fU}$.

One may verify that $\bar{f}^*: \mathbf{P}(Y) \to \mathbf{P}(X)$ so defined is a functor.

**4.6.3 Proposition** $\bar{f}^*: \mathbf{P}(Y) \to \mathbf{P}(X)$ is left adjoint to $\bar{f}_*: \mathbf{P}(X) \to \mathbf{P}(Y)$.

*Proof.* It is a formal verification that $\bar{f}^*$ in the present situation is precisely the functor constructed in 1.8.6 (Kan extension theorem); $\bar{f}^*$ assigns to a presheaf $Q$ on $Y$ a presheaf $\bar{f}^*Q$ on $X$, its Kan extension along $\bar{f}$.

**4.6.4 Definition** We define the functor $f^*: \mathbf{F}(Y) \to \mathbf{F}(X)$ by $f^* = S'\bar{f}^*T'$:

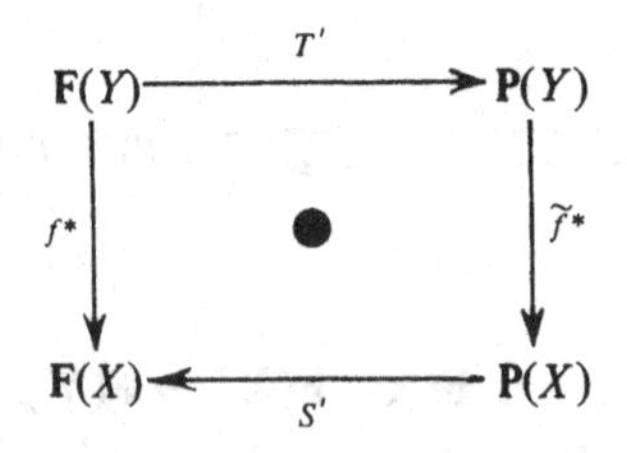


For $G \in \mathbf{F}(Y)$ the sheaf $f^*G$ on $X$ is called the *inverse image of G under f*.

**4.6.5 Proposition** $f^*: \mathbf{F}(Y) \to \mathbf{F}(X)$ is left adjoint to $f_*: \mathbf{F}(X) \to \mathbf{F}(Y)$.

*Proof.*
$$(f^*\text{-}, \text{-})_{\mathbf{F}(X)} = (S'\bar{f}^*T'\text{-}, \text{-})_{\mathbf{F}(X)} \simeq (\bar{f}^*T'\text{-}, T'\text{-},)_{\mathbf{P}(X)} \simeq (T'\text{-}, \bar{f}_*T'\text{-})_{\mathbf{P}(Y)}$$
$$= (T'\text{-}, T'f_*\text{-})_{\mathbf{P}(Y)} = (\text{-}, f_*\text{-})_{\mathbf{F}(Y)}.$$

**Exercises** (*a*) The functors $\bar{f}^*: \mathbf{P}(Y) \to \mathbf{P}(X)$ and $f^*: \mathbf{F}(Y) \to \mathbf{F}(X)$ preserve stalks in the following sense. For $x \in X$ and $Q \in \mathbf{P}(Y)$ there is an isomorphism $\sigma_x^Q: (\bar{f}^*Q)_x \rightrightarrows Q_{fx}$ and for $\eta: Q \to Q'$ the following diagram commutes:

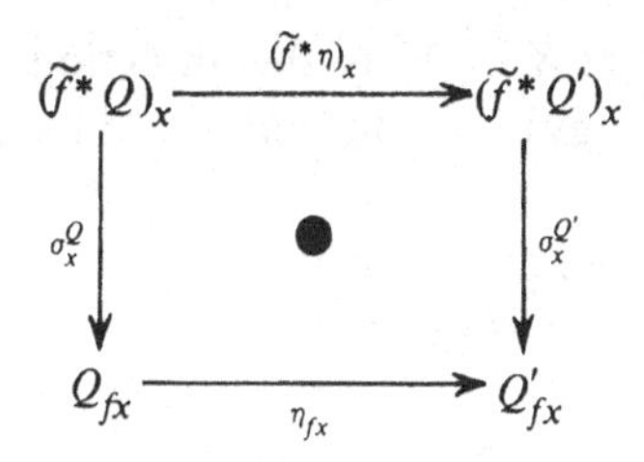


Similarly for the functor $f^*$ operating on sheaves.

(*b*) The functor $\tilde{f}^*$: $\mathbf{P}(Y) \to \mathbf{P}(X)$ preserves monic presheaves; i.e. if $Q \in \mathbf{P}(Y)$ satisfies S(i) then so does $\tilde{f}^*Q \in \mathbf{P}(X)$.

(*c*) However, in general $\tilde{f}^*$: $\mathbf{P}(Y) \to \mathbf{P}(X)$ does not preserve sheaves. consider $f$: $\mathbb{R} \to \mathbb{R}/\mathbb{Z}$. Take $F \in \mathbf{F}(\mathbb{R}/\mathbb{Z})$, the sheaf of continuous functions $g$: $V \to \mathbb{R}$ with $V$ open in $\mathbb{R}/\mathbb{Z}$. According to 4.5.4 (*b*) this is a sheaf on $\mathbb{R}/\mathbb{Z}$. Verify that $\tilde{f}^*F$ is not a sheaf on $\mathbb{R}$.

(*d*) Consider $d$: $\mathbb{R}_{\text{dis}} \to \mathbb{R}$ where $\mathbb{R}_{\text{dis}}$ denotes the set of real numbers with the discrete topology and $d$ is the identity mapping. Take for $F \in \mathbf{F}(\mathbb{R})$ the sheaf of continuous functions on open subsets of $\mathbb{R}$. Verify that $\tilde{f}^*F$ is not a sheaf on $\mathbb{R}_{\text{dis}}$.

(*e*) Let $Y$ be the set of real numbers with the topology defined by the open sets $\{V \backslash A \mid V$ is open in $\mathbb{R}$, $A$ is countable$\}$. Then the identity mapping $g$: $Y \to \mathbb{R}$ is continuous. For the sheaf $F \in \mathbf{F}(\mathbb{R})$ of continuous functions on open subsets of $\mathbb{R}$ we consider $G = \tilde{g}^*F \in \mathbf{P}(Y)$. Prove that $G(V \backslash A) = FV$. Prove that $G$ is a sheaf on $Y$ with the same stalks as $F$.

(*f*) Let $Y$ be as in (*e*) and let $X \subset Y$ be the subset of the rational numbers with the induced topology. Let $f$: $X \to Y$ be the embedding. Then $X$ has the discrete topology. Consider $\tilde{f}^*G \in \mathbf{P}(X)$, where $G$ is as defined in (*e*). Verify that $\tilde{f}^*G$ is not a sheaf on $X$. However $\tilde{f}^*G$ has the same stalks as $G$ and hence the same stalks as $F$. Since $X$ is discrete it is easy to see what the sheaf $f^*G \in \mathbf{F}(X)$ looks like:

$$(f^*G)A = \prod_{x \in A} F_x$$

for $A \subset X$. Now consider $f_*f^*G \in \mathbf{F}(Y)$. It turns out that

$$(f_*f^*G)(V \backslash A) = \prod_{x \in (V \backslash A) \cap X} F_x$$

for $V$ open and $A$ countable in $\mathbb{R}$. For $y \notin X$ we thus find that $(f_*f^*G)_y$ consists of just one element. For $x \in X \subset Y$ we have $(f_*f^*G)_x \simeq F_x$. Therefore $f_*f^*G \neq G$ in $\mathbf{F}(Y)$ for $f$: $X \hookrightarrow Y$.

In order to investigate the composite functors $\tilde{f}_*\tilde{f}^*$ and $\tilde{f}^*\tilde{f}_*$ (and similarly for their pendants operating on sheaves) we first establish the following.

**4.6.6 Proposition** Suppose $f: X \to Y$ is the injection of the subspace $X$ (with the induced topology) into the topological space $Y$. Then:

(i) $\tilde{\Psi}: \tilde{f}^* \tilde{f}_* \to I_{\mathbf{P}(X)}$ is an isomorphism of functors;

(ii) $\Psi: f^* f_* \to I_{\mathbf{F}(X)}$ is an isomorphism of functors;

(iii) $\tilde{f}_*$ is fully faithful;

(iv) $f_*$ is fully faithful.

*Proof.* Due to theorem 1.7.8 we have (i)$\Leftrightarrow$(iii) and (ii)$\Leftrightarrow$(iv). Since trivially (iii)$\Rightarrow$(iv) we only have to prove (i).

In 4.6.3 we constructed $\tilde{f}^* Q$ as the Kan extension of the presheaf $Q \in \mathbf{P}(Y)$ along $\tilde{f}: \mathbf{O}^\circ(Y) \to \mathbf{O}^\circ(X)$. Now $\tilde{f}$ is fully faithful, hence the remark after 1.8.6 proves (i).

**Exercises** (*a*) Prove that under the conditions of the last proposition $\tilde{f}_*$ and $f_*$ preserve stalks in the sense that $(\tilde{f}_* P)_{fx} \simeq P_x$ with commutativity similar to that in exercise 4.6.5(*a*), and the same for $f_*$. Relate this fact with (i) in 4.6.6. Let $F$ be a sheaf on $X$. Show that $(f_* F)_y \simeq pt$ for $y \in Y \backslash \bar{X}$ and that for $y \in \bar{X} \backslash X$ there are sheaves $F_1$ and $F_2$ on $X$ such that $(f_* F_1)_y = \varnothing$ and $(f_* F_2)_y \neq \varnothing$.

(*b*) Let $X$ be an open subset of $Y$ with the induced topology and let $f: X \to Y$ be the injection. Then we have the full embedding $\underset{\sim}{f}: \mathbf{O}^\circ(X) \to \mathbf{O}^\circ(Y) T$; show that $\tilde{f}^*$ is isomorphic to the functor $Q \mapsto Q\underset{\sim}{f}$. Show that $\tilde{f}^*$ preserves sheaves. Construct with 1.8.6 (Kan extension theorem) a left adjoint $\tilde{f}_!: \mathbf{P}(X) \to \mathbf{P}(Y)$ to $\tilde{f}^*$ and show that $f_! = S' \tilde{f}_! T': \mathbf{F}(X) \to \mathbf{F}(Y)$ is a left adjoint of $f^*: \mathbf{F}(Y) \to \mathbf{F}(X)$. Prove that $(f_! F)_y = \varnothing$ for $Y \backslash \bar{X}$ and hence $f_! \neq f_*$. Show that the adjunction $F \to f^* f_! F$ is an isomorphism and that $f_!$ is fully faithful and makes $\mathbf{F}(X)$ a full coreflective subcategory of $\mathbf{F}(Y)$.

**4.6.7 Proposition** Let $f: X \to Y$ be surjective. Then $f^*: \mathbf{F}(Y) \to \mathbf{F}(X)$ is faithful and hence the adjunction $\Phi: I \to f_* f^*$ is a monomorphism.

*Proof.* Consider morphisms $\eta, \eta': G_1 \to G_2$ in $\mathbf{F}(Y)$ with $f^* \eta = f^* \eta'$. For $y \in Y$ there is an $x \in X$ with $fx = y$ and, by exercise (*a*) in 4.6.5 it follows from $(f^* \eta)_x = (f^* \eta')_x$ that $\eta_y = \eta'_y$. So $\eta = \eta'$ which proves that $f^*$ is faithful. But then the adjunction $I \to f_* f^*$ is monic by the remark preceding 2.1.3.

Consider the continuous mapping $d: X_{\text{dis}} \to X$ in a topological space $X$ for which $X_{\text{dis}}$ is the same set with the discrete topology and where $d$ is the identity mapping. For a sheaf $F$ on $X$ the presheaf $\tilde{d}^* F$ is a presheaf with the same stalks as $F$. The sheafification of $\tilde{d}^*F$ yields the sheaf $d^*F$ with $(d^*F)A = \prod_{x\in A} F_x$ for $A \subset X$. Applying $d_*$ to $d^*F$ we obtain the sheaf $d_*d^*F$ on $X$ with $(d_*d^*F)U = \prod_{x\in U} F_x$. The stalks of $d_*d^*F$ are

$$(d_*d^*F)_{x_0} = \varinjlim_{U \ni x_0} \left( \prod_{x\in U} F_x \right).$$

In general the stalks of this sheaf are strictly larger than the stalks of $F$. This is expressed by saying that the functor $d_*d^*$ blows up the stalks.

**Example** Consider the constant concrete sheaf $F = \mathbb{R} \times D$ on $\mathbb{R}$ where $D$ is a set with the discrete topology. Then the stalks of $F$ are $F_x \simeq D$. Verify that the stalks of $d_*d^*F$ have the cardinality of the set of all mappings of $\mathbb{R}$ to $D$.

Concluding we can say that for an arbitrary continuous map $f: X \to Y$ in general $f^*f_* \not\simeq I_{\mathbf{F}(X)}$ and $f_*f^* \not\simeq I_{\mathbf{F}(Y)}$.

For $f: X \hookrightarrow Y$ one has $f^*f_* \simeq I_{\mathbf{F}(X)}$ but in general $f_*f^* \not\simeq I_{\mathbf{F}(Y)}$.

For a surjection $f: X \to Y$ one has $FU \rightarrowtail (f_*f^*F)U$ for any $F \in \mathbf{F}(Y)$ and $U$ open in $Y$.

**4.6.8 Proposition** Let $\ulcorner G, \rho, Y \urcorner \in \mathbf{C}(Y)$ and let $f: X \to Y$ be a morphism in **Top**. Consider the following cartesian square in **Top**:

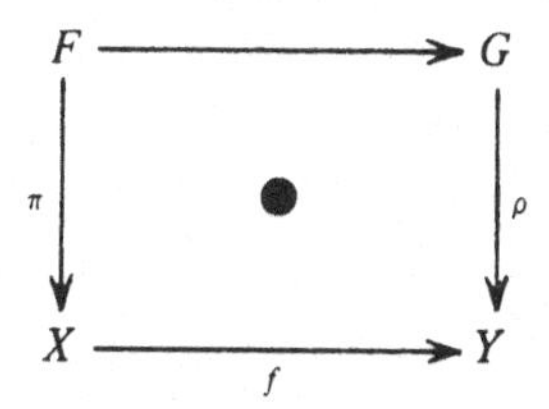


Then $\ulcorner F, \pi, X \urcorner \in \mathbf{C}(X)$ and $JF \simeq f^*JG$.

*Proof.* For the construction of the cartesian square in **Top** we refer to 1.5.5(*b*). A base for the topology of $X \times_Y G$ is given by the set $\{U \times_Y W\}_{U,W}$ with $U$ open in $X$ and $W$ open in $G$. The map $\pi$ is continuous by construction. We need to prove it is a local homeomorphism. Then we have

$$\begin{aligned}\pi(U \times_Y W) &= \{x \in U \mid \text{there exists an element } g \in W \text{ with } fx = \rho g\} \\ &= \{x \in U \mid fx \in \rho W\} \\ &= U \cap f^{-1}\rho W.\end{aligned}$$

Thus $\pi(U \times_Y W)$ is open, so $\pi$ is an open mapping.

Suppose $\rho|W$ is injective. We claim that then $\pi|(X \times_Y W)$ is injective. For suppose that $\pi\ulcorner x_1, g_1\urcorner = \pi\ulcorner x_2, g_2\urcorner$ for $\ulcorner x_i, g_i\urcorner \in X \times_Y W$ $(i = 1, 2)$. Then $x_1 = x_2$ and so $\rho g_1 = fx_1 = fx_2 = \rho g_2$. since $\rho|W$ is injective this implies $g_1 = g_2$. Thus $\pi$ is locally injective.

The projection $\pi: F \to X$ being continuous, open and locally injective we have shown that $\ulcorner F, \pi, X\urcorner$ is a concrete sheaf.

For the second part of the assertion, consider $\ulcorner M, \mu, X\urcorner = Sf^* TG$. Since the functors $T, f^*$ and $S$ preserve stalks we have $M_x \simeq G_{fx}$. As $M$ is the disjoint union of its stalks we may identify $M$ with

$$\{\ulcorner x, g\urcorner | x \in X, g \in G_{fx}\} = \{\ulcorner x, g\urcorner | fx = \rho g\}.$$

For open $U \subset X$ and $s \in (f^* TG)U = \varinjlim \Gamma(V, G)_{V \supset fU}$ we introduced the section $\tilde{s}: U \to M$ for the construction of $S(f^* TG)$. Going over the identifications made it turns out that $\tilde{s} = (x \in U) \mapsto \ulcorner x, tfx\urcorner$ for any $t \in \Gamma(V, G)$ with $V \supset fU$ and $s = \varinjlim (\text{res}\, {}^{V}_{V'})_{V' \supset fU} t$. But the collection of all $\tilde{s}U = \{\ulcorner x, tfx\urcorner | x \in U\}$ constitutes a base for the topology of $M$. Define the map $a: M \to G$ by $(\ulcorner x, g\urcorner \in M) \mapsto g$. Then $a$ is continuous. For let $\ulcorner x_0, g_0\urcorner \in M$ and let $W$ be an open neighbourhood of $g_0 \in G$. Then there is an open neighbourhood $V$ of $fx_0$ in $Y$ and a section $t \in \Gamma(V, G)$ such that $tfx_0 = g_0$ and $tV \subset W$. Then we have the open set $\{\ulcorner x, tfx\urcorner | x \in f^{-1}V\}$ containing $\ulcorner x_0, g_0\urcorner$ for which $a\{\ulcorner x, tfx\urcorner | x \in f^{-1}V\} \subset W$.

In the following diagram $\rho a = f\mu$.

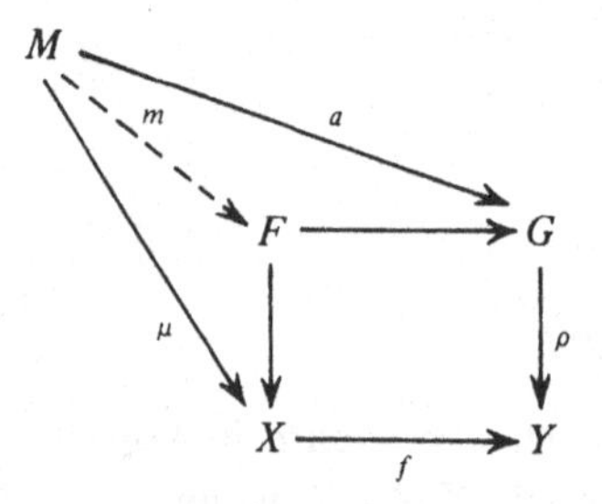


Hence there is a unique $m: M \to F$ making the diagram commutative. It follows that $m\ulcorner x, g\urcorner = \ulcorner x, g\urcorner$. Hence $m$ is bijective. It then follows from 4.2.6 that $m$ yields an ismorphism between the concrete sheaves $M$ and $F$: $F \simeq Sf^* TG$. Hence $JF \simeq JSf^* TG = S'f^* T'JG = f^* JG$.

**Corollary** Let $X$ be a subspace of $Y$ with the induced topology and let $f: X \to Y$ denote the injection. Let $\ulcorner G, \rho, Y\urcorner$ be a concrete sheaf on $Y$. Then $f^* G$ is isomorphic to the concrete sheaf $\ulcorner F, \pi, X\urcorner$ defined by $F = \rho^{-1}X \subset G$ with the induced topology and $\rho|F = \pi: F \to X$.

**Exercise** Consider $\ulcorner F_1, \pi_1, X \urcorner$ and $\ulcorner F_2, \pi_2, X \urcorner$ and the fibred product $F_1 \times_X F_2$ in **Top**. (We reserve our usual notation $\prod$, see 1.5.4, for the category $\mathbf{C}(X)$.)

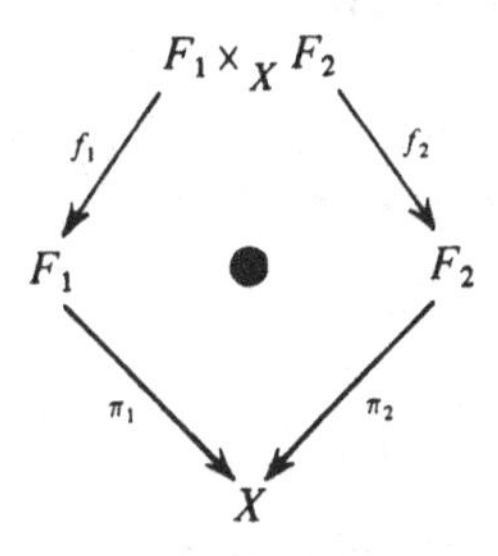


Let $\pi = \pi_1 f_1 = \pi_2 f_2$. Verify that $\ulcorner F_1 \times_X F_2, \pi, X \urcorner$ is a concrete sheaf and show that it is the product of the sheaves $F_1$ and $F_2$ in the category $\mathbf{C}(X)$.

We may therefore denote this sheaf by $F_1 \prod F_2$ (if no confusion can arise). Thus we find that $\mathbf{F}(X)$ has finite products. The stalks are the products of the stalks of the factors.

### 4.7 A pseudo-categorical survey

Certain functorial properties of our constructions are expressed in the following rules. For $1_X: X \to X$, $f: X \to Y$ and $g: Y \to Z$ in **Top** we have

(*a*) $(\tilde{1}_X)_* = I_{\mathbf{P}(X)}$ and $\tilde{g}_* \tilde{f}_* = (\widetilde{gf})_*$
(*b*) $(1_X)_* = I_{\mathbf{F}(X)}$ and $g_* f_* = (gf)_*$
(*c*) $(\tilde{1}_X)^* \simeq I_{\mathbf{P}(X)}$ and $\tilde{f}^* \tilde{g}^* \simeq (\widetilde{gf})^*$
(*d*) $(1_X)^* \simeq I_{\mathbf{F}(X)}$ and $f^* g^* \simeq (gf)^*$ (see 4.6.8).

The reader should prove these facts and convince himself that the isomorphisms cannot be replaced by identities. Using 1.7.5 one sees that both $f^* S'_Y$ and $S'_X \tilde{f}^*$ are left adjoints of $T'_Y f_* = \tilde{f}_* T'_X$ and hence isomorphic functors. Ignoring foundational difficulties from set theory, one can conceive of $\mathbf{P}$ and $\mathbf{F}$ as functors from **Top** to **Cat** (see 1.3.3), both in a covariant and a contravariant sense. Then $T': \mathbf{F} \to \mathbf{P}$ and $S': \mathbf{P} \to \mathbf{F}$ roughly play the role of natural transformations, $T'$ between the covariant functors, $S'$ between the contravariant ones.

### 4.8 Presheaves and sheaves of modules

**4.8.1 Definitions** For a topological space $X$ and an associative ring $R$ with identity we define ${}_R\mathbf{P}(X) = (\mathbf{O}^\circ(X), {}_R\mathbf{M})$. An object of ${}_R\mathbf{P}(X)$ will be called a *presheaf of modules on X*.

Remark. Since $\mathbf{O}^{\circ}(X)$ is small and since ${}_R\mathbf{M}$ is a Grothendieck category so is ${}_R\mathbf{P}(X)$ (compare 2.9.7). We shall frequently use the fact that the sequence $P_1 \to P_2 \to P_3$ is exact in ${}_R\mathbf{P}(X)$ if and only if the sequence $P_1U \to P_2U \to P_3U$ is exact in ${}_R\mathbf{M}$ for every open set $U \subset X$.

**4.8.2 Definition** A *concrete sheaf of modules on* $X$ is a concrete sheaf $\ulcorner F, \pi, X \urcorner$ for which the following properties hold.

(i) For every $x \in X$ the stalk $F_x \in {}_R\mathbf{M}$.

(ii) The addition $+$ defined on the stalks by

$$(\ulcorner f_1, f_2 \urcorner : \pi f_1 = \pi f_2) \mapsto f_1 + f_2$$

is continuous. In other words one has a continuous map $+ : F \times_X F \to F$ such that

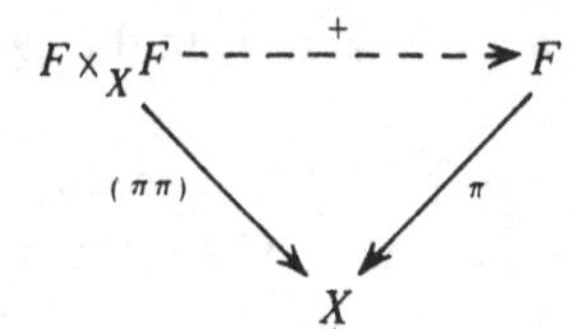


commutes in **Top**. In view of the exercise after 4.6.8 this means that the addition should be a morphism of concrete sheaves.

(iii) The scalar multiplication is continuous: for each $r \in R$ the mapping $r\cdot$ should be a morphism of concrete sheaves:

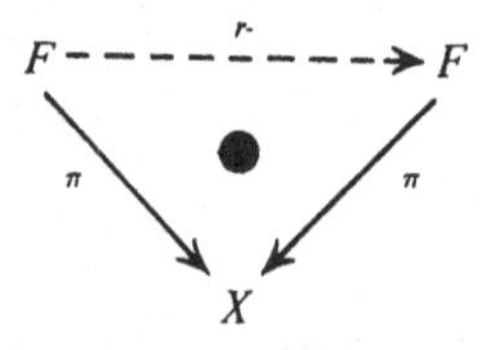


The category ${}_R\mathbf{C}(X)$ is the category whose objects are concrete sheaves of $R$-modules and whose morphisms are those morphisms in $\mathbf{C}(X)$ which induce $R$-homomorphisms on the stalks.

**4.8.3** The next items, which develop the theory, are given without full proofs.

(*a*) We want to show that there is a global null section $0 \in \Gamma(X, F)$. To see this, let $x$ be any point of $X$. Then there is a section $s_x \in \Gamma(U_x, F)$ for some open subset $U_x$ of $X$ containing $x$. Due to 4.8.2(iii) $-s_x \in \Gamma(U_x, F)$ also. From 4.8.2(ii) it follows that $s_x + (-s_x) \in \Gamma(U_x, F)$ being the composition of the two continuous mappings

$$\begin{pmatrix} s_x \\ -s_x \end{pmatrix} : U_x \to F \times_X F \quad \text{and} \quad F \times_X F \xrightarrow{+} F.$$

Since $(s_x+(-s_x))u=0_u\in F_u$ for all $u\in U_x$ we have shown that every point $x\in X$ has an open neighbourhood $U_x$ and a section $s_x\in(U_x, F)$ with the property that $s_xu=0$ for all $u\in U_x$. Patching all these local null sections together we obtain the global null section $0\in\Gamma(X, F)$.

(*b*) For any open subset $U$ of $X$ and any $\ulcorner F, \pi, X\urcorner\in {}_R\mathbf{C}(X)$ we may give $\Gamma(U, F)$ the structure of an $R$-module. By restriction of the global null section we always have $0|U\in\Gamma(U, F)$. Define $\Gamma(\varnothing, F)=0$. Verify that $\Gamma(\text{-}, F)$ becomes an object of ${}_R\mathbf{P}(X)$.

Next certain functors already constructed for sheaves and presheaves of topological spaces will be considered in the module case. While we do not mention this every time explicitly, they are all easily seen to be additive.

(*c*) Construct a functor ${}_RT$: ${}_R\mathbf{C}(X)\to{}_R\mathbf{P}(X)$ in the same way as the functor $T$ was constructed (4.3.1 example (*a*) and 4.3.2) such that the following diagram commutes:

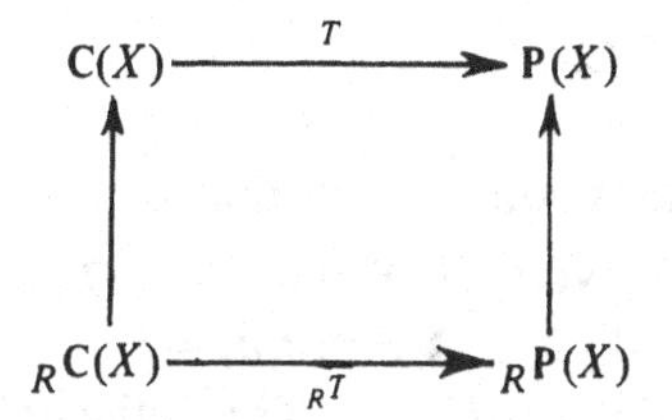


(*d*) Let $P\in{}_R\mathbf{P}(X)$. Give the stalks $P_x=\varinjlim PU_{U\ni x}$ the induced structure of an $R$-module. Verify that the object ${}_RSP$ defined this way is indeed an object of ${}_R\mathbf{C}(X)$ (see 4.4.2). Verify also that for $\nu$: $P\to P'$ in ${}_R\mathbf{P}(X)$ the morphism ${}_RS\nu$: ${}_RSP\to{}_RSP'$ is a morphism in ${}_R\mathbf{C}(X)$. Thus ${}_RS$: ${}_R\mathbf{P}(X)\to{}_R\mathbf{C}(X)$.

(*e*) Verify that the morphisms of functors $\Phi$ and $\Psi$ defined earlier may be considered as

$$\Phi: I_{{}_R\mathbf{P}(X)}\to{}_RT_RS \quad\text{and}\quad \Psi: {}_RS_RT\xrightarrow{\simeq} I_{{}_R\mathbf{C}(X)}.$$

Again ${}_RS$ is left adjoint to ${}_RT$ making ${}_R\mathbf{C}(X)$ a full reflective subcategory of ${}_R\mathbf{P}(X)$.

**4.8.4 Definition** The category ${}_R\mathbf{F}(X)$ is defined as the full subcategory of ${}_R\mathbf{P}(X)$ whose objects are those $P\in{}_R\mathbf{P}(X)$ such that there is an object $F\in{}_R\mathbf{C}(X)$ with $P\simeq TF$. Such an object is called a *sheaf of modules on X.*

One verifies easily that all of section 4.5 still holds in the case of presheaves and sheaves of modules. In particular $P\in{}_R\mathbf{P}(X)$ is a sheaf of $R$-modules if and only if $P$ satisfies S(i) and S(ii) of 4.5.3 or, equivalently, $d=\ker(d_1-d_2)$.

In the sequel we shall, unless stated otherwise, fix a ring $R$ and only consider presheaves and sheaves of $R$-modules and their morphisms. Therefore we shall usually omit the prefix $R$.

**4.8.5 Lemma** The functor $S'$: $\mathbf{P}(X) \to \mathbf{F}(X)$ preserves kernels.

*Proof.* Suppose $0 \to A \to B \to C$ is exact in $\mathbf{P}(X)$. That means that for every open set $U \subset X$ the sequence $0 \to AU \to BU \to CU$ is exact in ${}_R\mathbf{M}$. As ${}_R\mathbf{M}$ is a Grothendieck category it follows from AB5 that for every $x \in X$ the sequence $0 \to A_x \to B_x \to C_x$ is exact in ${}_R\mathbf{M}$.

We now consider $S'A$, $S'B$ and $S'C$ as concrete sheaves whose total space we will denote by $\tilde{A}$, $\tilde{B}$ and $\tilde{C}$ respectively. Since $A_x \to B_x$ is monic for each $x \in X$ the map $f: \tilde{A} \to \tilde{B}$ is injective and hence the morphism $S'A \to S'B$ is monic in $\mathbf{F}(X)$. Furthermore

$$S'A \to S'B \to S'C = S'A \overset{0}{\to} S'C.$$

Now suppose $F$ to be a concrete sheaf on $X$ such that $b: F \to S'B$ satisfies $0 = (F \overset{b}{\to} S'B \to S'C)$. Then, for every $x \in X$, $0 = (F_x \overset{b_x}{\to} B_x \to C_x)$ also. Hence there are unique $a_x: F_x \to A_x$ in ${}_R\mathbf{M}$ with $f_x a_x = b_x$.

Consider the mapping $a: F \to \tilde{A}$ given by $a|F_x = a_x$. This is the unique map satisfying $fa = b$. Since $f: \tilde{A} \to \tilde{B}$ is injective and according to 4.2.6 also open, it follows from the continuity of $b: F \to \tilde{B}$ that $a: F \to \tilde{A}$ is continuous. Hence $a$ is the unique morphism in $\mathbf{C}(X)$ such that $fa = b$, so $S'A \to S'B = \ker_{\mathbf{F}(X)}(S'B \to S'C)$.

Summarizing we thus far have an adjoint situation

$$\mathbf{P}(X) \underset{T'}{\overset{S'}{\rightleftarrows}} \mathbf{F}(X)$$

which makes $\mathbf{F}(X)$ a full reflective subcategory of $\mathbf{P}(X)$.

By theorem 1.9.5 $T'$ is left continuous and $S'$ right continuous. The last lemma tells us the latter is exact (see 2.7.5).

Furthermore the following statements hold.

**4.8.6** (*a*) The category $\mathbf{F}(X)$ is right complete. For if $D$ is a diagram in $\mathbf{F}(X)$, so is $T'D$ in $\mathbf{P}(X)$ where it has a supremum $L$. Then $S'L$ is a supremum of $S'T'D \simeq D$ in $\mathbf{F}(X)$.

(*b*) Since $\mathbf{P}(X)$ is also left complete (as ${}_R\mathbf{M}$ is left complete) it follows from theorem 1.9.6 that every diagram in $\mathbf{F}(X)$ has an infimum in $\mathbf{P}(X)$ which is also an infimum in $\mathbf{F}(X)$. Hence $\mathbf{F}(X)$ is left complete.

(*c*) Invoking theorem 2.7.7(iv) we may conclude from the fact that $\mathbf{P}(X)$ is abelian while $S'$ preserves kernels that $\mathbf{F}(X)$ is also abelian.

(*d*) $T'$ is left exact since it is left continuous.

(*e*) $\mathbf{P}(X)$ has a generator $G$. One proves easily that $S'G$ is a generator for $\mathbf{F}(X)$. Both categories therefore are locally small.

(*f*) Suppose $\{F'_\alpha\}_\alpha$, $\{F_\alpha\}_\alpha$ and $\{F''_\alpha\}_\alpha$ are inductive systems of sheaves in $\mathbf{F}(X)$ such that for every $\alpha$ the sequence $F''_\alpha \to F_\alpha \to F''_\alpha$ is short exact in $\mathbf{F}(X)$. Then $0 \to T'F'_\alpha \to T'F_\alpha \to T'F''_\alpha$ is exact. Since $\mathbf{P}(X)$ is a Grothendieck category we know that $0 \to \varinjlim T'F'_\alpha \to \varinjlim T'F_\alpha \to \varinjlim T'F''_\alpha$ is exact in $\mathbf{P}(X)$. Now $S'$ is exact and right continuous, therefore $0 \to \varinjlim S'T'F'_\alpha \to \varinjlim S'T'F_\alpha \to \varinjlim S'T'F''_\alpha$ is exact. But $\Psi: S'T' \rightrightarrows I_{\mathbf{F}(X)}$ yields exactness of $0 \to \varinjlim F'_\alpha \to \varinjlim F_\alpha \to \varinjlim F''_\alpha$. Hence $\varinjlim$ is exact so $\mathbf{F}(X)$ satisfies AB5.

**4.8.7 Theorem** The category $\mathbf{F}(X)$ is a Grothendieck category which is embedded by $T': \mathbf{F}(X) \to \mathbf{P}(X)$ as a full reflective subcategory of the Grothendieck category $\mathbf{P}(X)$. The left adjoint $S'$ of $T'$ is exact.

*Proof.* All the assertions have been proved in 4.8.6. We only mention the fact that it follows from 2.9.5 that $\mathbf{F}(X)$ has injectives and an injective cogenerator.

**Exercises** (*a*) Let $0 \to F' \to F \to F'' \to 0$ be a sequence in $\mathbf{F}(X)$. Then the following assertions are equivalent (we omit to write the embedding $T'$).

(i) The sequence is exact in $\mathbf{F}(X)$.

(ii) There is an object $P \in \mathbf{P}(X)$ with $S'P = 0$ such that $0 \to F' \to F \to F'' \to P \to 0$ is exact in $\mathbf{P}(X)$.

(iii) For every $x \in X$ the sequence $0 \to F'_x \to F_x \to F''_x \to 0$ is exact in ${}_R\mathbf{M}$.

(*b*) Let $M$ be a left $R$ module. Verify that $\tilde{M}$, as defined in 4.4.4 exercise (*d*), is in a natural way an object of ${}_R\mathbf{P}(X)$. Let $\hat{M} = S\tilde{M}$ be the constant sheaf with stalk $M$. Show that $\hat{M}$ is in a natural way an object of ${}_R\mathbf{C}(X)$ or ${}_R\mathbf{F}(X)$. Verify that the functors $M \to \tilde{M}$ and $M \to \hat{M}$ are exact functors ${}_R\mathbf{M} \to {}_R\mathbf{P}(X)$ and ${}_R\mathbf{M} \to {}_R\mathbf{F}(X)$ respectively.

All of the theory developed in 4.6 about functors between (pre)sheaves of sets induced by a continuous map $f: X \to Y$ carries over for (pre)sheaves of modules. All we have to do is introduce the obvious $R$-module structure from time to time and verify that the morphisms are $R$-homomorphisms. But now the base change functors enjoy additional exactness properties.

**4.8.8 Theorem** In the following commutative diagram attached to a morphism $f: X \to Y$ the functors $\tilde{f}_*$, $\tilde{f}^*$ and $f^*$ are exact while the functor $f_*$ is left exact.

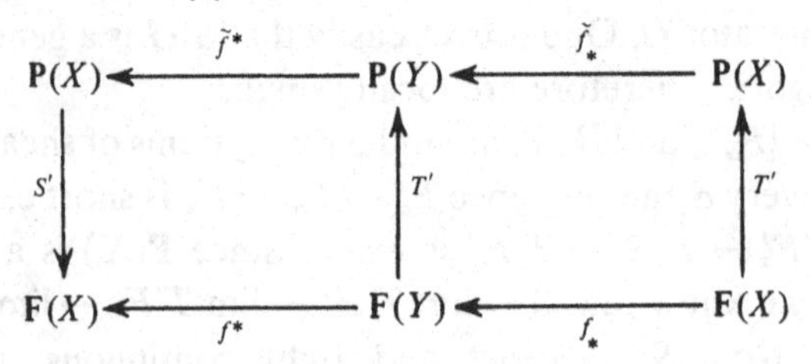


*Proof.* That the functor $\tilde{f}_*$ is exact follows directly from its definition: let $P_1 \to P_2 \to P_3$ be a short exact sequence in **P**($X$), then $\tilde{f}_*P_1 \to \tilde{f}_*P_2 \to \tilde{f}_*P_3$ is short exact since for every open $V$ in $Y$ the sequence $(\tilde{f}_*P_1)V \to (\tilde{f}_*P_2)V \to (\tilde{f}_*P_3)V$ or $P_1(f^{-1}V) \to P_2(f^{-1}V) \to P_3(f^{-1}V)$ is short exact in ${}_R\mathbf{M}$.

The exactness of $\tilde{f}^*$ follows from the fact that direct limits in ${}_R\mathbf{M}$ preserve exactness.

The exactness of $f^*$ follows from the relation $(f^*G)_x \simeq G_{fx}$ for $x \in X$, $G \in F(Y)$ and from exercise ($a$) after 4.8.7 ((i)⇔(iii)).

The left exactness of $f_*$ follows from the fact that $f_* \simeq S'\tilde{f}_*T'$, where $S'$ and $\tilde{f}_*$ are exact and $T'$ is left exact.

**Exercises** ($a$) We shall construct an example showing that $T'$ and $f_*$ are not exact in general.

Take $Y = \mathbb{R}^2$, $X = \mathbb{R}^2 \setminus \{0, 0)\}$ and $f: X \to Y$. For open $U \subset X$ let $FU$ denote the set of continuous functions $h: U \to \mathbb{R}$ and $GU$ the set of continuous mappings $g: U \to \mathbb{R}/\mathbb{Z}$. Consider $F$ and $G$ as sheaves on $X$ of $\mathbb{Z}$-modules in the obvious way. Then we have a morphism of sheaves $\nu: F \to G$ defined by

$$\nu_U: (h: U \to \mathbb{R}) \mapsto ((x \in U) \mapsto h(x) \bmod \mathbb{Z}).$$

One may verify easily that $\nu_x: F_x \to G_x$ is surjective for all $x \in X$ so $\nu: F \to G$ is epic in **F**($X$). One also verifies that $\nu_X: FX \to GX$ is not surjective so $T'\nu$ is not epic in **P**($X$). Considering the stalks of $f_*F$ and $f_*G$ in the origin it turns out that $f_*\nu$ is not epic in **F**($Y$). Therefore $T'$ and $f_*$ are not exact.

($b$) The forgetful functor $W: {}_R\mathbf{M} \to \mathbf{Sets}$ induces forgetful functors $\tilde{W}: {}_R\mathbf{P}(X) \to \mathbf{P}(X)$ and $\hat{W}: {}_R\mathbf{F}(X) \to \mathbf{F}(X)$. Verify that $T'$, $S'$, $\tilde{f}_*$, $\tilde{f}^*$, $f_*$ and $f^*$ commute with these forgetful functors in an obvious sense. The essential point here is that $W$ commutes with nonvoid direct limits (cf. 1.7.12) and that in the construction of $S'$ and $\tilde{f}^*$ no other suprema are used. Verify that most of exercise (*b*) after 4.6.6 can be adapted to sheaves of modules, but that $\tilde{f}_!$ and $f_!$ do not commute with $\tilde{W}$ and $\hat{W}$.

## 4.9 Subspaces and sheaves of modules

In the following we consider a topological space $X$, subsets $A \subset X$ with the induced topology and sheaves on $X$ and on $A$. For subsets $A$ and $B$ of $X$ we write as usual $A \backslash B = \{a \in A \mid a \notin B\}$.

**4.9.1 Definitions** The *restriction functor* $(-|A)$: $\mathbf{F}(X) \to \mathbf{F}(A)$ is defined by $(-|A) = i^*$ where $i: A \hookrightarrow X$ is the injection. Compare this with propositions 4.6.6 and 4.6.8; it is an exact functor by 4.8.8.

If $F_1$ is a sheaf on $A$ then $F$ on $X$ is called a *prolongation of* $F_1$ *to* $X$ if $F|A = F_1$ and $F|(X \backslash A) = 0$.

**4.9.2 Lemma** Let $\mu_1: F_1 \to G_1$ be a morphism in $\mathbf{F}(A)$. Assume $F$ and $G$ are prolongations to $X$ of $F_1$ and $G_1$ respectively. Then there is a unique $\mu: F \to G$ in $F(X)$ with $\mu|A = \mu_1$.

*Proof.* We consider $F$, $G$, $F_1$ and $G_1$ as concrete sheaves of modules. By proposition 4.6.8 we may assume $F_1 = \pi^{-1}A \subset F$ with the induced topology, and $G_1 = \rho^{-1}A \subset G$ likewise:

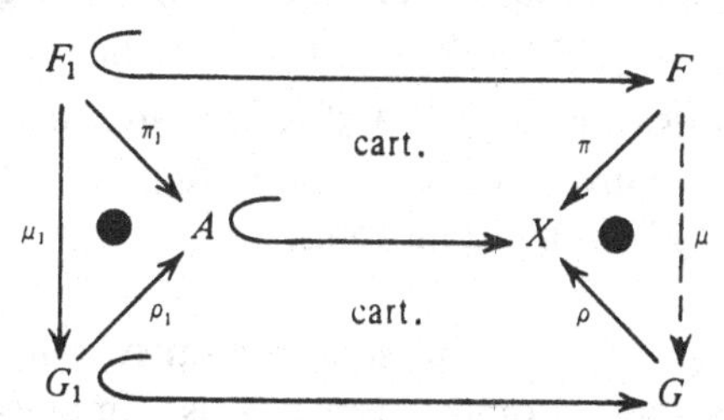


We denote the global null sections of $F$ and $G$ by $0_F$ and $0_G$ respectively. Then $F = F_1 \cup 0_F(X)$ and $G = G_1 \cup 0_G(X)$. So there is a unique mapping $\mu: F \to G$ with $\mu|F_1 = \mu_1$ and $\rho \circ \mu = \pi$. Since this mapping induces morphisms of $R$-modules in the stalks it suffices to prove that $\mu$ is continuous. As $0_F(X) \subset F$ is open and $\mu|0_F(X) = (0_G \circ \pi)|0_F(X)$ the mapping $\mu$ is continuous in the points of $0_F(X)$. Now consider a point $f \in F_1$ and an open neighbourhood $W$ of $\mu(f)$ in $G$. Since $\mu_1: F_1 \to G_1$ is continuous $f$ has an open neighbourhood $V_1 \subset F_1$ with $\mu(V_1) \subset W \cap G_1 \subset W$. But this means that there is an open neighbourhood $V \subset F$ of $f$ such that $V_1 = V \cap F_1$. Then $V \cap \pi^{-1}\rho W$ is an open neighbourhood of $f$ in $F$. We claim that $\mu(V \cap \pi^{-1}\rho W) \subset W$. This stems from the fact that $\mu(V \cap F_1) \subset W$ and $\mu(\pi^{-1}\rho W \backslash F_1) = W \backslash G_1 \subset W$. So now $\mu$ is also continuous in every point of $F_1$.

**Corollary** If $F_1 \in \mathbf{F}(A)$ has a prolongation to $X$ it is unique up to isomorphism.

**4.9.3 Definitions** If $F \in \mathbf{F}(A)$ has a prolongation to $X$ this prolongation is denoted by $F^X$. If $F \in \mathbf{F}(X)$ is such that $F|A$ has a prolongation to $X$ then this prolongation $(F|A)^X$ is also denoted by $F_A$.

A subset $A \subset X$ is called *locally closed* provided every $a \in A$ has an open neighbourhood $U \subset X$ such that $A \cap U$ is closed in $U$.

**Exercise** For any $A \subset X$ the following assertions are equivalent.

(i) $A$ is locally closed.

(ii) $A = U \cap C$ for an open $U \subset X$ and a closed $C \subset X$.

(iii) There is a closed $C \subset X$ such that $A \subset C$ is open for the induced topology on $C$.

**4.9.4 Proposition** Consider a closed subset $C \subset X$ and the open subset $U = (X \backslash C) \subset X$ with the injections $i: C \to X$ and $j: U \to X$. For any $F \in \mathbf{F}(X)$, $F_C$ and $F_U$ exist and there is a canonical exact sequence $0 \to F_U \to F \to F_C \to 0$ in $\mathbf{F}(X)$.

*Proof.* Use the adjunction morphism $\rho = \Phi(F): F \to i_* i^* F$ to get the exact sequence $0 \to K \to F \xrightarrow{\rho} i_* i^* F \to Q \to 0$. It now suffices to prove that $Q = 0$, $i_* i^* F$ is a prolongation of $F|C$ and $K$ is a prolongation of $F|U$. So all we have to do is to prove, in view of 4.8.7($a$),

(i) for $x \in C$, $\rho_x: F_x \to (i_* i^* F)_x$ is an isomorphism;

(ii) for $x \in U$, $(i_* i^* F)_x = 0$.

Since by 4.6.6 (ii) $\Psi: i^* i_* \to I_{\mathbf{F}(C)}$ is an isomorphism it follows that $\Psi(i^* F): i^* i_* i^* F \to i^* F$ is an isomorphism. This implies that $i^* \Phi(F): i^* F \to i^* i_* i^* F$ is an isomorphism and so $(i^* F)_x \to (i^* i_* i^* F)_x$ is an isomorphism. But $i^*$ preserves stalks which proves (i).

Since $j^*$ preserves stalks it suffices for (ii) to prove that $j^* i_* G = 0$ for $G = i^* F \in \mathbf{F}(C)$. Now it is easy to verify that $j^* i_* G = 0$ which establishes (ii).

**4.9.5 Theorem** For $A \subset X$ the following assertions are equivalent.

(i) $A$ is locally closed.

(ii) Every $F \in \mathbf{F}(A)$ has a prolongation to $X$.

(iii) $(\text{-})^X: \mathbf{F}(A) \to \mathbf{F}(X)$ can be made into a functor.

(iv) For every $F \in \mathbf{F}(X)$ the restriction $(F|A)$ has a prolongation to $X$.

(v) $(\text{-})_A: \mathbf{F}(X) \to \mathbf{F}(X)$ can be turned into a functor.

*Proof.* The equivalence of (ii) and (iii) follows from lemma 4.9.2. The same holds for the equivalence of (iv) and (v).

The implication (ii)$\Rightarrow$(iv) is trivial.

To prove (iv)$\Rightarrow$(ii) let $F \in \mathbf{F}(A)$. Then for $i: A \hookrightarrow X$ and $G = i_* F$ we have $F \simeq G|A$ by 4.6.6(ii).

To prove (i)$\Rightarrow$(ii) we first remark that, by 4.9.4, (iv) holds for $A$ open or closed. Hence (ii) holds for $A$ open or closed. In view of (iii) of the exercise opposite, (ii) then holds for $A$ locally closed.

Finally we prove (ii)$\Rightarrow$(i). Consider $R$ a left $R$-module and let $\hat{R}$ be the sheaf on $A$ as defined in exercise (b) after theorem 4.8.7. Thus $\hat{R} = A \times R \to A$ with discrete topology on $R$. By assumption $\hat{R}$ has a prolongation to $X$. Since $\{(a, 1) \in A \times R \mid a \in A\}$ is open in $\hat{R}$ and since $\hat{R} \subset \hat{R}^X$ has the induced topology, there is an open $W \subset \hat{R}^X$ such that $W \cap \hat{R} = \{(a, 1) \mid a \in A\}$. Now consider a fixed element $a \in A$. Since $(a, 1) \in W$ there is an open $U \subset X$ with $a \in U$ and $s \in \Gamma(U, \hat{R}^X)$ with $sU \subset W$. Then $A \cap U = \{x \in U \mid sx \neq 0(x)\}$ for $0 \in \Gamma(U, \hat{R}^X)$. So $A \cap U$ is closed in $U$, hence $A$ is locally closed.

**4.9.6 Proposition** Let $C \subset X$ be closed. Then $(\text{-})^X \simeq i_*: \mathbf{F}(C) \to \mathbf{F}(X)$ where $i: C \to X$ is the injection and $\Gamma(C, \text{-}) \simeq \Gamma(X, (\text{-})^X): \mathbf{F}(C) \to {}_R\mathbf{M}$.

*Proof.* By 4.6.6 (ii) we have, for $F \in \mathbf{F}(C)$, $F \simeq i^* i_* F = G|C$ if we denote $i_* F$ by $G$. Then by 4.9.4 the sheaf $G_C$ exists and is equal to $i_* i^* G = i_* i^* i_* F$. Since $G_C = (G|C)^X \simeq F^X$ we have $F^X \simeq i_* i^* i_* F \simeq i_* F$. Hence $(\text{-})^X \simeq i_*$. So $\Gamma(X, (\text{-})^X) \simeq \Gamma(X, i_* \text{-})$. Using lemma 4.9.2 in the following special case

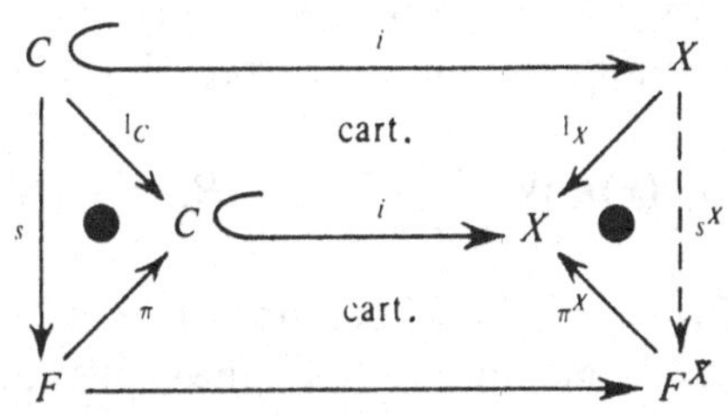


we find $\Gamma(X, i_* \text{-}) = \Gamma(C, \text{-})$.

**4.9.7 Exercises** (*a*) Verify that for $A \subset X$ locally closed the functors $(\text{-})_A: \mathbf{F}(X) \to \mathbf{F}(X)$ and $(\text{-})^X: \mathbf{F}(A) \to \mathbf{F}(X)$ are exact.

(*b*) Show that for $A_1$ and $A_2$ locally closed in $X$ it holds that $(\text{-})_{A_1} \circ (\text{-})_{A_2} = (\text{-})_{A_1 \cap A_2}$.

(*c*) For open $U \subset X$ prove that $\Gamma(U, (\text{-})|U)$ and $\Gamma(U, \text{-}): \mathbf{F}(X) \to {}_R\mathbf{M}$ are isomorphic.

($d$) Show that proposition 4.9.6 is not true for $A$ locally closed but not closed; for instance take $X$ connected and $A$ open in $X$, $\emptyset \neq A \neq X$.

($e$) Adapt exercise ($b$) after 4.6.6 and show that for an open subset $U$ with injection $U \hookrightarrow X$ the identity $i_! = (\text{-})^X: \mathbf{F}(U) \to \mathbf{F}(X)$ holds. Hence the prolongation functor $(\text{-})^X$ embeds $\mathbf{F}(U)$ as a full coreflective subcategory of $\mathbf{F}(X)$.

($f$) An alternative construction of $F_U$ for $F \in \mathbf{F}(X)$ and open $U \subset X$ runs as follows. Consider $F$ as a concrete sheaf $\ulcorner F, \pi, X \urcorner$. Define $F_U = \pi^{-1} U \cup 0_F(X)$ where $0_F$ is the null section in $\Gamma(X, F)$. Verify that $\ulcorner F_U, \pi | F_U, X \urcorner$ is a concrete sheaf and the prolongation of $F|U$ (cf. 4.9.2).

For use in the next section we need the following result.

**4.9.8 Proposition** Consider $R$ as a left $R$-module and define $\hat{R}$ on $X$ as in exercise ($b$) following 4.8.7. For open $U$, $V \subset X$ with $U \subset V$ we have $\hat{R}_U = (\hat{R}_V)_U \hookrightarrow \hat{R}_V$. So $\hat{R}_{\cdot}: \mathbf{O}(X) \to \mathbf{F}(X)$ is a functor. The functors $(\hat{R}_{\cdot}, \text{-})_{\mathbf{F}(X)}$ and $\Gamma(\text{-}, \text{-}): \mathbf{O}^{\circ}(X) \times \mathbf{F}(X) \to {}_R\mathbf{M}$ are isomorphic.

*Proof.* Let $q_{U,F}: (\hat{R}_U, F)_{\mathbf{F}(X)} \to \Gamma(U, F)$ be defined by $f \mapsto f s_U$ where the section $s_U \in \Gamma(U, \hat{R}_U)$ is given by $x \mapsto (1 \in \hat{R}_x = R)$. One verifies easily that $q$ is a morphism of bifunctors.

Let $t \in \Gamma(U, F)$. For every $f: \hat{R}_U \to F$ with $q_{U,F} f = t$ it follows that $f: (\ulcorner 1, x \urcorner \in \hat{R}_U) \mapsto tx$. Hence

$$f: (\ulcorner r, x \urcorner \in \hat{R}_U) \mapsto (r(tx) \quad \text{if } x \in U \quad \text{and} \quad 0(x) \in F_x \text{ otherwise}).$$

Therefore $q_{U,F}$ is injective.

To show that $q_{U,F}$ is also surjective let again $t \in \Gamma(U, F)$. Consider $f: \hat{R}_U \to F$ in $\mathbf{F}(X)$ defined by $f: \ulcorner r, x \urcorner \mapsto r(tx)$ for $x \in U$ and $0(x) \in F_x$ otherwise. This mapping is continuous since for open $W \subset F$ we have

$$f^{-1} W = \{\ulcorner 0, x \urcorner \in \hat{R}_U | 0(x) \in W\} \cup \bigcup_{r \neq 0} \{\ulcorner r, x \urcorner \in \hat{R}_U | x \in (rt)^{-1} W\}.$$

All these sets are open in $\hat{R}_U$ since all $rt$ are sections $U \to F$. Furthermore the mapping $f$ induces $R$-homomorphisms on the stalks and so $f$ is a morphism $\hat{R}_U \to F$.

Remark. Compare this with 4.5.4($a$).

## 4.10 Cohomology of sheaves

For a given topological space $X$ we consider the functor of the global sections $\Gamma(X, \text{-}): \mathbf{F}(X) \to {}_R\mathbf{M}$ and also the full embedding $T': \mathbf{F}(X) \to \mathbf{P}(X)$. Note that $T'F(X) = \Gamma(X, F)$ for $F \in \mathbf{F}(X)$. Both functors are left exact functors between abelian categories with injectives so their right derived functors $R^n\Gamma(X, \text{-})$ and $R^nT'$ respectively form exact sequences of right satellites. Calling these $H^n(X, \text{-})$, the *cohomology module functors*, and $\mathscr{H}^n(X, \text{-})$, the *cohomology presheaf functors*, respectively we will expect that $H^n(X, F) \simeq \mathscr{H}^n(X, F)$. Indeed this is a consequence of a more general result (see 4.10.3). In the sequel, however, the topological space $X$ being fixed, we will suppress the symbol $X$ and merely write $\mathscr{H}^n(\text{-})$.

The sheaf $F$ is called *acyclic* if $\mathscr{H}^n(F) = 0$ for all $n \geqslant 1$. Thus for an acyclic sheaf on $X$ the cohomology modules $H^n(X, F)$ are null for $n \geqslant 1$ (see the remark after 4.10.4).

A further remark is that, writing $\hat{R} = \hat{R}_X$ for the sheaf discussed in 4.9.8, we have $\Gamma(X, \text{-}) \simeq (\hat{R}, \text{-})_{\mathbf{F}(X)}$ which implies that $H^n(X, \text{-}) \simeq \mathrm{Ext}^n_{\mathbf{F}(X)}(\hat{R}, \text{-})$ for $n \geqslant 0$ (see [42, p. 51]).

**4.10.1 Lemma** If $Q$ is injective in $\mathbf{F}(X)$ then for open $U \subset X$ the restriction $Q|U$ is injective in $\mathbf{F}(U)$.

*Proof.* The functor $(\text{-})^X$ is exact and has a right adjoint $i^*$ (see 4.9.7 exercises $(a)$ and $(e)$). The functor $i^*$ preserves injectives in virtue of 2.8.9.

**4.10.2 Theorem** The cohomology presheaf functors $\mathscr{H}^n(\text{-})$ have the following relation with the cohomology module functors: for open $U \subset X$ there is a natural isomorphism $\mathscr{H}^n(\text{-})U \simeq H^n(U, \text{-}|U) \colon \mathbf{F}(X) \to {}_R\mathbf{M}$. So $\mathscr{H}^n(F)$ is, up to isomorphism, the presheaf assigning to an open subset $U$ the module $H^n(U, F|U)$.

*Proof.* The connected sequence of functors from $\mathbf{F}(X) \to {}_R\mathbf{M}$ given by $(H^n(U, \text{-}|U);\ \delta^n \circ (\text{-}|U))_{n \geqslant 0}$ is exact since $(\text{-}|U)$ is exact. With the aid of the previous lemma one sees easily that for $n \geqslant 1$ the functor $H^n(U, \text{-}|U)$ is effaceable. Thus by theorem 3.4.3 the above sequence must be the sequence of right satellites of

$$H^0(U, \text{-}|U) \simeq \Gamma(U, \text{-}|U) \simeq \Gamma(U, \text{-}) \colon \mathbf{F}(X) \to {}_R\mathbf{M}$$

(see 4.9.7 exercise $(c)$).

One also may verify that the sequence $(\mathscr{H}^n(\text{-})U;\ \delta^n_U)_{n \geqslant 0}$ is a sequence of right satellites of $\Gamma(U, \text{-})$. Therefore $\mathscr{H}^n(\text{-})U \simeq H^n(U, \text{-}|U)$.

**4.10.3 Corollary** $H^n(X, \text{-}) \simeq \mathscr{H}^n(\text{-})X$.

**4.10.4 Exercise** Verify that $S'\mathscr{H}^n(F) = 0$ for every sheaf $F$ on $X$ and for every $n \geqslant 1$.

Remark. For an acyclic sheaf $F \in \mathbf{F}(X)$ it follows that $H^n(X, F) = 0$ for $n \geqslant 1$.

According to Eilenberg and Steenrod a decent cohomology theory should satisfy certain axioms [12, p. 14]. We verify that the cohomology module functors $H^n(X, -)$ satisfy the so-called excision, exactness and dimension axioms.

**4.10.5 Definition** For $C_1$ and $C_2$ closed in $X$ we define the functors

$$H^n(C_1, C_2; -) = H^n(X, -) \circ (-)_{C_1 \setminus C_2} : \mathbf{F}(X) \to {}_R\mathbf{M}.$$

Remark that these definitions are possible since $C_1 \setminus C_2$ is locally closed in $X$; see the second definition of 4.9.3. For $F \in \mathbf{F}(X)$ we thus have $H^n(C_1, C_2; F) = H^n(X, F_{C_1 \setminus C_2})$ and $H^n(X, \emptyset; F) = H^n(X, F)$.

**4.10.6 Proposition** Let $C \subset X$ be closed. Then the following assertions hold.

(i) $H^n(C, -) \simeq H^n(X, (-)^X) : \mathbf{F}(C) \to {}_R\mathbf{M}$.

(ii) $H^n(C, -|C) \simeq H^n(X, (-)_C) : \mathbf{F}(X) \to {}_R\mathbf{M}$.

*Proof.* We claim that for $n \geqslant 1$ the functor $H^n(X, (-)^X) : \mathbf{F}(C) \to {}_R\mathbf{M}$ is effaceable. Since $\mathbf{F}(C)$ has injectives, it suffices to show by 3.4.2 (ii) that $H^n(X, Q^X) = 0$ for an injective sheaf $Q$ in $\mathbf{F}(C)$, $n \geqslant 1$. According to proposition 4.9.6, $(-)^X \simeq i_*$ where $i$ is the embedding $C \hookrightarrow X$. The left adjoint $i^*$ of $i_*$ is exact by theorem 4.8, hence preserves injectives by 2.8.9. Therefore $Q^X$ is injective in $\mathbf{F}(X)$ and $H^n(X, Q^X) = 0$ for $n \geqslant 1$ in virtue of 3.3.8 remark (iii), proving our contention.

Since $(-)^X$ is exact and since $(H^n(X, -); \delta^n)_{n \geqslant 0}$ is an exact connected sequence it follows that $(H^n(X, (-)^X); \delta^n \circ (-)^X)_{n \geqslant 0}$ is an exact connected sequence of functors from $\mathbf{F}(C)$ to ${}_R\mathbf{M}$ with $H^n(X, (-)^X)$ effaceable for $n \geqslant 1$. So according to theorem 3.4.3 this sequence must be the sequence of right satellites of $H^0(X, (-)^X) \simeq \Gamma(X, (-)^X) \simeq \Gamma(C, -)$. Therefore $H^n(C, -) \simeq H^n(X, (-)^X)$, which proves (i).

In view of the definition $(-)_C \simeq (-)^X \circ (-|C)$ (ii) follows directly from (i).

**4.10.7 Corollary** $H^n(C_1, C_2; F) \simeq H^n(C_1, C_2; F|C_1)$ for $F \in \mathbf{F}(X)$ and $C_1$, $C_2$ closed in $X$.

*Proof.*

$$\begin{aligned} H^n(C_1, C_2; F) &= H^n(X, F_{C_1\backslash C_2}) \\ &\simeq H^n(X, (F_{C_1\backslash C_2})_{C_1}) \\ &\simeq H^n(C_1, F_{C_1\backslash C_2}|C_1) \\ &\simeq H^n(C_1, (F|C_1)_{C_1\backslash C_2}) \\ &= H^n(C_1, C_2; F|C_1). \end{aligned}$$

**4.10.8 Corollary** (The excision axiom.) Let $X = C_1 \cup C_2$ with $C_1$ and $C_2$ closed subspaces of $X$. Then $H^n(X, C_2; F) \simeq H^n(C_1, C_1 \cap C_2; F|C_1)$.

*Proof.* This follows directly from the definition and 4.10.7, since $C_1 \cup C_2 \backslash C_2 = C_1 \backslash C_1 \cap C_2$.

**4.10.9 Proposition** (The exactness axiom.) For $C_1$ and $C_2$ closed in $X$ and for $F \in \mathbf{F}(X)$ the following sequence is exact: $0 \ldots \to H^n(X, C_1 \cup C_2; F) \to H^n(X, C_2; F) \to H^n(C_1, C_2; F) \to H^{n+1}(X, C_1 \cup C_2; F) \to \ldots, n \geq 0$.

*Proof.* In $\mathbf{F}(X\backslash C_2)$ we have the short exact sequence

$$0 \to (F|X\backslash C_2)_{X\backslash(C_1\cup C_2)} \to (F|X\backslash C_2) \to (F|X\backslash C_2)_{C_1\backslash C_2} \to 0$$

(compare 4.9.4). Since $(\text{-})^X$ is exact this sequence yields a short exact sequence in $\mathbf{F}(X)$:

$$0 \to F_{X\backslash C_1\cup C_2} \to F_{X\backslash C_2} \to F_{C_1\backslash C_2} \to 0.$$

This sequence in turn yields the long exact sequence in ${}_R\mathbf{M}$:

$$\begin{aligned} 0 \to H^0(X, C_1 \cup C_2; F) &\to H^0(X, C_2; F) \to H^0(C_1, C_2; F) \\ &\to H^1(X, C_1 \cup C_2; F) \to H^1(X, C_2; F) \to \ldots \end{aligned}$$

**4.10.10 Proposition** (The dimension axiom.) Suppose $X = \{x\}$. Then for every $F \in \mathbf{F}(X)$ we have $H^0(X, F) = F$ and $H^n(X, F) = 0$ for $n \geq 1$.

*Proof.* $\mathbf{F}(X) = \mathbf{P}(X)$. Hence $T'\colon \mathbf{F}(X) \to \mathbf{P}(X)$ and $\Gamma(X, \text{-})\colon \mathbf{F}(X) \to {}_R\mathbf{M}$ are both exact. Therefore $R^n\Gamma(X, \text{-}) = 0$ for $n \geq 1$.

**4.10.11 Exercise** Let $X$ be a $T_1$-space (see 4.12 ahead). Now, according to the exactness axiom and the dimension axiom there is an epimorphism $H^n(X, \{x\} \cup C; F) \to H^n(X, C; F)$ for $n \geq 1$ and for $x \in X$ whenever $C$ is closed in $X$ and $F \in \mathbf{F}(X)$. Conclude from this that $H^n(X, F) = 0$ for $n \geq 1$ for every finite discrete space $X$ and every sheaf $F$ on it.

## 4.11 Flabby sheaves and cohomology

Instead of injective sheaves one can work with so-called flabby sheaves with which there exists a canonical resolution.

**4.11.1 Definition** A sheaf $F$ on $X$ is called *flabby* provided for every open $U \subset X$ the restriction morphism $FX \to FU$ is epic. This means that every section on $U$ extends to a section on $X$.

**4.11.2 Proposition** If

$$0 \to F' \xrightarrow{k} F \xrightarrow{f} F'' \to 0$$

is exact in $\mathbf{F}(X)$ and $F'$ is flabby then the image of the sequence in $\mathbf{P}(X)$ is exact.

*Proof.* Since the sequence $0 \to F' \to F \to F''$ is already exact in $\mathbf{P}(X)$ (due to the left exactness of the functor $T'$: $\mathbf{F}(X) \to \mathbf{P}(X)$) all we have to prove is that $f: F \to F''$ is epic in $\mathbf{P}(X)$. This means that we have to prove that for each open $U_0 \subset X$ the morphism $f_{U_0}: \Gamma(U_0, F) \to \Gamma(U_0, F'')$ is epic in ${}_R\mathbf{M}$. Take $s \in \Gamma(U_0, F'')$ and consider the set $Y$ of all pairs $\ulcorner t, U \urcorner$ with open $U \subset U_0$ and $t \in \Gamma(U, F)$ such that $f_U t = \operatorname{res}{}^{U_0}_{U}\, s$. The set $Y$ is nonvoid since $\ulcorner 0, \varnothing \urcorner \in Y$. We order $Y$ by putting $\ulcorner t, U \urcorner \leq \ulcorner t', U' \urcorner$ provided $U \subset U'$ and $t = \operatorname{res}{}^{U'}_{U}\, t'$. Making use of the fact that $F$ and $F''$ are sheaves one verifies easily that any linearly ordered subset of $Y$ has an upper bound. Then $Y$ has a maximal element $\ulcorner t_1, U_1 \urcorner$ according to the lemma of Zorn. We claim that $U_1 = U_0$ so that $f_{U_0} t = s$. To prove our claim, let $x \in U_0 \backslash U_1$. The morphism $f_x: F_x \to F''_x$ is epic. This means that the element $x$ has an open neighbourhood $U_2 \subset U_0$ and there is a section $t_2 \in \Gamma(U_2, F)$ such that $f_{U_2} t_2 = \operatorname{res}{}^{U_0}_{U_2}\, s$. Now one verifies straightforwardly that

$$f_{U_1 \cap U_2}(\operatorname{res}{}^{U_1}_{U_1 \cap U_2}\, t_1) = f_{U_1 \cap U_2}(\operatorname{res}{}^{U_2}_{U_1 \cap U_2}\, t_2).$$

Since $\Gamma(U, \text{-})$ is left exact there is a section $t_3 \in \Gamma(U_1 \cap U_2, F')$ such that

$$k_{U_1 \cap U_2} t_3 = \operatorname{res}{}^{U_1}_{U_1 \cap U_2}\, t_1 - \operatorname{res}{}^{U_2}_{U_1 \cap U_2}\, t_2.$$

As $F'$ is flabby there is a section $t_4 \in \Gamma(U_2, F')$ with $t_3 = \mathrm{res}^{U_2}_{U_1^2 \cap U_2} t_4$. For $t_5 = k_{U_2} t_4 + t_2 \in \Gamma(U_2, F)$ one verifies immediately that

$$f_{U_2} t_5 = \mathrm{res}^{U_0}_{U_2^0} s \quad \text{and} \quad \mathrm{res}^{U_2}_{U_1^2 \cap U_2} t_5 = \mathrm{res}^{U_1}_{U_1^1 \cap U_2} t_1.$$

So, since $F$ is a sheaf, there is a section $t_6 \in \Gamma(U_1 \cup U_2, F)$ with

$$\mathrm{res}^{U_1 \cup U_2}_{U_1} t_6 = t_1 \quad \text{and} \quad \mathrm{res}^{U_1 \cup U_2}_{U_2} t_6 = t_5.$$

Since also $F''$ is a sheaf it follows that

$$f_{U_1 \cup U_2} t_6 = \mathrm{res}^{U_0}_{U_1^0 \cup U_2} s.$$

Hence $\ulcorner t_6, U_1 \cup U_2 \urcorner \geq \ulcorner t_1, U_1 \urcorner$. Since $U_1 \cup U_2 \neq U_1$, this contradicts the maximality of $\ulcorner t_1, U_1 \urcorner$. Hence $U_0 = U_1$.

**4.11.3 Proposition** If $0 \to F' \to F \to F'' \to 0$ is exact in $\mathbf{F}(X)$ and $F'$ and $F$ are flabby then $F''$ is also flabby.

*Proof.* Let $U$ be open in $X$. Due to our previous proposition and the flabbiness of $F'$ the following commutative diagram has exact rows:

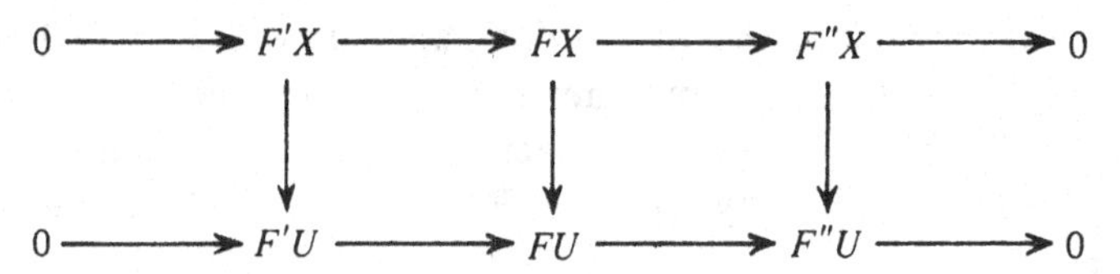


Since $F$ is also flabby the morphism $FX \to FU$ is epic in ${}_R\mathbf{M}$. It follows that $F''X \to F''U$ is epic, so $F''$ is flabby.

**4.11.4 Proposition** An injective sheaf is flabby.

*Proof.* Let $Q$ be an injective sheaf. Choose $U$ open in $X$. Then $\hat{R}_U \to \hat{R}_X$ is monic in $\mathbf{F}(X)$. But then $(\hat{R}_X, Q)_{\mathbf{F}(X)} \to (\hat{R}_U, Q)_{\mathbf{F}(X)}$ is epic. In view of proposition 4.9.8 $\Gamma(X, Q) \to \Gamma(U, Q)$ is epic.

**4.11.5 Proposition** A flabby sheaf is acyclic.

*Proof.* Choose an injective resolution $0 \to F \to Q^0 \to Q^1 \to \ldots$ of a flabby sheaf $F$. Then use propositions 4.11.3 and 4.11.4 to prove that $\ker(Q^i \to Q^{i+1})$ is flabby. Next use proposition 4.11.2 to prove that the image of the above sequence is exact in $\mathbf{P}(X)$ so the homology of the sequence is trivial.

**4.11.6 Proposition** For a topological space $X$ we define the *flabbifying functor* $\mathrm{Fl} = d_* d^*$ where $d: X_{\mathrm{dis}} \to X$. Then the adjunction $\Phi: I \to d_* d^*$ is a monomorphism; see 4.6.7.

The flabbifying functor $\mathrm{Fl}: \mathbf{F}(X) \to \mathbf{F}(X)$ is exact and makes every sheaf into a flabby sheaf.

*Proof.* Every sheaf on $X_{\text{dis}}$ is flabby. The functor $T': \mathbf{F}(X_{\text{dis}}) \to \mathbf{P}(X_{\text{dis}})$ is therefore exact. Because of the fact that $d_* \simeq S'\tilde{d}_*T': \mathbf{F}(X_{\text{dis}}) \to \mathbf{F}(X)$ it follows that $d_*$ is exact. As $d^*$ is exact we may conclude that Fl is exact.

For any $f: X \to Y$ the functor $f_*: \mathbf{F}(X) \to \mathbf{F}(Y)$ preserves flabbiness. Therefore if $F$ is a flabby sheaf on $X_{\text{dis}}$ the sheaf $d_*F$ on $X$ is also flabby.

**4.11.7 Exercises** (*a*) If $0 \to F' \to F \to F'' \to 0$ in $\mathbf{F}(X)$ and its image in $\mathbf{P}(X)$ are both exact sequences this need not imply that $F'$ is flabby (take $F'' = 0$).

(*b*) If $F \in \mathbf{F}(X)$ is flabby then for open $U \subset X$ the sheaf $F_U$ need not be flabby though $F|U$ patently is.

(*c*) For $F \in \mathbf{F}(X)$ and $x \in X$ the stalk $F_x$ is a direct summand of $(\mathrm{Fl}\, F)_x$ in ${}_R\mathbf{M}$.

**4.11.8 Definition** Let $F \in \mathbf{F}(X)$ and suppose $F^0 = F$. Then we define $F^{n+1}$ inductively by the requirement that the sequence $0 \to F^n \to \mathrm{Fl}\, F^n \to F^{n+1} \to 0$ be exact in $\mathbf{F}(X)$. Denote the composite $\mathrm{Fl}\, F^{n-1} \to F^n \to \mathrm{Fl}\, F^n$ by $d^{n-1}: \mathrm{Fl}\, F^{n-1} \to \mathrm{Fl}\, F^n$. Then the exact sequence

$$0 \to F \to \mathrm{Fl}\, F^0 \xrightarrow{d_0} \mathrm{Fl}\, F^1 \xrightarrow{d_1} \mathrm{Fl}\, F^2 \to \dots$$

is called the *canonical resolution* $\ulcorner F, \mathrm{Fl}\, F^+ \urcorner$ of $F$ in $\mathbf{F}(X)$.

Remark. If $F$ is flabby the image of the canonical resolution of $F$ is also exact in $\mathbf{P}(X)$ as follows from 4.11.2, 4.11.3.

**4.11.9 Theorem**

$$\mathscr{H}^n(F)X \simeq H^n(X, F) = H^n(T'\,\mathrm{Fl}\, F^+) = H^n(\Gamma(X, \mathrm{Fl}\, F^+)).$$

Indeed, every flabby sheaf is acyclic by 4.11.5 and therefore any flabby resolution yields sheaf cohomology in virtue of theorem 3.4.7. In particular the canonical resolution does this.

Remark. Alternatively it is possible to prove without the use of injectives the weaker result that the homology of the canonical resolution yields an exact connected sequence of functors, effaceable for $n \geqslant 1$. So this sequence must be the sequence of right satellites of the functor $T': \mathbf{F}(X) \to \mathbf{P}(X)$.

## 4.12 Soft and fine sheaves

In many applications it is convenient to work with two other classes of sheaves in our resolutions of which there are sufficiently many available. This development is mainly due to Godement [17, II. 3].

We first list, for further use, the separation axioms for a topological space $X$.

T1. Every set $\{x\}$ is closed, $x \in X$.

T2. Every two different points have disjoint neighbourhoods (Hausdorff space).

T3. For every $x \in X$ and for every closed $C \subset X$ with $x \notin C$ there are disjoint neighbourhoods $U$ and $V$ with $x \in U$ and $C \subset V$; besides $X$ should be T1 (regular space).

T4. For every two disjoint closed sets $C_1$ and $C_2$ in $X$ there are disjoint neighbourhoods $U_1$ and $U_2$; $X$ is also T1 (normal space).

One sees easily that T4⇒T3⇒T2⇒T1.

**4.12.1 Definition** A covering of a topological space is called *locally finite* if every point has a neighbourhood which intersects only finitely many elements of the covering.

A *refinement* of a covering of $X$ is a second covering each element of which is contained in an element of the first covering.

A topological space is called *paracompact* if it is Hausdorff and every open covering has a locally finite open refinement.

Without proof we mention that every paracompact space is normal; see [21, p. 68].

Also, if $\{U_\alpha\}_{\alpha \in I}$ is a (locally finite) open covering of a normal space $X$ then there exists an open (locally finite) covering $\{V_\alpha\}_{\alpha \in I}$ of $X$ with $\bar{V}_\alpha \subset U_\alpha$ for each $\alpha \in I$. The proof is not hard if one uses transfinite induction.

A compact space is paracompact (trivial).

One easily proves that a closed subspace of a paracompact space is again paracompact [21, p. 108].

**4.12.2 Definition** A topological space is called *fully paracompact* if every subspace is paracompact.

Again without proof we state that a metric space is fully paracompact [17, p. 149].

Now let $A \subset X$ be an arbitrary subset of the base space $X$ of a concrete sheaf of left $R$-modules $\ulcorner F, \pi, X \urcorner$. We define $\Gamma(A, F)$ as the set of all continuous mappings $s: A \to F$ such that $\pi s = \mathrm{id}_A$.

It is clear that $\Gamma(A, F)$ is a left $R$-module which is isomorphic to $\Gamma(A, F|A)$. In fact the functors $\Gamma(A, \text{-})$ and $\Gamma(A, \text{-}|A)$ from $\mathbf{F}(X)$ to ${}_R\mathbf{M}$ are isomorphic.

In accordance with 4.6.2 we define for all open $U \supset A$

$$\Gamma([A], F) = \varinjlim_{U \supset A} \Gamma(U, F)$$

for any $F \in \mathbf{F}(X)$ and any $A \subset X$. Then there is a morphism $\Gamma([A], F) \to \Gamma(A, F)$ which is easily seen to be monic.

**4.12.3 Lemma** For $F \in \mathbf{F}(X)$ with $X$ paracompact and $C$ closed in $X$ every $s \in \Gamma(C, F)$ can be extended to a section on an open neighbourhood of $C$.

*Proof.* For $x \in C$ the image $sx \in F$. Then $x$ has an open neighbourhood $M_x$ and a section $s'_x \in \Gamma(M_x, F)$ with $s'_x x = sx$. After restriction to the sheaf $F|C$ it turns out that $x$ has an open neighbourhood $N_x$ such that $N_x \subset M_x$ and $s'_x|(C \cap N_x) = s|(C \cap N_x)$. For $x \in C$ we write $s_x = (s'_x|N_x) \in \Gamma(N_x, F)$. For $x \in X \backslash C$ we set $N_x = X \backslash C$, so $N_x$ is open, and take $s_x = 0 \in \Gamma(N_x, F)$. Then every $x \in X$ has an open neighbourhood $N_x$ and a section $s_x \in \Gamma(N_x, F)$ such that $s_x|(C \cap N_x) = s|(C \cap N_x)$. The system $\{N_x\}_x$ is an open covering of $X$ which must have a locally finite open refinement $\{U_\alpha\}_{\alpha \in I}$ since $X$ is paracompact. For every $\alpha \in I$ we choose an element $x \in X$ with $U_\alpha \subset N_x$ and we put $s_\alpha = (s_x|U_\alpha) \in \Gamma(U_\alpha, F)$. Then $s|(C \cap U_\alpha) = s_\alpha|(C \cap U_\alpha)$.

Furthermore $X$ has an open covering $\{V_\alpha\}_{\alpha \in I}$ with $\bar{V}_\alpha \subset U_\alpha$.

Next we define two subsets of $X$:

$$W_1 = \{x \in X | s_\alpha x = s_\beta x \text{ whenever } x \in U_\alpha \cap U_\beta\}$$

and

$$W_2 = \{x \in X | s_\alpha x = s_\beta x \text{ whenever } x \in V_\alpha \cap V_\beta\}.$$

Then $C \subset W_1 \subset W_2$.

For the moment we assume that there exists an open subset $W \subset X$ with $W_1 \subset W \subset W_2$. Denote $W \cap V_\alpha$ by $W_\alpha$ and let $t_\alpha = (s_\alpha|W_\alpha) \in \Gamma(W_\alpha, F)$. Then $\{W_\alpha\}_{\alpha \in I}$ is an open covering of $W$ for which $\{t_\alpha\}_\alpha$ is a compatible system of sections. Thus there is a section $t \in \Gamma(W, F)$ with $t|W_\alpha = t_\alpha$. But then $t|C = s$.

To prove the existence of an open set $W$ with the above property, let $x \in W_1$. Since $\{V_\alpha\}_{\alpha \in I}$ is locally finite $x$ has an open neighbourhood $P'_1$ such that $P'_1$ has nonvoid intersections with only finitely many elements of the covering $\{V_\alpha\}_{\alpha\in I}$. But then one may shrink the open neighbourhood $P_1'$ to an open neighbourhood $P_1$ such that $x\in\bar V_\alpha$ if and only if $P_1\cap V_\alpha\neq\varnothing$. It follows that $P_2=\bigcap_{\alpha\in J}U_\alpha$ where $J=\{\alpha\in I|x\in U_\alpha\}$ is an open set. Consider the set $P_3$ consisting of all elements $y\in P_1\cap P_2$ such that $s_\alpha y=s_\beta y$ for $\alpha$ and $\beta\in J$. Since $J$ is finite $P_3$ is open. Since $x\in W_1$, the open set $P_3$ contains $x$ and thus is a neighbourhood of $x$. For $y\in P_3\cap V_\alpha$ it follows that $y\in P_1\cap V_\alpha$ and then $P_1\cap V_\alpha\neq\varnothing$ so that $x\in\bar V_\alpha\subset U_\alpha$. This means that $\alpha\in J$ and so $P_3\subset W_2$. Hence $W_2$ is a neighbourhood of $W_1$. From this it follows that there is an open subset $W\subset X$ such that $W_1\subset W\subset W_2$.

**4.12.4 Corollary** If $X$ is paracompact and $F$ is a sheaf on $X$ then $\Gamma([C],F)\to\Gamma(C,F)$ is an isomorphism for every closed subset $C\subset X$.

**4.12.5 Exercise** Verify that $\Gamma([X],G)\to\Gamma(X,G)$ is not epic for the sheaf $G$ on $Y$ as defined in exercise (*e*) of 4.6.5 where $X$ is the discrete subspace of all rational numbers. Conclude that $Y$ is not paracompact. Also prove this fact directly.

**4.12.6 Definition** A concrete sheaf $F$ on $X$ is called *soft* provided $\Gamma(X,F)\to\Gamma(C,F)$ is epic for all closed $C\subset X$.

Compare this with the definition of a flabby sheaf in 4.11.1. Here we have to work with concrete sheaves since the sheaf $F$ does not take values on closed sets. But again we shall not distinguish too closely between sheaves and concrete sheaves.

It follows immediately that a flabby sheaf on a paracompact space is soft.

Here is an example of a soft sheaf. Let $X$ be paracompact. We claim that the sheaf $F$ on $X$ defined by $FU=(U,\mathbb{R})_{\mathbf{Top}}$ is soft. To see this let $C\subset X$ be closed. Then

$$\Gamma(C,F)\simeq\Gamma([C],F)=\varinjlim_{U\supset C}\Gamma(U,F)\simeq\varinjlim_{U\supset C}(U,\mathbb{R})_{\mathbf{Top}}.$$

For $s\in\Gamma([C],F)$ there is an open neighbourhood $U_0$ of $C$ with $s_0\colon U_0\to\mathbb{R}$ such that $s=\mathrm{res}^{U_0}_{[C]}s_0$. As $X$ is normal, $C$ has an open neighbourhood $U_1$ with $\bar U_1\subset U_0$. For $s_1=(s_0|U_1)\colon U_1\to\mathbb{R}$ it holds that $s=\mathrm{res}^{U_1}_{[C]}s_1$. Now consider $t_1=(s_0|\bar U_1)\colon\bar U_1\to\mathbb{R}$. Since $X$ is normal and $\bar U_1$ is closed in $X$, the function $t_1$ has an extension $t\colon X\to\mathbb{R}$ such that $t_1=t|\bar U_1$ (Urysohn's lemma [21, p. 58]). Hence $t|U_1=s_1$ and $\mathrm{res}^X_{[C]}t=s$.

Remark. A soft sheaf need not be flabby since we may take $X=\mathbb{R}$ in the previous example. Then $F$ is soft but not flabby, for the continuous function $f: \mathbb{R}\backslash\{0\}\to\mathbb{R}$ defined by $f(x)=1/x$ cannot be extended to $\mathbb{R}$.

**Exercise** Verify that for a soft sheaf $F$ on $X$ and a closed subset $C\subset X$ the sheaf $F|C$ is again soft.

**4.12.7 Lemma** Let $F'$ be a soft sheaf on a regular space $X$ and suppose the sequence $0\to F'\xrightarrow{k} F\xrightarrow{f} F''\to 0$ is exact in $\mathbf{F}(X)$. Then for an open and paracompact subspace $U\subset X$ the sequence $0\to F'U\to FU\to F''U\to 0$ is exact in ${}_R\mathbf{M}$.

*Proof.* We only need to show that $k_U: FU\to F''U$ is epic. Let $s\in F''U=\Gamma(U,F'')$. For every $x\in X$ the morphism $f_x: F_x\to F''_x$ is epic. Since $X$ is a regular space it is possible to find for $x\in U$ an open neighbourhood $N_x$ such that $\bar{N}_x\subset U$ and a section $s_x\in\Gamma(N_x,F)$ such that $f\circ s_x=s|N_x$. The family $\{N_x\}_{x\in U}$ covers $U$ and $U$ is paracompact. So there is a locally finite open refinement $\{U_\alpha\}_{\alpha\in J}$ of $\{N_x\}_{x\in U}$ together with a collection $\{s_\alpha\in\Gamma(U_\alpha,F)\}_{\alpha\in J}$ such that $f\circ s_\alpha=s|U_\alpha$. As $U$ is normal there is an open refinement $\{V_\alpha\}_{\alpha\in J}$ of $\{U_\alpha\}_{\alpha\in J}$ such that $\bar{V}^U\subset U_\alpha$. Since $\bar{V}^U_\alpha=\bar{V}_\alpha\cap U$ and $\bar{V}_\alpha\subset\bar{U}_\alpha\subset\bigcup_x\bar{N}_x=U$ it follows that $\bar{V}^U_\alpha=\bar{V}_\alpha$. Therefore $\bar{V}_\alpha\subset U_\alpha$ for all $\alpha\in J$.

For $I\subset J$ we define $V(I)=\bigcup_{\alpha\in I}\bar{V}_\alpha$. Then $V(I)$ is always closed in $U$. For if $x\in U\backslash V(I)$ then $x$ has an open neighbourhood $P_0$ such that the set $\{\alpha|P_0\cap V_\alpha\neq\emptyset\}$ is finite so that particularly the set $\{\alpha|P_0\cap V_\alpha\neq\emptyset$ and $x\notin\bar{V}_\alpha\}$ is finite. This means that

$$P=\bigcap_{\alpha:x\notin\bar{V}_\alpha}(P_0\backslash\bar{V}_\alpha)$$

is an open neighbourhood of $x$. For all $\alpha$ we now have that if $P\cap V_\alpha\neq\emptyset$ then $x\in\bar{V}_\alpha$. Since $x\notin V(I)$ it follows that $P\cap\bar{V}_\alpha=\emptyset$ for all $\alpha\in I$. Hence $P\cap V(I)=\emptyset$, so $V(I)$ is closed in $U$.

Consider the set $Y$ of all pairs $\ulcorner t,I\urcorner$ such that $I\subset J$, $t\in\Gamma(V(I),F)$ and $ft=s|V(I)$. Define $\ulcorner t,I\urcorner\leq\ulcorner t',I'\urcorner$ by $I\subset I'$ and $t=t'|V(I)$. Then $Y$ is obviously nonvoid and, since $F$ and $F''$ are sheaves, one easily checks that every linearly ordered subset of $Y$ has an upper bound. According to Zorn's lemma $Y$ has a maximal element, say $\ulcorner t_0,I_0\urcorner$. We will prove in a minute that $V(I_0)=U$. Since $t_0\in\Gamma(U,F)$ satisfies $ft_0=s$ or also $f_U(t_0)=s$ we have proved that $f_U: FU\to F''U$ is epic.

To prove our claim, suppose $V(I_0)\neq U$. Then there is an element $\alpha_0\in J$ with $V_{\alpha_0}\backslash V(I_0)\neq\emptyset$. Since $V(I_0)$ is closed in the paracompact space $U$ we may extend $t_0$ to an open set $W \subset U$ with $V(I_0) \subset W$ such that there is a $t_1 \in \Gamma(W, F)$ with $t_1|V(I_0) = t_0$ and $ft_1 = s|W$. Then $f \circ (t_1|W \cap U_{\alpha_0}) = f \circ (s_{\alpha_0}|W \cap U_{\alpha_0})$. Since $0 \to F' \xrightarrow{k} F \xrightarrow{f} F''$ is exact in $\mathbf{P}(X)$ there exists an element $r_0 \in \Gamma(W \cap U_{\alpha_0}, F')$ with $kr_0 = (t_1|W \cap U_{\alpha_0}) - (s_{\alpha_0}|W \cap U_{\alpha_0})$. The set $\bar{V}_{\alpha_0}$ is contained in $U$ and closed in $X$. The set $V(I_0)$ is closed in $U$. Hence $V(I_0) \cap \bar{V}_{\alpha_0}$ is closed in $X$. Since $F'$ is soft we can find an element $r_1 \in \Gamma(\bar{V}_{\alpha_0}, F')$ with $r_1|V(I_0) \cap \bar{V}_{\alpha_0} = r_0|V(I_0) \cap \bar{V}_{\alpha_0}$. Putting $t_2 = kr_1 + (s_{\alpha_0}|\bar{V}_{\alpha_0}) \in \Gamma(\bar{V}_{\alpha_0}, F)$ we find that $t_2$ satisfies $t_2|\bar{V}_{\alpha_0} \cap V(I_0) = t_0|\bar{V}_{\alpha_0} \cap V(I_0)$ and $ft_2 = s|\bar{V}_{\alpha_0}$. Since both $\bar{V}_{\alpha_0}$ and $V(I_0)$ are closed in $U$ it is easy to see that there is an element $t_3 \in \Gamma(\bar{V}_{\alpha_0} \cup V(I_0), F)$ with $t_3|\bar{V}_{\alpha_0} = t_2$ and $t_3|V(I_0) = t_0$. Then also $ft_3 = s|\bar{V}_{\alpha_0} \cup V(I_0)$. Now for $I_3 = \{\alpha_0\} \cup I_0 \subset J$ the pair $\ulcorner t_3, I_3 \urcorner$ is strictly larger than $\ulcorner t_0, I_0 \urcorner$ in $Y$. This contradicts the maximality of $\ulcorner t_0, I_0 \urcorner$.

The following result does for soft sheaves what 4.11.3 does for flabby ones.

**4.12.8 Lemma** Let $X$ be paracompact and $0 \to F' \to F \to F'' \to 0$ an exact sequence in $\mathbf{F}(X)$ with $F'$ and $F$ soft. Then $F''$ is also soft.

*Proof.* Let $C$ be closed in $X$. Then $F'|C$ is a soft sheaf on $C$. The sequence $0 \to F'|C \to F|C \to F''|C \to 0$ is exact in $\mathbf{F}(C)$. Since $C$ is paracompact and hence regular, according to 4.12.7 the sequence $0 \to \Gamma(C, F'|C) \to \Gamma(C, F|C) \to \Gamma(C, F''|C) \to 0$ is exact in ${}_R\mathbf{M}$. Instead of $\Gamma(C, -|C)$ we take the isomorphic functor $\Gamma(C, -)$ and then consider the following commuting diagram

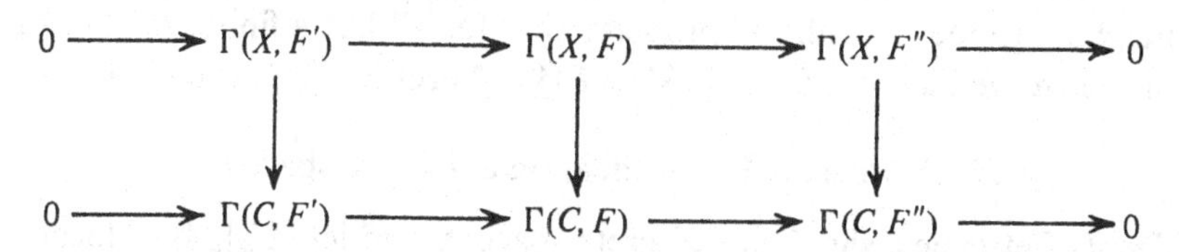

$$\begin{array}{ccccccccc} 0 & \to & \Gamma(X, F') & \to & \Gamma(X, F) & \to & \Gamma(X, F'') & \to & 0 \\ & & \downarrow & & \downarrow & & \downarrow & & \\ 0 & \to & \Gamma(C, F') & \to & \Gamma(C, F) & \to & \Gamma(C, F'') & \to & 0 \end{array}$$

for which the bottom row is exact. As $F$ is soft, $\Gamma(X, F'') \to \Gamma(C, F'')$ is epic.

**4.12.9 Lemma** Let $F$ be a soft sheaf on a paracompact space $X$. Then $H^n(X, F) = 0$ for $n \geq 1$. If $X$ is fully paracompact then $\mathscr{H}^n(F) = 0$ for $n \geq 1$ so a soft sheaf is then acyclic.

*Proof.* We first construct the canonical resolution of the sheaf $F$ as we did before, setting $F = F^0$ and making $0 \to F^n \to \mathrm{Fl}\, F^n \to F^{n+1} \to 0$ exact in $\mathbf{F}(X)$. Since $\mathrm{Fl}\, F^n$ is a flabby sheaf on a paracompact space it is a soft sheaf. But then all sheaves $F^n$ are soft due to 4.12.8. The long sequence

$$0 \to \Gamma(X, \mathrm{Fl}\, F^0) \to \Gamma(X, \mathrm{Fl}\, F^1) \to \Gamma(X, \mathrm{Fl}\, F^2)$$
$$\to \Gamma(X, \mathrm{Fl}\, F^3) \to \Gamma(X, \mathrm{Fl}\, F^4) \to \dots$$

is then exact according to 4.12.7 at all places $\Gamma(X, \mathrm{Fl}\, F^n)$ with $n \geq 1$. A flabby sheaf is acyclic and since $\mathscr{H}^n\,(\mathrm{Fl}\, F^i)X \simeq H^n\,(X, \mathrm{Fl}\, F^i)$ we have $H^n\,(X, \mathrm{Fl}\, F^i) = 0$ for $i \geq 0$. Now applying theorem 4.11.9 we find that $H^n(X, F) \simeq H^n(\Gamma(X, \mathrm{Fl}\, F^+)) = 0$.

If $X$ is fully paracompact then the long sequence

$$0 \to \mathrm{Fl}\, F^0 \to \mathrm{Fl}\, F^1 \to \mathrm{Fl}\, F^2 \to \mathrm{Fl}\, F^3 \to \dots$$

is exact in $\mathbf{P}(X)$ except perhaps at $\mathrm{Fl}\, F^0$ by 4.12.7. From this it follows again that $\mathscr{H}^n(F) = 0$ for $n \geq 1$.

**4.12.10 Corollary** On a paracompact space one can construct the cohomology modules from a soft resolution of the sheaf. On a fully paracompact space one can construct the cohomology presheaves from a soft resolution of the sheaf.

**4.12.11 Definition** A sheaf is called *fine* provided for every locally finite open covering $\{U_\alpha\}_\alpha$ of $X$ there is a collection of morphisms $i_\alpha: F \to F$ such that:

(i) $|i_\alpha| \subset \bar{U}_\alpha$ where $|i_\alpha|$ denotes the set $\{x \in X \mid (i_\alpha)_x \neq 0\}$ for all $\alpha$;

(ii) $\sum_\alpha i_\alpha = 1_F$.

Note: since the covering is locally finite every point $x \in X$ has an open neighbourhood $V$ with the property that for all but a finite number of indices $\alpha$ we have $i_\alpha | V = 0: F|V \to F|V$. Therefore $\sum_\alpha i_\alpha$ is defined.

**4.12.12 Lemma** Every sheaf on a discrete space is fine.

*Proof.* Let $F$ be a sheaf on a discrete space $X$ and let $\{U_\alpha\}_\alpha$ be a locally finite open covering of $X$. This implies that $X = \bigcup_\alpha U_\alpha$ and $\{\alpha \mid x \in U_\alpha\}$ is finite for every $x \in X$.

Choose for every $x \in X$ an index $\alpha(x)$ such that $x \in U_{\alpha(x)}$ and define $i_\alpha: F \to F$ by $i_\alpha | F_x =$ the embedding $F_x \hookrightarrow F$ if $\alpha = \alpha(x)$ and otherwise $0: F_x \to F$.

Since $X$ is discrete all the mappings $i_\alpha: F \to F$ are morphisms of sheaves on $X$. It is clear that $|i_\alpha| \subset U_\alpha$ for every $\alpha$ and also $\sum_\alpha i_\alpha = 1_F$.

**4.12.13 Lemma** If $F$ is a fine sheaf on $X$ and $f: X \to Y$ then $f_* F$ is a fine sheaf on $Y$.

*Proof.* Let $\{U_\alpha\}_\alpha$ be a locally finite open covering of $Y$. Since $f: X \to Y$ is continuous $\{f^{-1}U_\alpha\}_\alpha$ is a locally finite open covering of $X$. Hence there exists a collection $\{i_\alpha: F \to F\}_\alpha$ with $|i_\alpha| \subset \overline{f^{-1}U_\alpha}$ and $\sum_\alpha i_\alpha = 1$. For every index $\alpha$ the set $f^{-1}(Y\backslash\bar{U}_\alpha)$ is open in $X$ with $f^{-1}(Y\backslash\bar{U}_\alpha) \cap f^{-1}\bar{U}_\alpha = \varnothing$ and so $f^{-1}(Y\backslash\bar{U}_\alpha) \cap |i_\alpha| = \varnothing$. Thus $i_\alpha|f^{-1}(Y\backslash\bar{U}_\alpha) = 0$ and therefore $(f_*i_\alpha)|Y\backslash\bar{U}_\alpha = 0$. So $|f_*i_\alpha| \subset \bar{U}_\alpha$. Since $f_*$ is additive we have $\sum_\alpha f_*i_\alpha = 1_{f_*F}$.

**4.12.14 Corollary** For every sheaf $F$ on $X$ its flabbification $\mathrm{Fl}\, F$ is fine.

*Proof.* $\mathrm{Fl} = d_*d^*$ for $d: X_{\mathrm{dis}} \to X$ and every sheaf on $X_{\mathrm{dis}}$ is fine.

**4.12.15 Corollary** Every sheaf can be embedded in a fine sheaf.

*Proof.* The adjunction $\Phi: I \to d_*d^*$ is a monomorphism by proposition 4.6.7.

**4.12.16 Proposition** If $F$ is a direct summand of $G$ in $\mathbf{F}(X)$ and if $G$ is a fine sheaf then so is $F$.

*Proof.* By assumption there are morphisms $q: F \to G$ and $p: G \to F$ such that $1_F = pq$. Let $\{U_\alpha\}_\alpha$ be a locally finite covering of $X$. There is a system $\{j_\alpha: G \to G\}_\alpha$ satisfying $|j_\alpha| \subset \bar{U}_\alpha$ and $\sum_\alpha j_\alpha = 1_G$. But then the system $\{i_\alpha = pj_\alpha q: F \to F\}_\alpha$ satisfies $|i_\alpha| \subset \bar{U}_\alpha$ and $\sum_\alpha i_\alpha = 1_F$.

**4.12.17 Propositon** If $X$ is paracompact every fine sheaf on $X$ is soft.

*Proof.* Suppose $F \in \mathbf{F}(X)$ is fine and let $C \subset X$ be closed with $s \in \Gamma(C, F)$. Then, by 4.12.3, there exists an open $U_1 \supset C$ together with a section $s_1 \in \Gamma(U_1, F)$ such that $s_1|C = s$. Let $U_0 = X\backslash C$ and $s_0 = 0 \in \Gamma(U_0, F)$. Thus $\{U_0, U_1\}$ is an open covering of $X$. Since $X$ is normal there is an open covering $\{V_0, V_1\}$ of $X$ such that $\bar{V}_\alpha \subset U_\alpha$. Then there is a system $\{j_\alpha: F \to F\}_{\alpha=0,1}$ with $|j_\alpha| \subset \bar{V}_\alpha \subset U_\alpha$ and $\sum_\alpha j_\alpha = j_0 + j_1 = 1_F$. Finally one verifies easily that $t \in \Gamma(X, F)$ is well defined by $t = s_0 j_0 + s_1 j_1$ and that $t|C = s$.

**4.12.18 Corollary** A fine resolution of a sheaf on a fully paracompact space yields presheaf cohomology.

*Proof.* See 4.12.10.

We will end this section with an example of a fine sheaf. Let $X$ be a normal space and $F$ the sheaf defined by $FU = (U, \mathbb{R})_{\mathbf{Top}}$. We claim that this sheaf is fine. Let $\{U_\alpha\}_\alpha$ be a locally finite open covering. There is a second open covering $\{V_\alpha\}_\alpha$ with $\bar{V}_\alpha \subset U_\alpha$. By Urysohn's lemma there is a continuous mapping $f_\alpha: X \to \mathbb{R}$, for each $\alpha$, such that for all $x \in X$ we have $f_\alpha x \geq 0$, $f_\alpha x = 0$ for $x \notin U_\alpha$ and $f_\alpha x = 1$ for $x \in \bar{V}_\alpha$. We set $f = \sum_\alpha f_\alpha: X \to \mathbb{R}$. Then $fx \geq 1$ for all $x \in X$. We may therefore define $f^{-1}: X \to \mathbb{R}$ by $f^{-1}x = (fx)^{-1}$. Now give $i_\alpha: F \to F$ by $(i_\alpha)_U: (g: U \to \mathbb{R}) \mapsto (f^{-1}f_\alpha g: U \to \mathbb{R})$. Then the system $\{i_\alpha\}_\alpha$ satisfies $|i_\alpha| \subset \bar{U}_\alpha$ and $\sum_\alpha i_\alpha = 1_F$.

With the aid of 4.12.17 we see that the sheaf of continuous functions defined on open subsets is soft, as we already proved earlier.

For a general treatment of soft and fine sheaves and their role in sheaf cohomology the reader is invited to consult [17] or [6].

# References

[1] M. André, *Homologie des algèbres commutatives*, Springer, Berlin 1974.
[2] R. Baer, Erweiterung von Gruppen und ihre Isomorphismen, *Math. Z.* **38** (1934), 375–416.
[3] R. Baer, The subgroup of the elements of finite order of an abelian group, *Ann. Math. II* **37** (1936), 766–81.
[4] H. Bass, *Libération des modules projectifs sur certains anneaux de polynômes, Séminare Bourbaki 1973/74*, Lecture notes in math. 431, Springer, Berlin 1975, 228–54.
[5] N. Bourbaki, *Eléments de mathématique*, vol. II, ch. IV & V, 2nd edn., Hermann, Paris 1959.
[6] G. Bredon, *Sheaf theory*, McGraw-Hill, New York 1967.
[7] H.-B. Brinkmann, Baer addition of extensions, *Manuscr. Math.* **1** (1969), 99–109.
[8] W. Burgess, The meaning of mono and epi in some familiar categories, *Canad. Math. Bull.* **8** (1965), 759–70.
[9] H. Cartan and S. Eilenberg, *Homological algebra*, Princeton University Press, Princeton N.J. 1956.
[10] S. Eilenberg, *Automata, languages and machines*, vol. A and B, Academic Press, New York 1974 and 1976.
[11] S. Eilenberg and S. MacLane, General theory of natural equivalences, *Trans. Amer. Math. Soc.* **58** (1945), 231–94.
[12] S. Eilenberg and N. Steenrod, *Foundations of algebraic topology*, Princeton University Press, Princeton N.J. 1952.
[13] P. Freyd, *Abelian categories*, Harper & Row, New York 1964.
[14] P. Gabriel, Des catégories abéliennes, *Bull. Soc. Math. France* **90** (1962), 323–448.
[15] P. Gabriel, *Zentralblatt für Mathematik und ihre Grenzgebiete* **136** (1967), 6–7.
[16] P. Gabriel and N. Popescu, Caractérisation des catégories abéliennes avec générateurs et limites inductives exactes, *Comptes Rendus Acad. Sci. Paris série A* **258** (1964), 4188–91.
[17] R. Godement, *Topologie algébrique et théorie des faisceaux*, Hermann, Paris 1964.
[18] A. Grothendieck, Sur quelques points d'algèbre homologique, *Tôhoku Math. J. II* **9** (1957), 119–221.
[19] A. Grothendieck with J. A. Dieudonné, Eléments de géométrie algébrique, *Publ. Math. Inst. Ht. Etud. Scient.* **4**, **8**, **11**, **17**, **20**, **24**, **28**, **32**, Bures-s-Yvette 1960–67.
[20] P. J. Hilton and U. Stammbach, *A course in homological algebra*, Springer, Berlin 1971.
[21] S.-T. Hu, *Elements of general topology*, Holden-Day, San Francisco 1964.

[22] D. M. Kan, Adjoint functors, *Trans. Amer. Math. Soc.* **87** (1958), 294–329.
[23] J. Lambek, *Completions of categories*, Lecture notes in math. 24, Springer, Berlin 1966.
[24] J. Lambek, *Torsion theories, additive semantics and rings of quotients*, Lecture notes in math. 177, Springer, Berlin 1971.
[25] S. MacLane, *Homology*, Springer, Berlin 1963.
[26] S. MacLane, *Categories for the working mathematician*, Springer, Berlin 1971.
[27] W. Magnus, A. Karrass and D. Solitar, *Combinatorial group theory*, Wiley, New York 1966.
[28] C. R. F. Maunder, *Algebraic topology*, Van Nostrand, New York 1970.
[29] B. Mitchell, *Theory of categories*, Academic Press, New York 1965.
[30] D. G. Northcott, *An introduction to homological algebra*, Cambridge University Press, Cambridge 1960.
[31] F. Oort, On the definition of an abelian category, *Indag. Math.* **29** (1967), 83–95.
[32] F. Oort and J. R. Strooker, The category of finite bialgebras over a field, *Indag. Math.* **29** (1967), 163–9.
[33] N. Popescu, *Abelian categories with applications to rings and modules*, Academic Press, London 1973.
[34] D. Puppe, Über die Axiome für abelsche Kategorien, *Arch. d. Math.* **18** (1967), 217–22.
[35] D. G. Quillen, *Homotopical algebra*, Lecture notes in math. 43, Springer, Berlin 1967.
[36] D. Quillen, Projective modules over polynomial rings, *Invent. Math.* **36** (1976), 167–71.
[37] G. A. Reid, Epimorphisms and surjectivity, *Invent. Math.* **9** (1969/70), 295–302.
[38] D. W. Sharpe and P. Vámos, *Injective modules*, Cambridge Univ. Press, Cambridge 1972.
[39] E. H. Spanier, *Algebraic topology*, McGraw-Hill, New York 1966.
[40] B. Stenström, *Rings of quotients*, Springer, Berlin 1975.
[41] J. R. Strooker, Faithfully projective modules and clean algebras, Thesis, Groen, Leiden 1965.
[42] R. G. Swan, *The theory of sheaves*, Chicago University Press, Chicago 1964.
[43] M. Tierney, Axiomatic sheaf theory; some constructions and applications, in *Categories and commutative algebra, C.I.M.E. III 1971*, Cremonese, Rome 1973, 249–326.
[44] N. Yoneda, On the homology theory of modules, *J. Fac. Sci. Tokyo I* **7** (1954), 193–227.
[45] A. C. Zaanen, *Linear analysis*, North-Holland, Amsterdam 1953.

# Notation index

The standard notation employed for certain categories is listed here in alphabetical order.



In the next list, standard letters have been used in many cases to identify symbol notation for categories, morphisms, etc. Numerical suffices distinguish different individuals of the same type. The scheme is as follows: category **A** or **C**; equivalence class of short exact sequences $\mathscr{E}$; morphism $f$; object $A$ or $C$; short exact sequence $\dot{E}$.

# *Notation index*

# Index

# Index